A GUIDE TO
NORTH CAROLINA'S
FRESHWATER FISHES

A GUIDE TO

NORTH CAROLINA'S FRESHWATER FISHES

Bryn H. Tracy

Fred C. (Fritz) Rohde

Scott A. Smith

Jesse L. Bissette

Gabriela M. Hogue

THE UNIVERSITY OF NORTH CAROLINA PRESS | CHAPEL HILL

A Southern Gateways Guide

Designed by April Leidig
Set in Garamond by Copperline Book Services, Inc.

Manufactured in the United States of America

Cover photograph by Todd Pusser. *Foreground:* Crescent Shiner, *Luxilus cerasinus* (males in breeding color), atop a Bluehead Chub, *Nocomis leptocephalus,* nest in South Double Creek, Stokes County, NC. *Background:* Mountain Redbelly Dace, *Chrosomus oreas.*

Library of Congress Cataloging-in-Publication Data
Names: Tracy, Bryn H., author. | Rohde, Fred C., author. | Smith, Scott A. (Marine
    biologist), author. | Bissette, Jesse L., author. | Hogue, Gabriela M., author.
Title: A guide to North Carolina's freshwater fishes / Bryn H. Tracy, Fred C. (Fritz)
    Rohde, Scott A. Smith, Jesse L. Bissette, Gabriela M. Hogue.
Other titles: Southern gateways guide.
Description: Chapel Hill : The University of North Carolina Press, [2024] |
    Series: Southern gateways guide | Includes bibliographical references and index.
Identifiers: LCCN 2023040816 (print) | LCCN 2023040817 (ebook) |
    ISBN 9781469678115 (paperback) | ISBN 9781469678122 (ebook)
Subjects: LCSH: Freshwater fishes—North Carolina—Identification. |
    LCGFT: Field guides.
Classification: LCC QL628.N8 T73 2024 (print) | LCC QL628.N8 (ebook) |
    DDC 597.17609756—dc23/eng/20230920
LC record available at https://lccn.loc.gov/2023040816
LC ebook record available at https://lccn.loc.gov/2023040817

CONTENTS

PREFACE

North Carolina's fresh waters course through our state from streams origi-nating atop the highest peak east of the Mississippi River. Flowing westward and south to the Gulf of Mexico and eastward to the Atlantic Ocean, they create myriad and varied aquatic habitats. From our Mountain region the crystal-clear, cold, high-gradient, boulder-strewn streams give way to muddy and warm Piedmont streams and reservoirs. Our Sand Hills region contains unique tea-colored, tannin-stained, and sandy-bottomed streams. The dark, coffee-colored, and mucky-bottomed Coastal Plain streams find their way into the sounds and eventually the Atlantic Ocean.

Among all these areas of fresh water, the aquatic habitats are many and in-clude braided and channelized swamps, abandoned mill ponds, farm ponds, wetlands, bay lakes, and rivers and streams that drain our largest and densest metropolitan areas. In every one of these habitats, darting beneath the surface, can be found at least 1 of the 258 species of freshwater fishes. Ranging in size from under two inches to more than fourteen feet in length and with some as odd-shaped as science fiction–conjured images of alien animals, most species are unknown to the public and therefore live a life of obscurity. To us, this is dif-ficult to comprehend. Their diversity is astounding, and their beauty rivals that of their colorful tropical and marine counterparts. It is our hope that this book, along with "An Annotated Atlas of the Freshwater Fishes of North Carolina" (Tracy, Rohde, and Hogue 2020) and NCFishes.com, will ignite a spark and sustain your interest in the exploration and conservation of our rich freshwater fish fauna and their aquatic habitats.

From Wolf Creek, the westernmost community in Cherokee County, to the small Outer Banks town of Buxton in Dare County, North Carolina's fresh waters are home to forty families of fishes (table 1): thirty-one families whose species are primarily freshwater, five families whose species are primarily marine and estuarine, and four families whose species are almost evenly split between freshwater and marine. Also included are six families that are not indigenous (native) to North Carolina—Cichlidae, Cobitidae, Cyprinidae, Gasterosteidae, Loricariidae, and Xenocyprididae.

These forty families inhabiting North Carolina's fresh waters include 242 described species (three of which are extirpated species) and 16 undescribed species (tables 1 and 2). The two most speciose families are Leuciscidae (68 species) and Percidae (40 species). There are seventeen families that have only one freshwater species found in North Carolina (tables 1 and 2). The sixteen undescribed species that are currently known can be identified using the identification keys that were developed for each family found in the following chapters. There may be additional undescribed species within what are currently considered species complexes, such as Bluehead Chub, *Nocomis leptocephalus*; Mimic Shiner, *Paranotropis volucellus*; Mottled Sculpin, *Cottus bairdii*; and Fantail Darter, *Etheostoma flabellare* (Tracy, Rohde, and Hogue, 2020). These complexes are currently being studied and may add additional species to the already rich fauna. The full distributional picture of what we currently know about North Carolina's native (indigenous) and nonnative (nonindigenous) freshwater fish fauna can be extrapolated from table 2, which has been compiled via the labors of many ichthyologists over many decades.

TABLE 1. **North Carolina's freshwater fish fauna listed in phylogenetic order following Fricke, Eschmeyer, and van der Laan (2022)**

Family	Common Name	No. of Described and Undescribed Freshwater Species	No. of Additional Marine and Estuarine Species in North Carolina
Petromyzontidae	Lampreys	5	
Acipenseridae	Sturgeons	3	
Polyodontidae	Paddlefishes	1[a]	
Lepisosteidae	Gars	1	
Amiidae	Bowfins	1	
Anguillidae	Freshwater Eels	1	
Hiodontidae	Mooneyes	1	
Alosidae	Shads	5	2
Dorosomatidae	Thread Herrings	2	3
Engraulidae	Anchovies	1	5
Catostomidae	Suckers	29	
Cobitidae	Spined Loaches	1	
Cyprinidae	Carps	3	
Xenocyprididae	Sharpbellies	1	
Leuciscidae	Minnows	68	
Loricariidae	Suckermouth Armored Catfishes	1	
Ictaluridae	North American Catfishes	18	
Esocidae	Pikes	3	
Umbridae	Mudminnows	1	
Salmonidae	Trouts and Salmons	4	

TABLE 1. (*continued*)

Family	Common Name	No. of Described and Undescribed Freshwater Species	No. of Additional Marine and Estuarine Species in North Carolina
Aphredoderidae	Pirate Perches	1	
Amblyopsidae	Cavefishes	1	
Eleotridae	Sleepers	2	1
Gobiidae	Gobies	3	23
Paralichthyidae	Sand Flounders	1	19
Achiridae	American Soles	1	1
Cichlidae	Cichlids	2	
Atherinopsidae	New World Silversides	4	2
Fundulidae	Topminnows	10	2
Cyprinodontidae	Pupfishes	1	
Poeciliidae	Livebearers	4	
Belonidae	Needlefishes	1	4
Mugilidae	Mullets	2	1
Moronidae	Temperate Basses	3[b]	
Sciaenidae	Drums and Croakers	2	16
Percidae	Perches and Darters	40[c]	
Gasterosteidae	Sticklebacks	1	
Cottidae	Sculpins	3	
Centrarchidae	Sunfishes	23	
Elassomatidae	Pygmy Sunfishes	3	

[a] Currently extirpated from the state.
[b] A hybrid species is also stocked and not included in this number.
[c] Number does not include two extirpated species.

TABLE 2. Phylogenetic listing of freshwater fishes by river basin

I = Indigenous (native), IB = Indigenous but not in this basin, NI = Nonindigenous (introduced), and E = Extirpated

Family, Scientific Name	Mountain								Piedmont							Coastal						Total Basin Number of Occurrences[a]
	Hiwassee	Little Tennessee	Savannah	Pigeon	French Broad	Nolichucky	Watauga	New	Broad	Catawba	Yadkin	Cape Fear	Neuse	Tar	Roanoke	Chowan	Albemarle Sound Rivers	White Oak	Shallotte	Waccamaw	Lumber	
Petromyzontidae | Lampreys																						
Ichthyomyzon bdellium Ohio Lamprey	I	I				I																3
Ichthyomyzon greeleyi Mountain Brook Lamprey	I	I			I	I																4
Lampetra aepyptera Least Brook Lamprey													I	I								2
Lethenteron appendix American Brook Lamprey					I										I							2
Petromyzon marinus Sea Lamprey											I	I	I	I	I	I	I		I			8
Acipenseridae | Sturgeons																						
Acipenser brevirostrum Shortnose Sturgeon											I	I				I	I					4

[a] Extirpated and extant basin occurrences.

Species																#
Acipenser fulvescens — Lake Sturgeon	IB															1
Acipenser oxyrinchus — Atlantic Sturgeon			I							I					I	9
**Polyodontidae	Paddlefishes**																
Polyodon spathula — Paddlefish			E													1
**Lepisosteidae	Gars**																
Lepisosteus osseus — Longnose Gar	I		I			I	I	I	I	I	I		I	I	I	15
**Amiidae	Bowfins**																
Amia calva — Bowfin						I	I	I	I	I	I		I	I	I	11
**Anguillidae	Freshwater Eels**																
Anguilla rostrata — American Eel						I	I	I	I	I	I		I	I	I	12
**Hiodontidae	Mooneyes**																
Hiodon tergisus — Mooneye			I													1
**Alosidae	Shads**																
Alosa aestivalis — Blueback Herring	IB	IB	IB		IB	IB	I	I	I	I	I		I	I	I	14
Alosa mediocris — Hickory Shad						I	I	I	I	I	I		I	I	I	7

| 5

TABLE 2. (*continued*)

I = Indigenous (native), IB = Indigenous but not in this basin, NI = Nonindigenous (introduced), and E = Extirpated

Family, Scientific Name	Mountain								Piedmont							Coastal						Total Basin Number of Occurrences[a]
	Hiwassee	Little Tennessee	Savannah	Pigeon	French Broad	Nolichucky	Watauga	New	Broad	Catawba	Yadkin	Cape Fear	Neuse	Tar	Roanoke	Chowan	Albemarle Sound Rivers	White Oak	Shallotte	Waccamaw	Lumber	
Alosa pseudoharengus Alewife									I	IB		I	I	I	I	I	I	I	I			9
Alosa sapidissima American Shad											I	I	I	I	I	I	I	I	I	I	I	11
Brevoortia tyrannus Atlantic Menhaden												I	I	I	I	I	I	I	I			8
**Dorosomatidae	Thread Herrings**																						
Dorosoma cepedianum Gizzard Shad	I	I		I	I	I			NI	NI	NI	NI	NI	NI	NI	NI	NI	NI	NI	I	I	18
Dorosoma petenense Threadfin Shad	NI	NI			NI				NI	NI	NI	NI	NI	NI	NI	NI	NI	NI	NI			14
**Engraulidae	Anchovies**																						
Anchoa mitchilli Bay Anchovy												I	I	I	I	I	I	I	I			8

[a] Extirpated and extant basin occurrences.

Catostomidae	Suckers		

Catostomidae | Suckers

Species		Count
Carpiodes carpio River Carpsucker		2
Carpiodes cyprinus Quillback		4
Carpiodes sp. "Atlantic" Highfin Carpsucker		3
Carpiodes sp. "Carolina" Quillback		3
Catostomus commersonii White Sucker		15
Erimyzon oblongus Eastern Creek Chubsucker		13
Erimyzon sucetta Lake Chubsucker		11
Hypentelium nigricans Northern Hog Sucker		14
Hypentelium roanokense Roanoke Hog Sucker		2
Ictiobus bubalus Smallmouth Buffalo		4
Ictiobus cyprinellus Bigmouth Buffalo		2
Ictiobus niger Black Buffalo		1

TABLE 2. (continued)

I = Indigenous (native), IB = Indigenous but not in this basin, NI = Nonindigenous (introduced), and E = Extirpated

Family, Scientific Name	Hiwassee	Little Tennessee	Savannah	Pigeon	French Broad	Nolichucky	Watauga	New	Broad	Catawba	Yadkin	Cape Fear	Neuse	Tar	Roanoke	Chowan	Albemarle Sound Rivers	White Oak	Shallotte	Waccamaw	Lumber	Total Basin Number of Occurrences[a]
Minytrema melanops Spotted Sucker	I	I																I	I	I	I	6
Moxostoma anisurum Silver Redhorse	I	I			I	I																4
Moxostoma ariommum Bigeye Jumprock															I							1
Moxostoma breviceps Smallmouth Redhorse	I	I		I	I	I																5
Moxostoma carinatum River Redhorse	I	I		I	I																	4
Moxostoma cervinum Blacktip Jumprock													I	I	I							3
Moxostoma collapsum Notchlip Redhorse									I	I	I	I	I	I	I	I						8

[a] Extirpated and extant basin occurrences.

8 |

Species														Count
Moxostoma duquesnei Black Redhorse	I	I						I						7
Moxostoma erythrurum Golden Redhorse	I	I	I								I			6
Moxostoma macrolepidotum Shorthead Redhorse										I	I			8
Moxostoma pappillosum V-lip Redhorse					I				I	I	I	I	I	8
Moxostoma robustum Robust Redhorse								E						2
Moxostoma rupiscartes Striped Jumprock			I		I		IB	I		I	I			4
Moxostoma sp. "Brassy" Jumprock					I		I	I		I	I			4
Moxostoma sp. "Carolina" Redhorse							I	I						2
Moxostoma sp. "Sicklefin" Redhorse	I													2
Thoburnia hamiltoni Rustyside Sucker									I					1
**Cobitidae	Spined Loaches**														
Misgurnus anguillicaudatus Pond Loach							NI							1

TABLE 2. *(continued)*

I = Indigenous (native), **IB** = Indigenous but not in this basin, **NI** = Nonindigenous (introduced), and **E** = Extirpated

Family, Scientific Name	Mountain								Piedmont								Coastal					Total Basin Number of Occurrences[a]
	Hiwassee	Little Tennessee	Savannah	Pigeon	French Broad	Nolichucky	Watauga	New	Broad	Catawba	Yadkin	Cape Fear	Neuse	Tar	Roanoke	Chowan	Albemarle Sound Rivers	White Oak	Shallotte	Waccamaw	Lumber	
Cyprinidae | Carps																						
Carassius auratus Goldfish	NI	NI		NI	NI				NI	NI	NI	NI	NI	NI	NI	NI		NI			NI	14
Cyprinus carpio Common Carp	NI	NI	NI	NI	NI	NI	NI	NI	NI	NI	NI	NI	NI	NI	NI	NI	NI	NI	NI	NI	NI	21
Cyprinus rubrofuscus Koi												NI	NI		NI							3
Xenocyprididae | Sharpbellies																						
Ctenopharyngodon idella Grass Carp			NI	NI	NI			NI	NI	NI	NI	NI	NI	NI	NI	NI	NI	NI	NI	NI	NI	17
Leuciscidae | Minnows																						
Alburnops chalybaeus Ironcolor Shiner												I	I	I	I	I	I	I	I	I	I	10

[a] Extirpated and extant basin occurrences.

10

Species															Total
Alburnops petersoni Coastal Shiner	I	I	I	I	I	I	I	I	I				I	I	I	7
Campostoma anomalum Central Stoneroller	I	I	I	I	I	I	I	NI	I	I						12
Chrosomus oreas Mountain Redbelly Dace	IB			IB	IB	IB	IB	IB	IB	I	I					9
Clinostomus funduloides Rosyside Dace	IB	I	IB	IB	IB	I	I	IB	IB	I	I	I				12
Clinostomus sp. "Hiwassee" Dace	I															1
Clinostomus sp. "Smoky" Dace		I														1
Coccotis coccogenis Warpaint Shiner	I	I	I	I	I	IB	IB	IB	I/IB[b]	I	I					11
Cyprinella analostana Satinfin Shiner									I	I	I	I	I			7
Cyprinella chloristia Greenfin Shiner						I	I	I	IB							3
Cyprinella galactura Whitetail Shiner		I	I	I	IB	I	I									9
Cyprinella labrosa Thicklip Chub		I	I	I	I	I		I								3
Cyprinella lutrensis Red Shiner							NI	NI	NI	NI						3

[b] *Coccotis coccogenis* is indigenous (I) in the Linville River watershed, but nonindigenous (IB) in other parts of the Catawba basin.

TABLE 2. (continued)

I = Indigenous (native), IB = Indigenous but not in this basin, NI = Nonindigenous (introduced), and E = Extirpated

Family, Scientific Name	Mountain								Piedmont								Coastal					Total Basin Number of Occurrences[a]
	Hiwassee	Little Tennessee	Savannah	Pigeon	French Broad	Nolichucky	Watauga	New	Broad	Catawba	Yadkin	Cape Fear	Neuse	Tar	Roanoke	Chowan	Albemarle Sound Rivers	White Oak	Shallotte	Waccamaw	Lumber	
Cyprinella nivea Whitefin Shiner			I						I	I	I	I										5
Cyprinella pyrrhomelas Fieryblack Shiner									I	I	I											3
Cyprinella spiloptera Spotfin Shiner	E			I	I	I		I														5
Cyprinella zanema Santee Chub									I	I												2
Cyprinella sp. "Thinlip" Chub									I		I	I										3
Erimonax monachus Spotfin Chub		I			E																	2
Erimystax insignis Blotched Chub	I				I	I																3

[a] Extirpated and extant basin occurrences.

12

Species														Total
Exoglossum laurae Tonguetied Minnow								I						1
Exoglossum maxillingua Cutlip Minnow					I									1
Hudsonius altipinnis Highfin Shiner	I	I	I	I	I	I	I	I						8
Hudsonius hudsonius Spottail Shiner		I	I	I	I	I	I	I	I	I				9
Hybognathus regius Eastern Silvery Minnow	E	I	I	I	I	I	I	I	I	I				10
Hybopsis amblops Bigeye Chub										I	I	I	I	4
Hybopsis hypsinotus Highback Chub						IB	I	I	IB					4
Hybopsis rubrifrons Rosyface Chub											I			1
Hydrophlox chiliticus Redlip Shiner						IB	IB	IB	IB	I	I			6
Hydrophlox chlorocephalus Greenhead Shiner								I						1
Hydrophlox lutipinnis Yellowfin Shiner											IB	I		2
Hydrophlox rubricroceus Saffron Shiner						I	I	IB	I	IB	I	IB	I	8

TABLE 2. (continued)

I = Indigenous (native), **IB** = Indigenous but not in this basin, **NI** = Nonindigenous (introduced), and **E** = Extirpated

Family, Scientific Name	Hiwassee	Little Tennessee	Savannah	Pigeon	French Broad	Nolichucky	Watauga	New	Broad	Catawba	Yadkin	Cape Fear	Neuse	Tar	Roanoke	Chowan	Albemarle Sound Rivers	White Oak	Shallotte	Waccamaw	Lumber	Total Basin Number of Occurrences[a]
Hydrophlox sp. "Piedmont" Shiner									I													1
Luxilus albeolus White Shiner								I				I	I	I	I	I						6
Luxilus cerasinus Crescent Shiner												IB			I	I						3
Luxilus chrysocephalus Striped Shiner				I	I	I																4
Lythrurus ardens Rosefin Shiner											IB	IB			I							3
Lythrurus matutinus Pinewoods Shiner											I	I	I	I								2
Miniellus alborus Whitemouth Shiner										IB	I	I			I							4

a Extirpated and extant basin occurrences.

14

Species	Count
Miniellus mekistocholas Cape Fear Shiner	1
Miniellus procne Swallowtail Shiner	9
Miniellus scabriceps New River Shiner	1
Nocomis leptocephalus Bluehead Chub	14
Nocomis micropogon River Chub	8
Nocomis platyrhynchus Bigmouth Chub	1
Nocomis raneyi Bull Chub	4
Notemigonus crysoleucas Golden Shiner	21
Notropis amoenus Comely Shiner	8
Notropis bifrenatus Bridle Shiner	2
Notropis maculatus Taillight Shiner	4
Notropis micropteryx Highland Shiner	5

TABLE 2. (continued)

I = Indigenous (native), IB = Indigenous but not in this basin, NI = Nonindigenous (introduced), and E = Extirpated

Family, Scientific Name	Mountain								Piedmont							Coastal						Total Basin Number of Occurrences[a]
	Hiwassee	Little Tennessee	Savannah	Pigeon	French Broad	Nolichucky	Watauga	New	Broad	Catawba	Yadkin	Cape Fear	Neuse	Tar	Roanoke	Chowan	Albemarle Sound Rivers	White Oak	Shallotte	Waccamaw	Lumber	
Notropis photogenis Silver Shiner	I	I		I	I	I		I														6
Notropis scepticus Sandbar Shiner									I	I	I	I										4
Notropis telescopus Telescope Shiner	I	I		I	I	I				I												6
Notropis sp. "Kanawha" Rosyface Shiner								I														1
Paranotropis leuciodus Tennessee Shiner	I	I	I	I	I	I	I	IB		I												9
Paranotropis spectrunculus Mirror Shiner	I	I		I	I	I	I		IB	I												8
Paranotropis volucellus Mimic Shiner		IB			I	I		I					I	I	I							7

[a] Extirpated and extant basin occurrences.

Species		...	Count
Phenacobius crassilabrum Fatlips Minnow	NI	I	3
Phenacobius teretulus Kanawha Minnow		I	1
Pimephales notatus Bluntnose Minnow	I	I	4
Pimephales promelas Fathead Minnow	NI	NI	13
Pteronotropis cummingsae Dusky Shiner			9
Rhinichthys atratulus Blacknose Dace			1
Rhinichthys cataractae Longnose Dace	I		9
Rhinichthys obtusus Western Blacknose Dace	I		11
Semotilus atromaculatus Creek Chub	I		16
Semotilus lumbee Sandhills Chub			3
Loricariidae | Suckermouth Armored Catfishes			
Pterygoplichthys pardalis Amazon Sailfin Catfish		NI	1

TABLE 2. (continued)

I = Indigenous (native), **IB** = Indigenous but not in this basin, **NI** = Nonindigenous (introduced), and **E** = Extirpated

Family, Scientific Name	Mountain								Piedmont							Coastal						Total Basin Number of Occurrences[a]
	Hiwassee	Little Tennessee	Savannah	Pigeon	French Broad	Nolichucky	Watauga	New	Broad	Catawba	Yadkin	Cape Fear	Neuse	Tar	Roanoke	Chowan	Albemarle Sound Rivers	White Oak	Shallotte	Waccamaw	Lumber	
Ictaluridae | North American Catfishes																						
Ameiurus brunneus Snail Bullhead	IB	IB	I		IB				I	I	I	I	I	I	IB	I	I	I			I	11
Ameiurus catus White Catfish				IB	IB			IB	I	I	I	I	I	I	I	I	I	I	I	I	I	16
Ameiurus melas Black Bullhead										NI	NI				NI							3
Ameiurus natalis Yellow Bullhead	IB								I	I	I	I	I	I	I	I	I	I	I	I	I	13
Ameiurus nebulosus Brown Bullhead	I	I		I	I	I	IB	IB	I	I	I	I	I	I	I	I	I	I	I	I	I	19
Ameiurus platycephalus Flat Bullhead	IB	IB		IB	IB	IB	IB		I	I	I	I	I	I	I	I	I	I	I	I	I	17

[a] Extirpated and extant basin occurrences.

18

Ictalurus furcatus Blue Catfish		**I**								**NI**	**NI**	**NI**	**NI**	**NI**	**NI**	**NI**	**NI**	**NI**		**NI**	**NI**	**12**
Ictalurus punctatus Channel Catfish	**I**	**I**		**I**	**I**	**I**		**I**	**IB**	**IB**	**IB**	**IB**	**IB**	**IB**	**IB**	**IB**	**IB**			**IB**	**IB**	**17**
Noturus eleutherus Mountain Madtom					**I**																	**1**
Noturus flavus Stonecat		**I**			**I**	**I**																**3**
Noturus furiosus Carolina Madtom													**I**	**I**								**2**
Noturus gilberti Orangefin Madtom															**I**							**1**
Noturus gyrinus Tadpole Madtom											**I**	**I**	**I**	**I**	**I**	**I**	**I**	**I**	**I**	**I**	**I**	**11**
Noturus insignis Margined Madtom			**I**				**IB**	**IB**	**I**	**I**	**I**	**I**	**I**	**I**	**I**	**I**	**I**	**I**	**I**	**I**	**I**	**16**
Noturus sp. "Cape Fear Broadtail" Madtom												**I**										**1**
Noturus sp. "Lake Waccamaw Broadtail" Madtom																				**I**		**1**
Noturus sp. "Pee Dee Broadtail" Madtom																				**I**	**I**	**2**
Pylodictis olivaris Flathead Catfish	**I**	**I**		**I**	**I**	**I**		**I**	**IB**	**IB**	**IB**	**IB**	**IB**	**IB**	**IB**			**IB**		**IB**	**IB**	**16**

TABLE 2. (*continued*)

I = Indigenous (native), **IB** = Indigenous but not in this basin, **NI** = Nonindigenous (introduced), and **E** = Extirpated

Family, Scientific Name	Mountain									Piedmont						Coastal						Total Basin Number of Occurrences[a]
	Hiwassee	Little Tennessee	Savannah	Pigeon	French Broad	Nolichucky	Watauga	New	Broad	Catawba	Yadkin	Cape Fear	Neuse	Tar	Roanoke	Chowan	Albemarle Sound Rivers	White Oak	Shallotte	Waccamaw	Lumber	
Esocidae | Pikes																						
Esox americanus Redfin Pickerel										I	I	I	I	I	I	I	I	I	I	I	I	12
Esox masquinongy Muskellunge	I	I			I	I		IB	IB													6
Esox niger Chain Pickerel					IB						I	I	I	I	I	I	I	I	I	I	I	12
Umbridae | Mudminnows																						
Umbra pygmaea Eastern Mudminnow											I	I	I	I	I	I	I	I	I	I	I	11
Salmonidae | Trouts and Salmons																						
Oncorhynchus mykiss Rainbow Trout	NI	NI	NI	NI	NI	NI	NI	NI	NI	NI	NI				NI							12

[a] Extirpated and extant basin occurrences.

Species		Total
Oncorhynchus nerka Sockeye Salmon	NI	1
Salmo trutta Brown Trout	NI NI NI NI NI NI NI NI NI NI	12
Salvelinus fontinalis Brook Trout	I I I I I IB IB I I I	12
Aphredoderidae | Pirate Perches		
Aphredoderus sayanus Pirate Perch	I I I I I I I I	12
Amblyopsidae | Cavefishes		
Chologaster cornuta Swampfish	I I I I I I I	10
Eleotridae | Sleepers		
Dormitator maculatus Fat Sleeper	I I I I	4
Eleotris amblyopsis Largescaled Spinycheek Sleeper	I I I I	4
Gobiidae | Gobies		
Awaous banana River Goby	NI NI	2
Ctenogobius shufeldti Freshwater Goby	I I I I I	5
Evorthodus lyricus Lyre Goby	I I I I	4

TABLE 2. (continued)

I = Indigenous (native), IB = Indigenous but not in this basin, NI = Nonindigenous (introduced), and E = Extirpated

Family, Scientific Name	Hiwassee	Little Tennessee	Savannah	Pigeon	French Broad	Nolichucky	Watauga	New	Broad	Catawba	Yadkin	Cape Fear	Neuse	Tar	Roanoke	Chowan	Albemarle Sound Rivers	White Oak	Shallotte	Waccamaw	Lumber	Total Basin Number of Occurrences[a]
**Paralichthyidae	Sand Flounders**																						
Paralichthys lethostigma Southern Flounder												I	I	I	I	I	I	I	I			8
**Achiridae	American Soles**																						
Trinectes maculatus Hogchoker											I	I	I	I	I	I	I	I	I		I	10
**Cichlidae	Cichlids**																						
Coptodon zillii Redbelly Tilapia												NI			NI							2
Oreochromis aureus Blue Tilapia					NI					NI					NI							3

[a] Extirpated and extant basin occurrences.

Atherinopsidae | New World Silversides

Species									
Labidesthes sicculus Brook Silverside	–								1
Labidesthes vanhyningi Green Silverside							–	–	4
Menidia beryllina Inland Silverside		IB	–	–	–	–	–		9
Menidia extensa Waccamaw Silverside							–		1

Fundulidae | Topminnows

Fundulus chrysotus Golden Topminnow							–		1
Fundulus confluentus Marsh Killifish				–	–	–			5
Fundulus diaphanus Banded Killifish				–	–	–			6
Fundulus heteroclitus Mummichog			–	–	–	–	IB		8
Fundulus lineolatus Lined Topminnow				–	–	–	–	–	11
Fundulus rathbuni Speckled Killifish		IB	–	–					5
Fundulus waccamensis Waccamaw Killifish							–		1

TABLE 2. (continued)

I = Indigenous (native), IB = Indigenous but not in this basin, NI = Nonindigenous (introduced), and E = Extirpated

Family, Scientific Name	Hiwassee	Little Tennessee	Savannah	Pigeon	French Broad	Nolichucky	Watauga	New	Broad	Catawba	Yadkin	Cape Fear	Neuse	Tar	Roanoke	Chowan	Albemarle Sound Rivers	White Oak	Shallotte	Waccamaw	Lumber	Total Basin Number of Occurrences[a]
												Piedmont							Coastal			
	Mountain								Piedmont							Coastal						
Fundulus sp. "Lake Phelps" Killifish												IB					I					2
Lucania goodei Bluefin Killifish												NI										1
Lucania parva Rainwater Killifish												I	I	I		I		I	I			7
**Cyprinodontidae	Pupfishes**																						
Cyprinodon variegatus Sheepshead Minnow												I	I	I		I		I	I			7
**Poeciliidae	Livebearers**																						
Gambusia affinis Western Mosquitofish	I	I			I																	3
Gambusia holbrooki Eastern Mosquitofish					IB				I	I	I	I	I	I	I	I		I	I	I	I	14

[a] Extirpated and extant basin occurrences.

Species		Value
Heterandria formosa	Least Killifish	2
Poecilia latipinna	Sailfin Molly	3
Belonidae | Needlefishes		
Strongylura marina	Atlantic Needlefish	9
Mugilidae | Mullets		
Dajaus monticola	Mountain Mullet	2
Mugil cephalus	Striped Mullet	9
Moronidae | Temperate Basses		
Morone americana	White Perch	11
Morone chrysops	White Bass	9
Morone saxatilis	Striped Bass	12
Sciaenidae | Drums and Croakers		
Aplodinotus grunniens	Freshwater Drum	2
Leiostomus xanthurus	Spot	7

TABLE 2. *(continued)*

I = Indigenous (native), IB = Indigenous but not in this basin, NI = Nonindigenous (introduced), and E = Extirpated

Family, Scientific Name	Mountain								Piedmont							Coastal						Total Basin Number of Occurrences[a]
	Hiwassee	Little Tennessee	Savannah	Pigeon	French Broad	Nolichucky	Watauga	New	Broad	Catawba	Yadkin	Cape Fear	Neuse	Tar	Roanoke	Chowan	Albemarle Sound Rivers	White Oak	Shallotte	Waccamaw	Lumber	
**Percidae	Perches and Darters**																						
Etheostoma blennioides Greenside Darter	I	I	I		I	I		I														6
Etheostoma brevispinum Carolina Fantail Darter									I	I	I											3
Etheostoma collis Carolina Darter										I	I	I	I	I	I							6
Etheostoma flabellare Fantail Darter				IB	IB			I	I	I	I	I	I	I	I							8
Etheostoma fusiforme Swamp Darter								I	I	I	I	I	I	I	I	I	I	I	I	I	I	14
Etheostoma gutselli Tuckasegee Darter		I		I																		2

a Extirpated and extant basin occurrences.

Species	Common Name																							Total
Etheostoma inscriptum	Turquoise Darter	I																						1
Etheostoma jessiae	Blueside Darter		E																					1
Etheostoma kanawhae	Kanawha Darter										I													1
Etheostoma maculaticeps	Southern Tessellated Darter				IB	I	I	I					I	I	I							8
Etheostoma mariae	Pinewoods Darter																				I			1
Etheostoma perlongum	Waccamaw Darter																					I		1
Etheostoma podostemone	Riverweed Darter									I														1
Etheostoma serrifer	Sawcheek Darter						I	I	I				I	I	I	I			I	I	I			11
Etheostoma simoterum	Snubnose Darter			I	I																			2
Etheostoma swannanoa	Swannanoa Darter			I	I	I																		3
Etheostoma thalassinum	Seagreen Darter									I														2
Etheostoma vitreum	Glassy Darter												I		I		I							4

TABLE 2. (continued)

I = Indigenous (native), IB = Indigenous but not in this basin, NI = Nonindigenous (introduced), and E = Extirpated

Family, Scientific Name	Mountain								Piedmont							Coastal						Total Basin Number of Occurrences[a]
	Hiwassee	Little Tennessee	Savannah	Pigeon	French Broad	Nolichucky	Watauga	New	Broad	Catawba	Yadkin	Cape Fear	Neuse	Tar	Roanoke	Chowan	Albemarle Sound Rivers	White Oak	Shallotte	Waccamaw	Lumber	
Etheostoma zonale Banded Darter	I	I		I	I	I																5
Etheostoma sp. "Tessellated" Darter													I	I	I	I	I	I				6
Nothonotus acuticeps Sharphead Darter						I																1
Nothonotus chlorobranchius Greenfin Darter		I		I	I	I	I															5
Nothonotus ruflineatus Redline Darter	I	IB		I	I																	4
Nothonotus vulneratus Wounded Darter		I			E																	2

[a] Extirpated and extant basin occurrences.

Species (common name)																		Total
Perca flavescens Yellow Perch	IB	IB		IB	IB	—	—	—	—	—	—	—	—	—	—	—	—	18
Percina aurantiaca Tangerine Darter	—	—	—	—	—	—												6
Percina burtoni Blotchside Logperch			E	—														2
Percina caprodes Logperch		—	—	—														3
Percina crassa Piedmont Darter							—	—	—	—	—							5
Percina evides Gilt Darter	—	—	—	—														5
Percina gymnocephala Appalachia Darter					IB													1
Percina nevisensis Chainback Darter							—	—	—	—								4
Percina oxyrhynchus Sharpnose Darter					IB													1
Percina rex Roanoke Logperch								—										1
Percina roanoka Roanoke Darter							—	—	—	—								4
Percina squamata Olive Darter	—	—	—	—												—		5

TABLE 2. (continued)

I = Indigenous (native), IB = Indigenous but not in this basin, NI = Nonindigenous (introduced), and E = Extirpated

Family, Scientific Name	Mountain								Piedmont							Coastal						Total Basin Number of Occurrences[a]
	Hiwassee	Little Tennessee	Savannah	Pigeon	French Broad	Nolichucky	Watauga	New	Broad	Catawba	Yadkin	Cape Fear	Neuse	Tar	Roanoke	Chowan	Albemarle Sound Rivers	White Oak	Shallotte	Waccamaw	Lumber	
Percina westfalli Sooty-banded Darter			I																			1
Percina williamsi Sickle Darter					E																	1
Sander canadensis Sauger	I			I	I					IB												4
Sander vitreus Walleye	I	I		I	I				IB	IB	IB		E		I	I	I					11
**Gasterosteidae	Sticklebacks**																						
Apeltes quadracus Fourspine Stickleback													NI	NI			NI					3

[a] Extirpated and extant basin occurrences.

Family / Species	Common name	Distribution	No.
Cottidae | Sculpins			
Cottus bairdii	Mottled Sculpin	I I I I I I I IB	9
Cottus caeruleomentum	Blue Ridge Sculpin	I	1
Cottus carolinae	Banded Sculpin	I I	2
Centrarchidae | Sunfishes			
Acantharchus pomotis	Mud Sunfish	I I I IB	11
Ambloplites cavifrons	Roanoke Bass	IB IB I I	5
Ambloplites rupestris	Rock Bass	I IB IB IB IB IB	12
Centrarchus macropterus	Flier	I I	11
Enneacanthus chaetodon	Blackbanded Sunfish	I	8
Enneacanthus gloriosus	Bluespotted Sunfish	I	11
Enneacanthus obesus	Banded Sunfish	I	10
Lepomis auritus	Redbreast Sunfish	IB IB I IB IB I IB I	21

TABLE 2. (continued)

I = Indigenous (native), **IB** = Indigenous but not in this basin, **NI** = Nonindigenous (introduced), and **E** = Extirpated

Family, Scientific Name	Mountain								Piedmont							Coastal						Total Basin Number of Occurrences[a]
	Hiwassee	Little Tennessee	Savannah	Pigeon	French Broad	Nolichucky	Watauga	New	Broad	Catawba	Yadkin	Cape Fear	Neuse	Tar	Roanoke	Chowan	Albemarle Sound Rivers	White Oak	Shallotte	Waccamaw	Lumber	
Lepomis cyanellus Green Sunfish	NI	NI	—	NI	NI	NI	NI	NI	NI	NI	NI	NI	NI	NI	NI	NI	NI	—	—	—	—	16
Lepomis gibbosus Pumpkinseed	IB	IB	IB	IB	IB	IB	—	IB	I	I	I	I	I	I	I	I	NI	—	—	NI	—	17
Lepomis gulosus Warmouth	IB	IB	IB	IB	IB	I	—	—	I	I	I	I	I	I	I	I	—	I	I	I	I	18
Lepomis macrochirus Bluegill	I	I	I	I	I	I	I	IB	I	I	I	I	IB	IB	IB	IB	IB	I	I	I	I	21
Lepomis marginatus Dollar Sunfish											—	—	I	I	I	I		I	I	I	NI	8
Lepomis microlophus Redear Sunfish	NI	NI	—	—	NI	—	—	NI	NI	NI	NI	NI	NI	NI	NI	NI	NI	NI	NI	NI	NI	17
Lepomis punctatus Spotted Sunfish											—	I						I	I	I	I	5

[a] Extirpated and extant basin occurrences.

32

Micropterus coosae Redeye Bass	NI													1
Micropterus dolomieu Smallmouth Bass	I	IB				NI		I		IB		IB		12
Micropterus henshalli Alabama Bass	NI	NI		NI		I		NI		NI		NI	NI	9
Micropterus punctulatus Spotted Bass	I			I		IB		IB	NI	IB		IB	IB	11
Micropterus salmoides Largemouth Bass	I		IB	I	IB	I	I	I	I	I	I	I	I	21
Micropterus sp. "Barttram's" Bass			I		IB	IB								2
Pomoxis annularis White Crappie	I			I		IB	IB	IB	IB	IB	IB	IB		14
Pomoxis nigromaculatus Black Crappie	I			I	IB	I	I	I	I	I	I	I	I	18
Elassomatidae | Pygmy Sunfishes														
Elassoma boehlkei Carolina Pygmy Sunfish												I	I	1
Elassoma evergladei Everglades Pygmy Sunfish					I							I	I	4
Elassoma zonatum Banded Pygmy Sunfish					I							I	I	8

TABLE 2. (*continued*)

I = Indigenous (native), **IB** = Indigenous but not in this basin, **NI** = Nonindigenous (introduced), and **E** = Extirpated

Family, Scientific Name	Mountain								Piedmont								Coastal					Total Basin Number of Occurrences[a]
	Hiwassee	Little Tennessee	Savannah	Pigeon	French Broad	Nolichucky	Watauga	New	Broad	Catawba	Yadkin	Cape Fear	Neuse	Tar	Roanoke	Chowan	Albemarle Sound Rivers	White Oak	Shallotte	Waccamaw	Lumber	
Total Number of Species	70	77	38	63	103	65	30	56	71	99[b]	117	127	109	100	123	85	77	73	64	67	65	
No. of Indigenous Species (= I + E)	51	51	28	49	76	54	19	28	42	65	82	101	93	87	96	75	67	65	60	57	54	
No. of Nonindigenous Species (= IB + NI)	19	26	10	14	27	11	11	28	29	35	35	26	16	13	27	10	10	8	4	10	11	

[a] Extirpated and extant basin occurrences.

[b] *Coccotis coccogenis* is indigenous (I) in the Linville River watershed, but nonindigenous (IB) in other parts of the Catawba basin.

Note: Additional estuarine species found within identifications keys.

34

A Chronological History of Ichthyological Surveys and Collections in North Carolina

Our state's rich ichthyological history harks back to 1682 when Thomas Ash wrote a general description of the fish fauna of "Carolina," which referred to all the coastal lands between Florida and Virginia (Ash, 1682; Tracy, Rohde, and Hogue, 2020). In the roughly 340 years since then, ichthyologists from near and far have surveyed, collected, and published on our diverse freshwater and estuarine fauna (table 3). The first state-specific checklist was provided by John Lawson (1709, 152–60), and it was received in such high regard that it was later plagiarized extensively by Brickell (1737) when he wrote *The Natural History of North-Carolina*. For the past 150 years, checklists and publications detailing the fishes of the state have appeared regularly, beginning with Cope (1870b) and continuing with Jordan (1889b), Jordan and Evermann (1896–1900), Smith (1907), Jordan and Clark (1930), Fowler (1945), Louder (1962a), and Ratledge, Carnes, and Collins (1966). Menhinick, Burton, and Bailey (1974) published an annotated checklist, which relied heavily upon Randall (1957), the North Carolina Wildlife Resources Commission's (NCWRC) 1960s stream survey data (Starnes and Hogue, 2011), and Joseph R. Bailey's (Duke University) unpublished survey data from 1947 and 1949 of the Hiwassee, Little Tennessee, Savannah, Pigeon, French Broad, Nolichucky, Watauga, New, and Yadkin basins. Later, Menhinick published *The Freshwater Fishes of North Carolina* (1991) using datasets from the NCWRC and Bailey, distributional maps from Lee et al. (1980), the 1974 checklist, unpublished manuscripts archived at the North Carolina Museum of Natural Sciences, and his and other researchers' personal collections. Although the distributional maps and some of the taxonomic nomenclature are now outdated, the book is still popular and widely in use. More recent field guides, including identifying characteristics, illustrations, photographs, and distributional maps, have appeared (for example, Page and Burr, 2011; Rohde et al., 1994). In 2020, Tracy, Rohde, and Hogue (2020) published an updated and annotated atlas of the state's indigenous and nonindigenous freshwater fish fauna, which utilized the tremendous amount of accessible data from museums and state agencies to create the most up-to-date distributional picture of North Carolina's freshwater fish fauna.

Smith's 1907 publication *The Fishes of North Carolina* was truly the first publication that included identification keys for 345 fresh, brackish, and saltwater species and their abundances, distributions, habitats, migrations, spawning, and food value. In the book's preface, Joseph Hyde Pratt, state geologist, stated that the goal of Smith's book was to create:

a deeper interest in the welfare of both fishes and fishermen, and a better understanding of the condition and needs of the fishing industry, with a view to placing this important branch on a permanent basis and making it yield an increasing revenue to both State and people.... It is hoped that this volume will be the means of creating such an interest in the fisheries that suitable laws for their protection may be enacted as needed, and that the State officers charged with the administration of the fisheries may have the sympathy and cooperation of all citizens. (iv)

In this publication, Smith described a new species, a goby, *Microgobius holmesi*, and gave it the common name of Holmes' Goby (figure 1). An interesting fact was that the goby was named after Professor J. A. Holmes, a former state geologist and director of the North Carolina Geological and Economic Survey, who had requested that Smith produce this guide to the fishes of North Carolina. The species was known from a single specimen collected in 1904 from Uncle Israel Shoal in Beaufort Harbor. Unfortunately, this species and the other species of *Microgobius*, *M. eulepis* Eigenmann and Eigenmann, which Smith listed as occurring in Beaufort Harbor at the same shoal, were later synonymized by Birdsong (1981) with Green Goby, *M. thalassinus*.

Prior to Smith (1907), twenty species had been described from North Carolina, including one catfish, five darters, nine minnows, four suckers, and one topminnow (table 4). The first species described from North Carolina was *Clinostomus carolinus* (Girard, 1856, p. 212). However, because this species was later synonymized with the Rosyside Dace, *Clinostomus funduloides* Girard, 1856, by Lachner and Deubler (1960), the Bluehead Chub, *Nocomis leptocephalus*

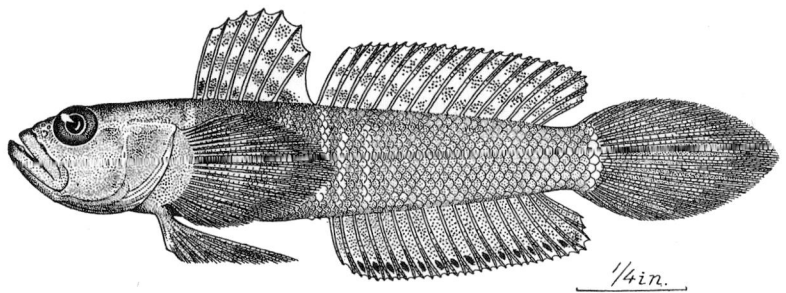

FIGURE 1. *Microgobius holmesi* Smith 1907, Holmes' Goby.
Illustration adapted from Smith (1907).

TABLE 3. Important milestones in the ichthyological history of North Carolina since the 1850s

Time Period	Ichthyological Contribution
1856	Descriptions of the first freshwater fish species (Rosyside Dace, *Clinostomus carolinus* [later considered a synonym of *Clinostomus funduloides*], and Bluehead Chub, *Nocomis leptocephalus*) from NC are made (Girard 1856)
1870	Edward Drinker Cope visits NC's streams and surveys their fauna (Cope 1870a, 1870b; Tracy and Jenkins 2021)
1889	David Starr Jordan and companions visit NC's streams and survey their fauna (Jordan 1889a, 1889b)
1896	"The Fishes of the Neuse River Basin" (Evermann and Cox 1896)
1907	*The Fishes of North Carolina* (Smith 1907)—the first book on all the known species of fishes across NC with a strong emphasis on the commercial fisheries resources of the Albemarle–Pamlico estuaries
1916	"Notes on the Fishes of the Lumbee River" (Evermann 1916)
1932	"On a Collection of Fishes from the Tuckaseegee and Upper Catawba River Basins, N.C., with a Description of a New Darter" (Hildebrand 1932)
1940s	The Department of Conservation and Development, Division of Game and Inland Fisheries (predecessor of the NCWRC), conducts surveys in the Mountain river basins (e.g., Engels 1941a, 1941b, 1942; King and Grove 1942a, 1942b) and by Joseph R. Bailey (Duke University) in 1947 and 1949 (Bailey 1949)
1940s	Henry W. Fowler examines NC specimens and records at the North Carolina Museum of Natural Sciences and reviews collections made by others in the 1930s–early 1940s (Fowler 1945)
1940s	A series of annual spring break field trips to survey NC's fish fauna are led by Edward C. Raney with his graduate students at Cornell University, often resulting in new species descriptions; these trips, led by his graduate students, continue into the 1960s–70s
1940s–50s	Bartholomew B. Brandt, William W. Hassler, and graduate students at North Carolina State University conduct surveys of the Roanoke and Cape Fear Rivers to determine the impacts of paper mill effluents on the fishery resources and a survey of the Crabtree Creek watershed in Wake County

TABLE 3. (*continued*)

Time Period	Ichthyological Contribution
1940s–60s	Earl E. Deubler Jr., William E. Fahy, Austin B. Williams, and many others of the University of North Carolina's Institute of Fisheries Research (predecessor of Institute of Marine Sciences) in Morehead City conduct surveys of Albemarle–Pamlico Sounds and Neuse River estuary
1940s–present	Studies begin on the impacts of pulp and paper mill effluents on the fish community and aquatic life of the Pigeon River (Haywood County, Pigeon basin)
1945–80s	Collections made by Lora M. Outten, Mars Hill College, primarily within a < ten-mile radius of Mars Hill, NC, in Madison, Buncombe, and Yancey Counties result in life history studies on Warpaint Shiner, *Coccotis coccogenis*; Saffron Shiner, *Hydrophlox rubricroceus*; and Whitetail Shiner, *Cyprinella galactura* (Outten 1956, 1957, and 1958)
1946	"Endemic Fish Fauna of Lake Waccamaw, North Carolina" (Hubbs and Raney 1946)
1950s–70s	Joseph R. Bailey and students at Duke University conduct surveys primarily in the Raleigh–Durham–Chapel Hill area and Lake Waccamaw (Waccamaw basin) and pre-impoundment surveys of New Hope Creek (Cape Fear basin) and Eno and Little Rivers (Neuse basin) before inundation by Jordan and Falls reservoirs
1951	"The Fishes of North Carolina's Bay Lakes and Their Intraspecific Variation" (Frey 1951)
1957	"The Distribution of the Fishes of the Catawba-Wateree River Drainage Basin, North Carolina and South Carolina" (Randall 1957)
1960s	The NCWRC conducts surveys across the state whose purpose was to provide a checklist of the species present in each drainage and classify the streams based on their faunal, physical, and chemical characteristics. This was seminal work as it was the first statewide survey completed in NC from the Mountains to the Coast that assessed the impacts of water quality on the fauna of a stream prior to implementation of the Clean Water Act in 1972 (e.g., Baker and Smith 1965a, 1965b; Bayless 1963; Bayless and Shannon 1965; Bayless and Smith 1963; Carnes, 1965; Carnes, Davis, and Tatum 1964; Crowell 1965; Davis and McCoy 1965; Louder 1962b, 1963, 1964; Messer 1964a, 1964b, 1965; Messer and Ratledge 1963; Messer et al. 1965; Ratledge, Carnes, and Collins 1966; Richardson 1964; Richardson and Carnes 1964; Richardson, Messer, and Ratledge 1963; Smith 1963; Smith and Bayless, 1964; Starnes and Hogue 2011; Tatum, Carnes, and Richardson 1963)

TABLE 3. (*continued*)

Time Period	Ichthyological Contribution
1960s	The NCWRC conducts surveys and checklists of North Carolina bay lakes, natural lakes, and reservoirs (e.g., Bayless 1966; Crowell 1966; Louder 1962a; Messer 1966; NCWRC 1961; Ratledge, Carnes, and Collins 1966)
1960s–80s	William W. Hassler and colleagues at North Carolina State University conduct a long-term study of the lower Neuse River to assess the potential impacts of paper mill effluents on the fish community; Hassler and colleagues initiated the Striped Bass, *Morone saxatilis*, juvenile abundance index survey in the Albemarle Sound
1960s–2000s	Edward F. Menhinick (University of North Carolina at Charlotte) established a collection to aid in his studies on the distribution of North Carolina fishes
1960s–present	Bartholomew B. Brandt, Jeffrey Buckel, William W. Hassler, F. Eugene Hester, Joseph Hightower, Thomas Kwak, John Miller, Richard Noble, and James Rice, along with graduate students in the North Carolina Cooperative Fish and Wildlife Research Unit and Zoology Department at North Carolina State University, conducted master's and doctoral studies on Atlantic Sturgeon, American Shad, Cape Fear Shiner, Robust Redhorse, "Sicklefin" Redhorse, *Moxostoma* sp.; Carolina Madtom, Flathead Catfish, and many other freshwater species
1965	"Zoogeographic Studies of the Freshwater Fish Fauna of Rivers Draining the Southern Appalachian Region" (Ramsey 1965)
1965–2003	Jerry West and students at Western Carolina University conduct surveys in the vicinity of the university and take other field trips to the coast of North Carolina
1970s	B. J. Copeland, Ron Hodson, William S. Birkhead, R. Wilson Laney, and E. C. Pendelton, and many technicians at North Carolina State University, survey the lower Cape Fear River estuary to assess impacts of the Brunswick Nuclear Plant in Southport, NC (Cape Fear basin)
1970s	Garland B. Pardue, Melvin T. Huish, and graduate students at North Carolina State University's Cooperative Fisheries Research Unit conduct pre-channelization study of the fish communities of Hoggard Mill and Duke Swamp (Gates and Bertie Counties, Chowan basin)
1970s–80s	John Lundberg and graduate students at Duke University conduct collections at many of the same locales where Bailey had previously collected and conducted taxonomic research on chubsuckers, *Erimyzon* spp., and bullheads, *Ameiurus* spp.

TABLE 3. *(continued)*

Time Period	Ichthyological Contribution
1970s–90s	Frank J. Schwartz and graduate students at the University of North Carolina's Institute of Marine Sciences conduct faunal surveys primarily across eastern North Carolina and the Sand Hills; preconstruction assessments of the Brunswick Nuclear Plant in Southport, NC (Cape Fear basin); and surveys of the freshwater and brackish ponds in the Cape Hatteras and Cape Lookout National Seashores
1970s–present	Carolina Power & Light Company and Duke Energy initiate long-term times series studies of the impacts of thermal discharges, impingement/entrainment, fish passage, and minimum flows on aquatic life at their coal-fired, nuclear, and hydroelectric plants
1974	"An Annotated Checklist of the Freshwater Fishes of North Carolina" (Menhinick, Burton, and Bailey 1974)
1978	*The Fishes of Mecklenburg County, North Carolina* (Cloutman and Olmsted 1978)
1979	"Freshwater Fishes of Croatan National Forest, North Carolina, with Comments on the Zoogeography of Coastal Plain Fishes" (Rohde, Burgess, and Link 1979)
1980	*Atlas of North American Freshwater Fishes* (Lee et al. 1980)
1980s–2000s	Rudy G. Arndt and students at Stockton State University (NJ) conduct a series of annual spring break field trips to survey North Carolina's fish fauna
1980s–present	Roger Rulifson and graduate students at East Carolina University conduct studies on the ecology and evolution of anadromous and catadromous species using otolith chemistry and on long-distance migration patterns and population demographics of Blueback Herring, *Alosa aestivalis*; Hickory Shad, *A. mediocris*; Alewife, *A. pseudoharengus*; American Shad; and Striped Bass along the Eastern Seaboard of North America
1981	"Fishes of the Waccamaw River Drainage" (Shute, Shute, and Lindquist 1981)
1984	*Fisherman's Guide: Fishes of the Southeastern United States* (Manooch 1984)
1991	*The Freshwater Fishes of North Carolina* (Menhinick 1991)
1990s–present	North Carolina Division of Water Resources (formerly North Carolina Division of Water Quality) conducts basin-wide surveys by using fish community structure as a biological and water quality assessment tool

TABLE 3. *(continued)*

Time Period	Ichthyological Contribution
1990s–present	The NCWRC conducts statewide surveys of nongame fish species
1994	*Freshwater Fishes of the Carolinas, Virginia, Maryland and Delaware* (Rohde et al. 1994)
1995–present	The first Curator of Fishes (1995) and Collection Manager of Fishes (1996) are hired at the North Carolina Museum of Natural Sciences, firmly establishing the longevity and sustainability of the collection. Orphaned collections and voucher specimens are received from Charleston Museum, Duke University, Duke University Marine Lab, Mars Hill College, North Carolina Division of Marine Fisheries (NCDMF), North Carolina Division of Water Resources, North Carolina State University, North Carolina Wildlife Resources Commission, Fred C. Rohde, Steve W. Ross, Wayne C. Starnes, Stockton State University, University of North Carolina at Charlotte, University of North Carolina Institute of Marine Sciences, US Army Corps of Engineers, US Geological Survey, Western Carolina University, and West Virginia Department of Natural Resources. Collection currently includes 1.4 million specimens (including over 900,000 specimens from North Carolina)
ca. 1996	*Studies of Streams of the Uwharrie National Forest* (Menhinick, n.d.)
1998	"Distribution and Status of Selected Fishes in North Carolina with a New State Record" (Rohde, Moser, and Arndt 1998)
2001	"Longitudinal Succession of Fishes in the Dan River in Virginia and North Carolina (Blue Ridge/Piedmont Provinces)" (Rohde, Arndt, and Smith 2001)
2003	"An Annotated List of the Fishes Known from the Dan River in Virginia and North Carolina (Blue Ridge/Piedmont Provinces)" (Rohde et al. 2003)
2006–present	The North Carolina Museum of Natural Sciences fishes database is globally accessible
2010	*Fishes of Gaston County, Fishes by Species* (Menhinick 2010)
1990s–present	The Robust Redhorse Conservation Committee is created and subsequent surveys and research are conducted by the NCWRC, North Carolina State University, Carolina Power & Light Company, Duke Energy, US Fish and Wildlife Service, US Geological Survey, and North Carolina Museum of Natural Sciences

TABLE 3. (*continued*)

Time Period	Ichthyological Contribution
1990s–present	The NCWRC and NCDMF conduct long-term time series studies of diadromous fish species (American Eel, American Shad; Striped Bass; and river herring) and Atlantic Sturgeon
2013	"History of Fish Investigations in the Yadkin–Pee Dee River Drainage of North Carolina and Virginia with an Analysis of Nonindigenous Species and Invasion Dynamics of Three Species Suckers (Catostomidae)" (Tracy, Jenkins, and Starnes 2013)
2013	NCFishes.com is created and launched
2020	"An Annotated Atlas of the Freshwater Fishes of North Carolina" (Tracy, Rohde, and Hogue 2020)

(Girard, 1856, p. 213) became the first species to be described from North Carolina that has not been synonymized with any other species (Tracy, 2013; Tracy, Rohde, and Hogue, 2020). Since Smith's 1907 work, another seventeen species have been described from North Carolina (table 4). Professor Edward Drinker Cope was the most prolific ichthyologist in species descriptions, describing 46 of our 242 known species between 1865 and 1871 (Tracy, Rohde, and Hogue, 2020; Tracy and Jenkins, 2021). The most recently described species is the Carolina Pygmy Sunfish, *Elassoma boehlkei*, described by Rohde and Arndt in 1987. Currently, there are sixteen species awaiting formal taxonomic descriptions. Those species are within the families Leuciscidae (5), Catostomidae (5), Ictaluridae (3), Fundulidae (1), Percidae (1), and Centrarchidae (1) (table 2).

More than 340 years have passed since Thomas Ash's publication, and today's aquatic environments and fish communities would be unrecognizable to him. There has been a dramatic change in the landscape of North Carolina as environments have been altered by dams, channelization, pollution, terrestrial modifications, population growth, and urbanization. Our fish communities have been overexploited and have suffered the deleterious effects from pollution, habitat fragmentation, species homogenization, and introductions of nonnative species. Extirpations of native species have occurred, and many species may face the same fate in the near future.

Since Lawson's publication in 1709 (pp. 152–60), three species have been extirpated from our waters: the Paddlefish, Blueside Darter, and Sickle Darter. To our knowledge, the Paddlefish has not been seen in North Carolina since the late 1870s, Sickle Darter since 1940, and the Blueside Darter since 1950 (Tracy, Rohde, and Hogue, 2020). The Paddlefish was known anecdotally and historically (in 1869, 1873, and 1874) only from the French Broad River near Asheville and Brevard (Cope, 1870b; Anonymous, 1873; Anonymous, 1874; Luke Etchison, NCWRC, personal communication) and has long been considered extirpated from the state. The following species that we believed to have been extirpated have been rediscovered or reintroduced: the Lake Sturgeon (French Broad basin); Bridle Shiner (Neuse and Chowan basins), Rustyside Sucker (Roanoke basin), Robust Redhorse (Yadkin basin), Mountain Madtom (French Broad basin), Banded Sculpin (Pigeon and French Broad basins), and Snubnose Darter (French Broad and Nolichucky basins) (Tracy, Rohde, and Hogue, 2020).

There are twenty-three indigenous species that are restricted to one specific drainage (basin) in North Carolina and surrounding states—a term referred to as "hydrologic endemism" (tables 2 and 5). The families containing the greatest number of hydrologic endemic species are Leuciscidae (nine species) and Percidae (six species). There are five species found in only one North Carolina river basin and nowhere else in the world: the Cape Fear Shiner, "Lake Waccamaw Broadtail" Madtom, "Cape Fear Broadtail" Madtom; Waccamaw Killifish, and Waccamaw Silverside.

Several nonindigenous species that have been infrequently reported as single occurrences (or hoaxes) but that do not have established, reproducing populations in North Carolina are not covered in this book. These species include but may not be limited to the Florida Gar, *Lepisosteus platyrhincus* (family Lepisosteidae); Clown Knifefish, *Chitala ornate* (family Notopteridae); Bighead Carp, *Hypophthalmichthys nobilis* (family Cyprinidae); Tench, *Tinca tinca* (family Tincidae); various species of piranhas and pacus (family Serassalmidae); Snakehead, *Channa* spp. (family Channidae); and Oscar, *Astronotus ocellatus* (family Cichlidae) (USGS, 2021). Further information on these fishes may be found in USGS (2021) and Fuller, Nico, and Williams (1999).

Format of Identification Keys

The identification keys are dichotomous keys (see chapter 2) and include the scientific and common name of every species along with the year in which the species was described. Phylogenetic order and family and scientific names follow *Eschmeyer's Catalog of Fishes* (Fricke, Eschmeyer, and van der Laan, 2022). The author (the describer of the species) is placed in parentheses when the species is currently assigned to a genus other than that in which it was first

TABLE 4. Species described from North Carolina and their type localities

Scientific Name, Author, Year Described	Common Name	Type Locality	Collector(s)
Family Catostomidae			
Moxostoma collapsum (Cope, 1870)	Notchlip Redhorse	Neuse, Yadkin, and Catawba Rivers, unknown counties	E. D. Cope
Moxostoma pappillosum (Cope, 1870)	V-lip Redhorse	Catawba and Yadkin Rivers, unknown counties	E. D. Cope
Moxostoma robustum (Cope, 1870)	Robust Redhorse	Inhabiting the Yadkin River, unknown county	E. D. Cope
Moxostoma rupiscartes Jordan and Jenkins, 1889	Striped Jumprock	Catawba River, near Marion, McDowell Co.	D. S. Jordan, O. P. Jenkins, and S. E. Meek
Family Leuciscidae			
Alburnops petersoni (Fowler, 1942)	Coastal Shiner	Crane Creek below US 1, Moore Co.	G. A. Coventry, C. B. Peterson, and F. A. Ulmer Jr.
Cyprinella labrosa (Cope, 1870)	Thicklip Chub	Tributaries of the upper Catawba River, McDowell and Burke Cos.	E. D. Cope
Cyprinella nivea (Cope, 1870)	Whitefin Shiner	Upper waters of the Catawba River, unknown counties	E. D. Cope
Cyprinella pyrrhomelas (Cope, 1870)	Fieryblack Shiner	Tributaries of the upper Catawba River, unknown counties	E. D. Cope
Hudsonius altipinnis (Cope, 1870)	Highfin Shiner	Yadkin River, Rowan Co.	E. D. Cope
Hybopsis hypsinotus (Cope, 1870)	Highback Chub	Creeks heading the Catawba River, McDowell Co., or tributary to the Yadkin River, Rowan Co.	E. D. Cope

Species	Common name	Locality	Authority
Hydrophlox chiliticus (Cope, 1870)	Redlip Shiner	Tributaries of the Yadkin River, Rowan Co.	E. D. Cope
Hydrophlox chlorocephalus (Cope, 1870)	Greenhead Shiner	Tributaries of the Catawba River	E. D. Cope
Lythrurus matutinus (Cope, 1870)	Pinewoods Shiner	Neuse River, Wake Co.	E. D. Cope
Miniellus alborus (Hubbs and Raney, 1947)	Whitemouth Shiner	Brush Creek, near Siler City, Randolph Co.	E. C. Raney and E. A. Lachner
Miniellus mekistocholas (Snelson, 1971)	Cape Fear Shiner	Rocky River at NC 902, Chatham Co.	F. F. Snelson and W. M. Palmer
Nocomis leptocephalus (Girard, 1856)	Bluehead Chub	Salem, Forsyth Co.	J. T. Lineback and students
Pteronotropis cummingsae (Myers, 1925)	Dusky Shiner	Upper Burnt Mill Creek, New Hanover Co.	G. S. Meyers
Semotilus lumbee Snelson and Suttkus, 1978	Sandhills Chub	Tributary of Aberdeen Creek at culvert on US 1, Moore Co.	W. M. Palmer, A. L. Braswell, and J. E. Cooper
Family Ictaluridae			
Noturus furiosus Jordan and Meek, 1889	Carolina Madtom	Neuse River, at Millburnie, Wake Co.	O. P. Jenkins and S. E. Meek
Family Atherinopsidae			
Menidia extensa Hubbs and Raney, 1946	Waccamaw Silverside	Lake Waccamaw, Columbus Co.	E. C. Raney, E. A. Lachner, and R. A. Pfeiffer

TABLE 4. (continued)

Scientific Name, Author, Year Described	Common Name	Type Locality	Collector(s)
Family Fundulidae			
Fundulus rathbuni Jordan and Meek, 1889	Speckled Killifish	Reedy Fork, at Fulk's Mill, South Buffalo Creek, near Greensboro; Little Alamance Creek, near Greensboro; and from an unnamed small, very clear brook or spring-run, near Fulk's Mill, Guilford Co.	D. S. Jordan, O. P. Jenkins, and S. E. Meek
Fundulus waccamensis Hubbs and Raney, 1946	Waccamaw Killifish	Lake Waccamaw, Columbus Co.	E. C. Raney, E. A. Lachner, and R. A. Pfeiffer
Family Percidae			
Etheostoma brevispinum (Coker, 1926)	Carolina Fantail Darter	Paddys Creek, near Lake James, Burke Co.	R. E. Coker
Etheostoma gutselli (Hildebrand, 1932)	Tuckasegee Darter	Tuckasegee River, near Ela, Swain Co.	J. S. Gutsell
Etheostoma kanawhae (Raney, 1941)	Kanawha Darter	North Fork of the New River, at Crumpler, Ashe Co.	E. C. Raney, E. A. Lachner, and L. J. Kezer
Etheostoma maculaticeps (Cope 1870)	Southern Tessel-lated Darter	Upper waters of the Catawba River, unknown counties	E. D. Cope
Etheostoma mariae (Fowler, 1947)	Pinewoods Darter	Outlet of Watson's Lake, near Southern Pines, Moore Co.	H. W. Fowler
Etheostoma perlongum (Hubbs and Raney, 1946)	Waccamaw Darter	Lake Waccamaw, Columbus Co.	E. C. Raney, E. A. Lachner, and R. A. Pfeiffer
Etheostoma serrifer (Hubbs and Cannon, 1935)	Sawcheek Darter	Buffalo Creek, near Wendell, Wake Co.	C. S. Brimley and J. A. Harris

Etheostoma vitreum (Cope, 1870)	Glassy Darter	Walnut Creek, at Raleigh, Wake Co.	E. D. Cope
Nothonotus chlorobranchius (Zorach, 1972)	Greenfin Darter	Cullasaja River, near Franklin, Macon Co.	H. E. Winn and C. F. Powers
Nothonotus rufilineatus (Cope, 1870)	Redline Darter	Spring Creek, at Hot Springs, Madison Co.	E. D. Cope
Nothonotus vulneratus (Cope, 1870)	Wounded Darter	Spring Creek, at Hot Springs, Madison Co.	E. D. Cope
Percina burtoni Fowler, 1945	Blotchside Logperch	Swannanoa River, near Oteen, Buncombe Co.	E. M. Burton
Percina gymnocephala Beckham, 1980	Appalachia Darter	South Fork New River, near West Jefferson, Ashe Co.	E. C. Beckham and E. B. Beckham
Percina nevisensis (Cope, 1870)	Chainback Darter	Falls of the Neuse River, near Raleigh, Wake Co.	E. D. Cope
Family Elassomatidae			
Elassoma boehlkei Rohde and Arndt, 1987	Carolina Pygmy Sunfish	Juniper Creek, at SR 1340, Brunswick Co.	F. C. Rohde

TABLE 5. North Carolina's hydrologic endemic species

Family, Scientific Name	Common Name
Catostomidae	
Moxostoma ariommum	Bigeye Jumprock
Thoburnia hamiltoni	Rustyside Sucker
Leuciscidae	
Clinostomus sp. "Hiwassee" Dace	
Clinostomus sp. "Smoky" Dace	
Hydrophlox chlorocephalus	Greenhead Shiner
Hydrophlox sp. "Piedmont" Shiner	
Miniellus mekistocholas	Cape Fear Shiner
Miniellus scabriceps	New River Shiner
Nocomis platyrhynchus	Bigmouth Chub
Notropis sp. "Kanawha" Rosyface Shiner	
Phenacobius teretulus	Kanawha Minnow
Ictaluridae	
Noturus gilberti	Orangefin Madtom
Noturus sp. "Cape Fear Broadtail" Madtom	
Noturus sp. "Lake Waccamaw Broadtail" Madtom	
Atherinopsidae	
Menidia extensa	Waccamaw Silverside
Fundulidae	
Fundulus waccamensis	Waccamaw Killifish
Percidae	
Etheostoma kanawhae	Kanawha Darter
Etheostoma mariae	Pinewoods Darter
Etheostoma perlongum	Waccamaw Darter
Etheostoma podostemone	Riverweed Darter
Percina gymnocephala	Appalachia Darter
Percina rex	Roanoke Logperch
Elassomatidae	
Elassoma boehlkei	Carolina Pygmy Sunfish

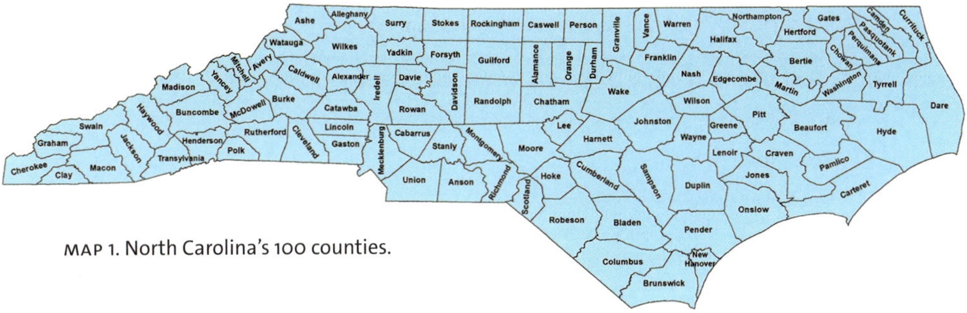

MAP 1. North Carolina's 100 counties.

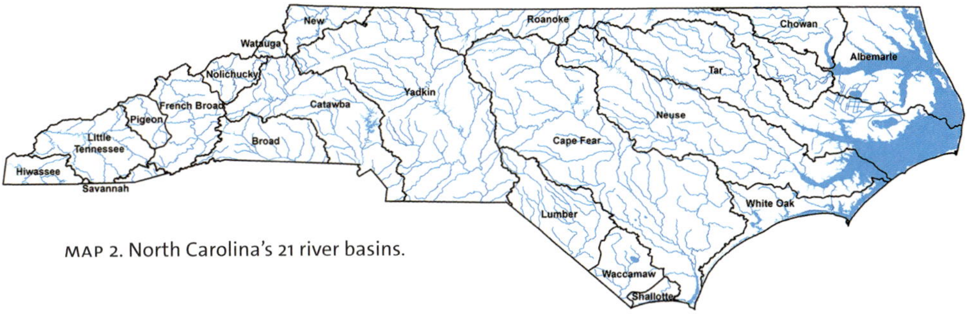

MAP 2. North Carolina's 21 river basins.

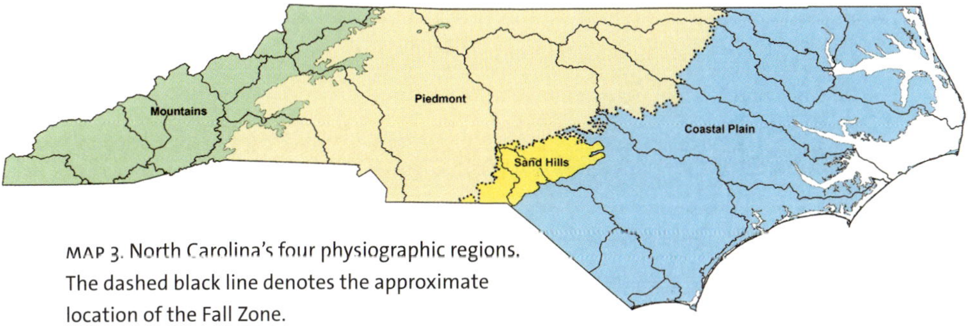

MAP 3. North Carolina's four physiographic regions. The dashed black line denotes the approximate location of the Fall Zone.

described. For example, within the family Leuciscidae, *Photogenis pyrrhomelas*, Cope, 1870 is now *Cyprinella pyrrhomelas* (Cope, 1870). The common name is an American Fisheries Society–accepted common name (Page et al., 2023). Additional regional, vernacular, and colloquial common names found herein were obtained primarily from Cloutman and Olmsted (1983), Manooch (1984), and Smith (1907). Common names for undescribed species are used strictly for convenience and may or may not be the accepted common name once the species is scientifically described. The meaning of each of the scientific names is presented in the appendix. Imperilment and management status was determined from NCAC (2023), NCDMF (2020), NCNHP (2022), and NCWRC (2020, 2021, 2022).

Distributional data totaling more than 276,000 records were compiled from material found in Tracy, Rohde, and Hogue (2020) and recent captures. Species distributional maps were produced using ArcGIS Version 10.5, projection GCS_WGS_1984, and WGS 84 datum. Red dots indicate species localities based upon the dataset in Tracy, Rohde, and Hogue (2020) and recent captures. When a species was known to occur within a basin but voucher material was not present, the occurrence was designated with a gray diamond in the approximate middle of the basin. Coordinates for type localities containing specific locality information were added by the authors, and these localities are plotted as yellow stars. Several species may be underrepresented in the dataset due to various reasons including imperilment status and possibly their large size, which increases the time and cost of preservation and archival. Maps were not produced for several species that are included in the dichotomous keys because they are primarily found in estuarine or marine environments.

Supplemental maps (maps 1–3) are included, which delineate the counties, basins, and physiographic regions of North Carolina. River basin designations for the twenty-one basins within North Carolina's 100 counties and four physiographic regions generally follow those of the NCWRC's 1960s surveys (Starnes and Hogue, 2011) and Menhinick (1991), with the exceptions of using the Nolichucky basin instead of the Toe basin and the inclusion of the Pamlico Sound within the Neuse and Tar basins. Three extremely small river basins, the Ocoee, Tallulah, and Lynches, are combined with the Hiwassee, Savannah, and Yadkin basins, respectively. Even though the Pee Dee River originates at the confluence of the Yadkin and the Uwharrie Rivers, for simplicity the entire basin is referred to as the Yadkin basin. When basins are listed within the following text they are ordered by physiographic region from the mountains to the coast (see table 2).

Aids for Identification

The correct identification of any species of fish requires a knowledge and understanding of external anatomy and how certain measurements are determined. In 1964 Hubbs and Lagler proposed methods to accurately measure and count the parts of a fish, and these methods are still being used. Another practice that Hubbs and Lagler proposed and that is commonly followed is, when possible, making measurements and taking photographs on the left side of the fish. Observing certain anatomical structures or taking certain measurements often requires the aid of a dissecting microscope or a high-quality hand lens.

Definitions of the more common terms are presented below, and a glossary is provided for other scientific terms used within this book. "Ca." is an abbreviation of the Latin word "circa," meaning "approximately." The following illustrations (figures 2 and 3) and definitions have been adapted primarily from Jenkins and Burkhead (1994) and Rohde et al. (2009).

Measurements

Measurements are taken on a straight line using dividers, calipers, or a ruler. Measurements commonly made are body depth, eye diameter, gape width, predorsal length, standard length, and total length. Dimensions are often expressed as a percentage of total or standard length.

- **Body depth**—The greatest vertical distance across the body, exclusive of the fins.
- **Eye diameter**—The greatest distance across the soft margins of the eye.
- **Gape width**—In the case of sturgeons (family Acipenseridae), distance between the inside corner of the lips.

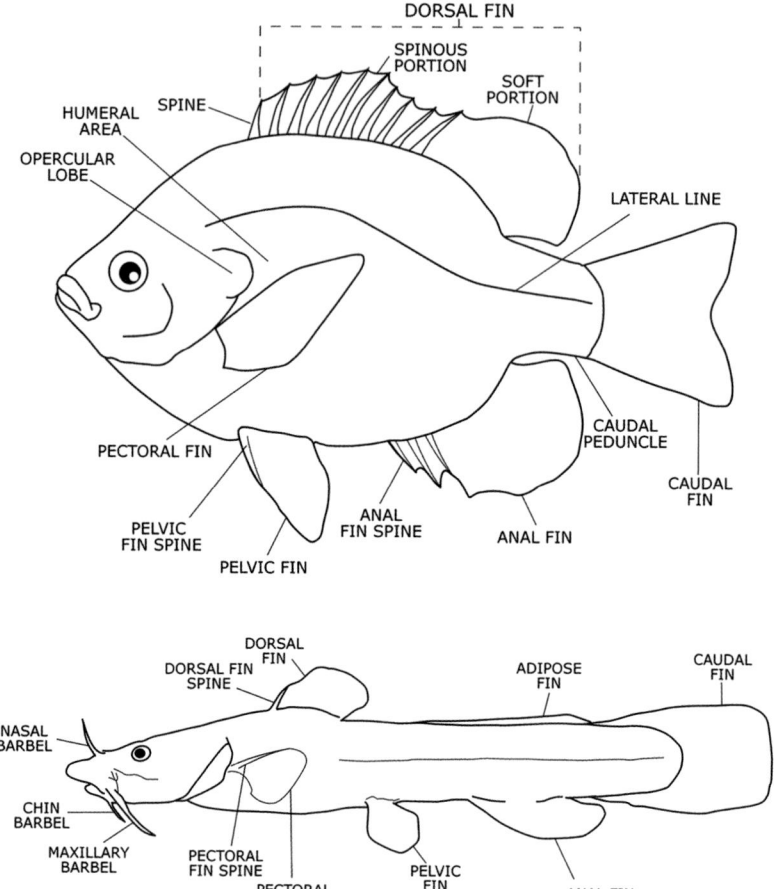

FIGURE 2. General terminology of the anatomical features of fishes.
Illustration adapted from Rohde et al. (2009).

- Predorsal length—The distance from the tip of the snout to the structural base of the first dorsal fin, ray, or spine.
- Standard length (SL)—Straight-line distance from the anteriormost point on the head to the posterior end on the caudal fin base (hypural plate).
- Total length (TL)—Straight-line distance from the anteriormost part of the head to the posteriormost part of the caudal fin with the caudal fin tips compressed together.

A

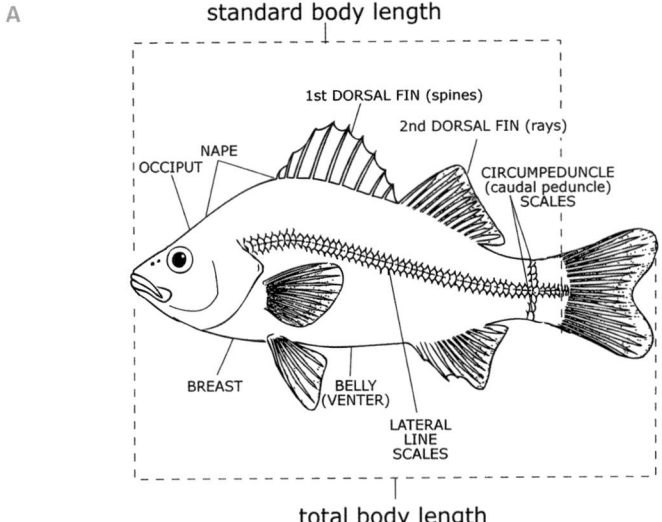

standard body length

1st DORSAL FIN (spines)

2nd DORSAL FIN (rays)

NAPE

OCCIPUT

CIRCUMPEDUNCLE
(caudal peduncle)
SCALES

BREAST

BELLY
(VENTER)

LATERAL
LINE
SCALES

total body length

B

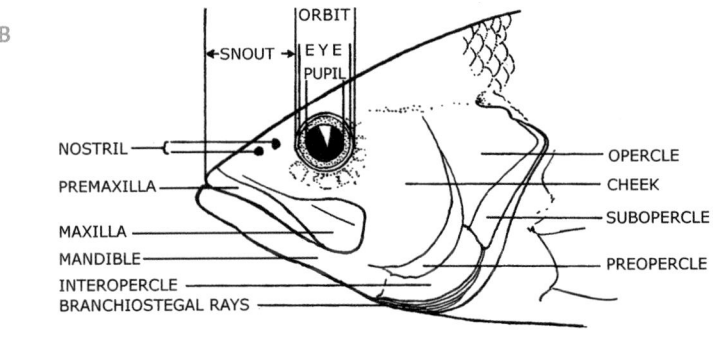

ORBIT

←SNOUT→ EYE
PUPIL

NOSTRIL

OPERCLE

PREMAXILLA

CHEEK

MAXILLA

SUBOPERCLE

MANDIBLE

PREOPERCLE

INTEROPERCLE
BRANCHIOSTEGAL RAYS

FIGURE 3. Body measurements, scale counts, and anatomical features of the head. Illustrations adapted from Rohde et al. (2009) (A) and Eddy (1969) (B).

Meristic Data

Meristic data refer to characteristics that are discrete and can be counted, such as the number of fin rays and lateral line scales. For example, "lateral line scales (35) 37–40 (42)" indicates that 80–90 percent of the specimens have thirty-seven to forty scales and that the full range (that is, minimum to maximum) is thirty-five to forty-two scales. Ranges without parenthetical values indicate the full ranges (Jenkins and Burkhead, 1994; Rohde et al., 2009).

Scale Counts

- Circumpeduncle scale rows (caudal peduncle scale rows)—The number of scales around the narrowest part of the caudal peduncle.
- Lateral line scales—The number of scales beginning with the first pored lateral scale on the shoulder (pectoral girdle) to the base of the caudal fin. Scales posterior to the caudal fin base are not counted. The caudal fin base is where the vertebral column ends and can be seen as a crease when the caudal fin is flexed from side to side. In fishes with incomplete pored lateral lines, the number of pored and unpored scales is usually noted. In the case of fishes that do not have a developed lateral line, scales along the midline of the body (midlateral) are counted.
- Predorsal circumferential scale rows—The number of scales encircling the body anterior to the dorsal fin.

Fin Ray and Spine Counts

- Caudal fin rays—Counts may be given as branched rays only or as principal rays (number of branched rays plus the two unbranched rays at the dorsal and ventral edges).
- Dorsal fin and anal fin rays—Most fins are bilateral, flexible, segmented, and usually branched. In most families, only the principal rays (those in which the tip reaches the distal margin of the fin) are counted. If the bases of the last two rays are closer together than those of more anterior rays, they are counted as one because they share a common basal element. In catfishes (family Ictaluridae) and trouts and salmons (family Salmonidae), all rays, including the rudimentary anterior ones, are counted. With catfishes it may be necessary to make a small incision along the base of the fin through the fatty tissue to expose all the rays.
- Dorsal fin and anal fin spines—All spines, which are unsegmented and usually hard and sharp, are counted, regardless of their size.
- Pectoral fin and pelvic fin rays—All rays are counted.

Branchiostegal Ray and Gill Raker Counts

- Branchiostegal rays—These are the long and thin bones that support the gill membranes on the ventral surface of the head. All rays on one side are counted.

• Gill rakers—All rakers, including the rudimentary ones, are counted on the first gill arch. This is accomplished by lifting the opercular (gill) flap. If the rakers are numerous and crowded, it may be necessary to remove the arch.

Pharyngeal Tooth Count

• Pharyngeal teeth—The fifth gill arch is modified in the families Catostomidae, Cyprinidae, Leuciscidae, and Xenocyprididae by being strengthened and bearing inward-facing teeth. A pharyngeal arch can bear one, two, or three rows of teeth, and all are counted. Counts are usually given for both sides, beginning with the shorter row on the left arch and concluding with the shorter row on the right arch. For example, a specimen with two teeth in the shorter row and four in the longer row on both arches would be stated as 2,4–4,2. One with only a single row of four teeth on either arch would be 4–4. To count the teeth, it is necessary to carefully cut out the arch and remove the flesh. See Menhinick (1991) and Jenkins and Burkhead (1994) for excellent instructions and illustrations on how to extract and clean the arch.

Mouth Position

• Inferior mouth—Mouth position distinctly on underside of head.
• Subterminal mouth—Mouth position in which the anterior end of the mouth is just below the tip of the snout.
• Superior mouth—Mouth position in which the anterior end of the mouth is distinctly above the tip of the snout.
• Supraterminal mouth—Mouth position in which the anterior end of the mouth is slightly above the tip of the snout.
• Terminal mouth—Mouth position in which the anterior end of the mouth is even with the tip of the snout.

2

Identification Key to the Families of Freshwater Fishes in North Carolina

How to Use a Dichotomous Key

The identification keys within this book are formatted as dichotomous keys. Each key is represented as a series of paired statements (couplets) laid out in sequential fashion. Each half of the couplet presents the user with one or more different diagnostic characters. If each half of the couplet contains multiple characters, then the first character is generally the most diagnostic. With any dichotomous key, one should always begin with the first couplet.

Using the following key to the families as an example, one would decide if the fish that is being identified fits the diagnostic characters in Couplet 1a or in Couplet 1b: elongate and snakelike with a long dorsal fin (Couplet 1a), or not elongate and snakelike with a variable dorsal fin (Couplet 1b). If the fish satisfies the features in Couplet 1a, one would proceed to Couplet 2; if not, one would then proceed to Couplet 3. Choices continue to be presented within the couplets until one reaches a final determination—for example, Lampreys, Petromyzontidae (Couplet 2a) or Freshwater Eels, Anguillidae (Couplet 2b).

Key to the Families of Freshwater Fishes of North Carolina

1a. Body elongate and snakelike. Dorsal fin long..............................2
1b. Body not elongate and snakelike. Dorsal fin variable....................3

2a. Jaws absent; mouth a disk-shaped funnel. Seven pairs of gill openings. Pectoral fins absent (figure 4)...............Lampreys, Petromyzontidae
2b. Jaws present. One pair of gill openings. Pectoral fins present (figure 5)...
....Freshwater Eels, Anguillidae

FIGURE 4. Least Brook Lamprey, *Lampetra aepyptera*, family Petromyzontidae.

FIGURE 5. American Eel, *Anguilla rostrata*, family Anguillidae; elver stage.

3a. Caudal fin strongly heterocercal. Snout long; teeth absent or very small. Body covered with five rows of bony plates or else almost scaleless......4
3b. Caudal fin not strongly heterocercal. If snout is long, then teeth are well developed. Body usually scaled, but without bony plates................5

4a. Body well scaled with five rows of bony plates. Snout flat and triangular (figure 6).......................................Sturgeons, Acipenseridae
4b. Body almost scaleless with no plates. Snout long and shaped like a canoe paddle (figure 7)..............................Paddlefishes, Polyodontidae

FIGURE 6. Atlantic Sturgeon, *Acipenser oxyrinchus*, family Acipenseridae. National Marine Fisheries Service Endangered Species Act Permit 21198.

FIGURE 7. Paddlefish, *Polyodon spathula*, family Polyodontidae. Arkansas. Photograph courtesy of David A. Neely.

5a. Large gular plate found on the chin between the two jaws. Dorsal fin long, greater than one-half TL (figure 8) .... Bowfins, Amiidae
5b. No large gular plate found on the chin between the two jaws. Dorsal fin length variable.. 6

FIGURE 8. Bowfin, *Amia calva*, family Amiidae.

6a. Jaws elongate, slender, and needlelike.................................... 7
6b. Jaws not elongate, slender, or needlelike................................. 8

7a. Ganoid scales large and rhomboid shaped. Lateral line scales ca. 57–63. Caudal fin abbreviate heterocercal (figure 9)........Gars, Lepisosteidae

7b. Cycloid scales very small. Lateral line scales ca. 300. Caudal fin homocercal (figure 10)................................Needlefishes, Belonidae

FIGURE 9. Longnose Gar, *Lepisosteus osseus*, family Lepisosteidae.

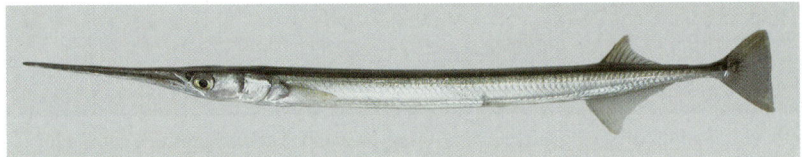

FIGURE 10. Atlantic Needlefish, *Strongylura marina*, family Belonidae.

8a. Both eyes on same side of head; eyed side with most pigment. Dorsal and anal fins long, extending from head to near caudal fin...................9

8b. Both eyes not on same side of head. Both sides of body equally pigmented. Dorsal and anal fins not extending from head to caudal fin............10

9a. Eyes on right side of head. Pectoral fins absent. Mouth small and inconspicuous (figure 11)...........................American Soles, Achiridae

9b. Eyes on left side of head. Pectoral fins well developed. Mouth prominent (figure 12)...............................Sand Flounders, Paralichthyidae

FIGURE 11. Hogchoker, *Trinectes maculatus*, family Achiridae.

FIGURE 12. Southern Flounder, *Paralichthys lethostigma*, family Paralichthyidae.

10a. Anus in throat region between the gills in early to late juveniles and adults. Caudal fin rounded..11

10b. Anus in "normal" position, anterior to anal fin. Caudal fin shape variable 12

11a. Pelvic fins present. Dorsal and anal fins with spines. Body thick, not elongate. Scales visible. Eyes not covered with translucent skin (figure 13).....
....Pirate Perches, Aphredoderidae

11b. Pelvic fins absent. Dorsal and anal fin spines absent. Body elongate. Scales embedded, not visible. Eyes covered with translucent skin (figure 14).....
....Cavefishes, Amblyopsidae

FIGURE 13. Pirate Perch, *Aphredoderus sayanus*, family Aphredoderidae.

FIGURE 14. Swampfish, *Chologaster cornuta*, family Amblyopsidae.

FIGURE 15. Margined Madtom, *Noturus insignis*, family Ictaluridae.

FIGURE 16. Amazon Sailfin Catfish, *Pterygoplichthys pardalis*, family Loricariidae. Peru.

FIGURE 17. Brown Trout, *Salmo trutta*, family Salmonidae.

16a. Subterminal mouth surrounded by ten to twelve barbels. Caudal fin rounded. Lateral line present (figure 18)......Spined Loaches, Cobitidae
16b. No barbels surrounding the mouth. Caudal fin variable. Lateral line present or absent...17

FIGURE 18. Pond Loach, *Misgurnus anguillicaudatus*, family Cobitidae.

17a. Caudal fin rounded or truncate. Lateral line absent. Scales present on top of head..18
17b. Caudal fin forked. Lateral line variable. Scales absent on top of head.....
....21

18a. Mouth terminal and not protractile; upper jaw broadly bound to snout with a wide frenum (figure 19)...................Mudminnows, Umbridae
18b. Mouth superior and protractile; upper jaw without a frenum..........19

FIGURE 19. Eastern Mudminnow, *Umbra pygmaea*, family Umbridae.

19a. Third anal fin ray unbranched. Third anal fin ray of male slender and rod-like (figure 20) . . . . Livebearers, Poeciliidae

19b. Third anal fin ray branched. Third anal fin ray of male not slender and rodlike 20

FIGURE 20. Western Mosquitofish, *Gambusia affinis*, family Poeciliidae. Ohio.

20a. Body slender. Teeth in jaws small and conical (figure 21) . . . . Topminnows, Fundulidae

20b. Body deep. Teeth in jaws tricuspid and incisor-like (figure 22) . . . . Pupfishes, Cyprinodontidae

FIGURE 21. Golden Topminnow, *Fundulus chrysotus*, family Fundulidae.

FIGURE 22. Sheepshead Minnow, *Cyprinodon variegatus*, family Cyprinodontidae.

21a. Jaws extending far behind eye; mouth strongly inferior (figure 23)........
....Anchovies, Engraulidae
21b. Jaws not extending very far behind eye; mouth position variable.......22

FIGURE 23.
Bay Anchovy,
Anchoa mitchilli,
family Engraulidae.
Florida. Photo-
graph courtesy of
Dan Marotta.

22a. Belly with a keel. Adipose eyelids present................................23
22b. Belly rounded. Adipose eyelids absent..................................25

23a. Lateral line present. Breast without a sawtooth margin (figure 24).......
....Mooneyes, Hiodontidae
23b. Lateral line absent. Midline of breast with a sawtooth line of scales...24

FIGURE 24.
Mooneye, *Hiodon
tergisus*, family
Hiodontidae.
Ohio. Photograph
courtesy of Brian
Zimmerman.

24a. Last dorsal fin ray filamentous (figure 25) ....
....Thread Herrings, Dorosomatidae
24b. Last dorsal fin ray not filamentous........................Shads, Alosidae

FIGURE 25.
Gizzard Shad,
*Dorosoma cepe-
dianum*, family
Dorosomatidae.

25a. Jaws elongate and wide, forming a duck-like beak (figure 26) . . . .
. . . . Pikes, Esocidae
25b. Jaws not elongate, not forming a duck-like beak 26

FIGURE 26. Redfin Pickerel,
Esox americanus, family Esocidae.

26a. Lips thick and fleshy with plicae or papillae. Dorsal fin without a thick
spine (figure 27) . . . . .Suckers, Catostomidae
26b. Lips usually thin and smooth. Dorsal fin with or without a thick spine . . .
. 27

FIGURE 27. Smallmouth Redhorse, *Moxostoma
breviceps*, family Catostomidae.

27a. Dorsal fin long with a stout, saw-toothed spine, followed by thirteen or
more branched rays. Anal fin with a stout spine (figure 28) . . . .
. . . . .Carps, Cyprinidae

FIGURE 28. Common Carp, *Cyprinus
carpio*, family Cyprinidae.

27b. Dorsal fin without a stout, saw-toothed spine, followed by twelve or fewer branched rays. Anal fin without a stout spine............................28

28a. Distance from anal fin origin to tip of snout three or more times as long as the distance from the anal fin origin to the caudal fin base (figure 29) ....Sharpbellies, Xenocyprididae
28b. Distance from anal fin origin to tip of snout less than three times as long as the distance from the anal fin origin to the caudal fin base (figure 30) ....Minnows, Leuciscidae

FIGURE 29. Grass Carp, *Ctenopharyngodon idella*, family Xenocyprididae. South Carolina.

FIGURE 30. Swallowtail Shiner, *Miniellus procne*, family Leuciscidae.

29a. Body without scales or scales greatly reduced in size or body with bony plates...30
29b. Body with scales..31

30a. Scaleless body or with bony plates. Three or four spines in first dorsal fin; when four, angled alternately left to right. Caudal peduncle extremely narrowed (figure 31) . . . . .Sticklebacks, Gasterosteidae

30b. Scales greatly reduced in size. Six to nine dorsal fin spines not angled left to right. Caudal peduncle not extremely narrowed (figure 32) . . . . .Sculpins, Cottidae

FIGURE 31. Fourspine Stickleback, *Apeltes quadracus*, family Gasterosteidae

FIGURE 32. Blue Ridge Sculpin, *Cottus caeruleomentum*, family Cottidae.

31a. Pelvic fins abdominal. Dorsal fins widely separated .32
31b. Pelvic fins thoracic. Dorsal fins joined or slightly separated 33

32a. Anal fin with one thin spine. First dorsal fin with five or six thin spines. Adipose eyelids absent (figure 33) . . . . .New World Silversides, Atherinopsidae

32b. Anal fin with two or three thick spines. First dorsal fin with four thick spines. Adipose eyelids present or absent (figure 34) . . . . .Mullets, Mugilidae

FIGURE 33. Green Silverside, *Labidesthes vanhyningi*, family Atherinopsidae.

FIGURE 34. Striped Mullet, *Mugil cephalus*, family Mugilidae.

33a. Anal fin with two spines. Dorsal fins joined (figure 35).....................
....Drums and Croakers, Sciaenidae
33b. Anal fin with one or two spines and dorsal fins separated or anal fin with
three or more spines and dorsal fins usually joined......................34

FIGURE 35. Freshwater Drum, *Aplodinotus grunniens*, family Sciaenidae. Photograph courtesy of Luke Etchison.

34a. Anal fin with one or two spines. Dorsal fins separated...................35
34b. Anal fin with three or more spines. Dorsal fins usually joined..........37

35a. Lateral line present; may be incomplete. Caudal fin usually emarginate or
truncate. Dorsal fin spines stiff (figure 36)....................................
....Perches and Darters, Percidae
35b. Lateral line absent. Caudal fin rounded or pointed. Dorsal fin spines flexible.
....36

FIGURE 36. Turquoise Darter, *Etheostoma inscriptum*, family Percidae.

36a. Pelvic fins united, forming a disk (figure 37)............Gobies, Gobiidae

36b. Pelvic fins separate, not united into a disk (figure 38).....................
....Sleepers, Eleotridae

FIGURE 37.
Freshwater
Goby, *Ctenogo-
bius shufeldti*,
family Gobiidae.

FIGURE 38.
Fat Sleeper,
*Dormitator
maculatus*,
family
Eleotridae.

37a. Lateral line absent (figure 39)............Pygmy Sunfishes, Elassomatidae

37b. Lateral line present, may be interrupted.................................38

FIGURE 39.
Carolina
Pygmy Sunfish,
*Elassoma
boehlkei*, family
Elassomatidae.

38a. Lateral line interrupted under soft dorsal fin with anterior series of pored scales higher than posterior series. Each nostril with a single opening (figure 40)...Cichlids, Cichlidae

38b. Lateral line not interrupted. Each nostril with two openings...........39

FIGURE 40. Blue Tilapia, *Oreochromis aureus*, family Cichlidae. Florida. Photograph courtesy of Tim Aldridge.

39a. Dorsal fins separate or slightly joined. Sharp spine on rear of opercle. Posterior edge of preopercle strongly serrated (figure 41) . . . . Temperate Basses, Moronidae

39b. Dorsal fins joined; a notch may be present. No spine on rear of opercle. Posterior edge of preopercle smooth (figure 42) . . . . Sunfishes, Centrarchidae

FIGURE 41. White Perch, *Morone americana*, family Moronidae.

FIGURE 42. Mud Sunfish, *Acantharchus pomotis*, family Centrarchidae.

Note to Table 6, facing page. Only species discussed in this chapter are included in this table. Family Xenocyprididae, which has one species in North Carolina, is included in chapter 8, "Minnows." Family Paralichthyidae, which has one species that is a seasonal inhabitant of fresh water, is found in chapter 14, "Sand Flounders," along with two other species that may be encountered in fresh water.

3

Monospecific Families

There are fifteen families that are monospecific (meaning, having only one species or having only one species inhabiting fresh water) in North Carolina (tables 2 and 6; Tracy, Rohde, and Hogue, 2020). The family Xenocyprididae is monospecific but is not included in this chapter because, for the purpose of identification, it is combined with the families Cyprinidae and Leuciscidae as the "minnows." Some families listed in table 6 may be represented by only one species in fresh water but are more diverse in estuarine or marine waters (for example, families Engraulidae, Belonidae, and Achiridae) (Kells and Carpenter, 2011).

TABLE 6. Families with only one species in fresh water in North Carolina

Family	Scientific Name	Common Name
Polyodontidae	*Polyodon spathula*	Paddlefish
Lepisosteidae	*Lepisosteus osseus*	Longnose Gar
Amiidae	*Amia calva*	Bowfin
Anguillidae	*Anguilla rostrata*	American Eel
Hiodontidae	*Hiodon tergisus*	Mooneye
Engraulidae	*Anchoa mitchilli*	Bay Anchovy
Cobitidae	*Misgurnus anguillicaudatus*	Pond Loach
Loricariidae	*Pterygoplichthys pardalis*	Amazon Sailfin Catfish
Umbridae	*Umbra pygmaea*	Eastern Mudminnow
Aphredoderidae	*Aphredoderus sayanus*	Pirate Perch
Amblyopsidae	*Chologaster cornuta*	Swampfish
Achiridae	*Trinectes maculatus*	Hogchoker
Cyprinodontidae	*Cyprinodon variegatus*	Sheepshead Minnow
Belonidae	*Strongylura marina*	Atlantic Needlefish
Gasterosteidae	*Apeltes quadracus*	Fourspine Stickleback

Polyodontidae—Paddlefish, *Polyodon spathula* (Walbaum, 1792)

- **Unique characteristics:** Long, canoe paddle–shaped snout that is ca. one-third TL. Huge mouth. Heterocercal caudal fin. Large, fleshy, pointed flap on rear edge of gill cover. Tiny eyes. Toothless jaws as adults (figure 43).
- **Most likely to be confused with (in fresh water):** No other species.
- **Remarks:** Paddlefishes were known anecdotally and historically (1869, 1873, and 1874) only from the French Broad River near Asheville and Brevard (Buncombe and Transylvania Counties) (Cope, 1870b; Anonymous, 1873; Anonymous, 1874; Luke Etchison, NCWRC, personal communication) and have been long considered extirpated from the state. Range restricted to French Broad basin.
- **Colloquial names:** spoonbill cat, duck-billed cat.

FIGURE 43. Paddlefish, *Polyodon spathula*. Arkansas. Photograph courtesy of David A. Neely.

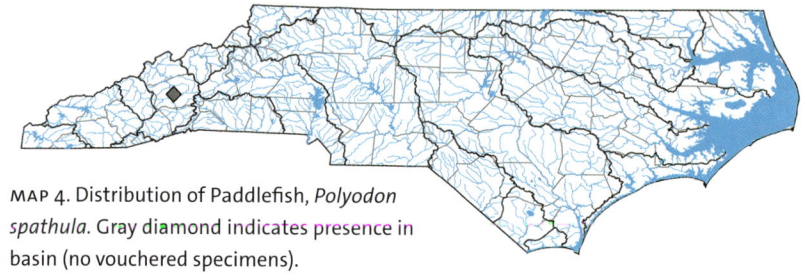

MAP 4. Distribution of Paddlefish, *Polyodon spathula*. Gray diamond indicates presence in basin (no vouchered specimens).

Lepisosteidae—Longnose Gar, *Lepisosteus osseus* (Linnaeus, 1758)

- **Unique characteristics:** Elongate, slender jaws with many needlelike teeth. Long, slender body covered with large, armor-like ganoid scales that are rhomboid. Lateral line scales ca. 57–63. Dorsal fin short, less than half TL, located near caudal fin. Caudal fin is abbreviate heterocercal (figure 44).

- **Most likely to be confused with (in fresh water):** Atlantic Needlefish (family Belonidae). The Atlantic Needlefish has scales that are very small and cycloid; lateral line scales are ca. 300; and the caudal fin is homocercal.
- **Colloquial names:** needlenosed gar, gar-pike.

FIGURE 44. Longnose Gar, *Lepisosteus osseus*. A—adult; B—young-of-year.

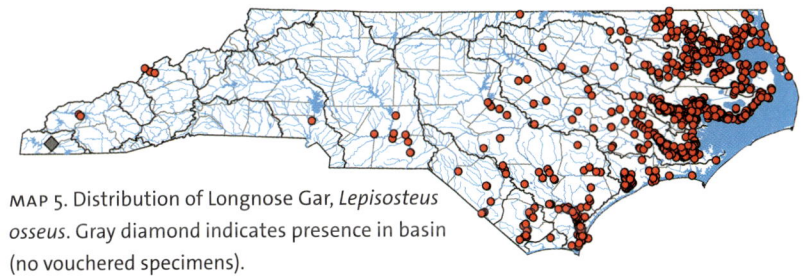

MAP 5. Distribution of Longnose Gar, *Lepisosteus osseus*. Gray diamond indicates presence in basin (no vouchered specimens).

Amiidae—Bowfin, *Amia calva* Linnaeus, 1766

- **Unique characteristics:** Large gular plate found on the throat between the two jaws. Caudal fin is abbreviate heterocercal. Long dorsal fin that extends for more than half TL (Brownstein et al., 2022) (figure 45).
- **Most likely to be confused with (in fresh water):** Juvenile Eastern Mudminnow and Snakehead (family Chanidae) (Robins et al., 2018). The Eastern Mudminnow (family Umbridae) has a short dorsal fin less than half TL, and there is no large gular plate found on the throat between the two jaws. The Snakehead does not have a gular plate, and the anal fin is elongated.
- **Colloquial names:** blackfish, cypress trout, grinnel, grindle, choupique.

FIGURE 45. Bowfin, *Amia calva*. A—adult; B—young-of-year.

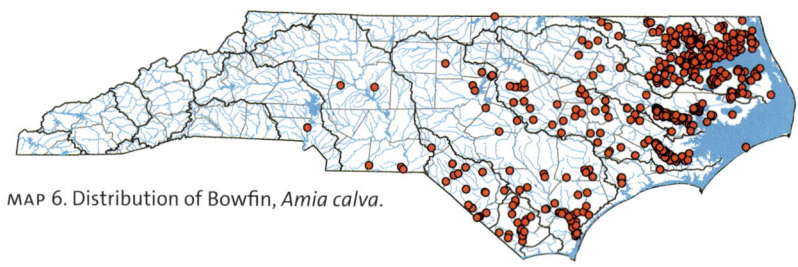

MAP 6. Distribution of Bowfin, *Amia calva*.

Anguillidae—American Eel, *Anguilla rostrata* (Lesueur, 1817)

- **Unique characteristics:** Body elongate and serpentine-like. Dorsal fin long. Jaws present. One pair of gill openings. Pectoral fins present; pelvic fins absent. Tiny, embedded scales (figure 46).
- **Most likely to be confused with (in fresh water):** Lampreys (family Petromyzontidae). Lampreys are lacking jaws, pectoral fins, and scales and have seven pairs of gill openings.
- **Colloquial names:** slippery eel, silver eel, freshwater eel.

FIGURE 46. American Eel, *Anguilla rostrata*.

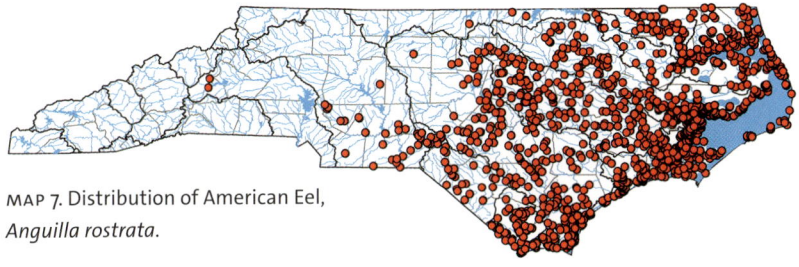

MAP 7. Distribution of American Eel, *Anguilla rostrata*.

Hiodontidae—Mooneye, *Hiodon tergisus* Lesueur, 1818

- **Unique characteristics:** Lateral line present. Breast without a sawtooth margin. Untoothed keel along the belly from the pelvic fin bases to the anus. Strongly laterally compressed body. Scales absent from atop the head. A single dorsal fin; fins without spines. Anal fin origin behind dorsal fin origin. Large eyes with adipose eyelids (figure 47).
- **Most likely to be confused with (in fresh water):** Gizzard Shad and Threadfin Shad (family Dorosomatidae). For both species, the lateral line is absent, midline of the breast has a sawtooth line of scales, and tongue is devoid of teeth.
- **Remarks:** Range restricted to French Broad basin.
- **Colloquial names:** slicker, toothed herring, moon-eye.

FIGURE 47. Mooneye, *Hiodon tergisus*. Ohio. Photograph courtesy of Brian Zimmerman.

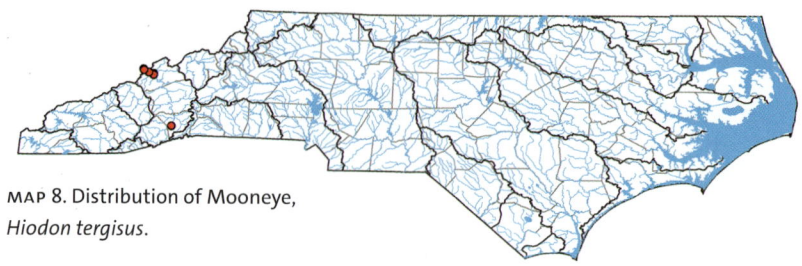

MAP 8. Distribution of Mooneye,
Hiodon tergisus.

Engraulidae—Bay Anchovy, *Anchoa mitchilli* (Valenciennes, 1848)

- **Unique characteristics:** Jaws very long, extending far behind eye; maxilla long, tip pointed, reaching onto or beyond preoperculum; mouth strongly inferior. Dorsal fin origin above anal fin origin (figure 48).
- **Most likely to be confused with (in fresh water):** No other species.
- **Colloquial names:** anchovy.

FIGURE 48. Bay Anchovy, *Anchoa mitchilli*. Florida. Photograph courtesy of Dan Marotta.

MAP 9. Distribution of Bay Anchovy,
Anchoa mitchilli.

Cobitidae—Pond Loach, *Misgurnus anguillicaudatus* (Cantor, 1842)

- **Unique characteristics:** Elongate body. Subterminal mouth surrounded by ten to twelve barbels. Caudal fin rounded. Stout spine on pectoral fin. Dorsal fin origin above pelvic fin origin. Tiny scales (figure 49).
- **Most likely to be confused with (in fresh water):** No other species.
- **Remarks:** Nonindigenous; range currently restricted to upper Cape Fear basin.
- **Colloquial names:** dojo, dojo loach, weather loach, Japanese weatherfish, Amur weatherfish, Oriental weatherfish.

FIGURE 49. Pond Loach, *Misgurnus anguillicaudatus*.

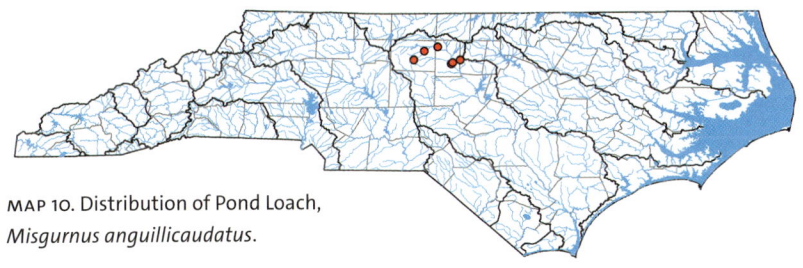

MAP 10. Distribution of Pond Loach, *Misgurnus anguillicaudatus*.

Loricariidae—Amazon Sailfin Catfish, *Pterygoplichthys pardalis* (Castelnau, 1855)

- **Unique characteristics:** Body covered with flexible bony plates. One pair of barbels on a large, subterminal mouth with papillose lips. Adipose fin with anterior spine. Dorsal fin with one spine and ten to fourteen rays (figure 50).
- **Most likely to be confused with (in fresh water):** No other species.
- **Remarks:** Illegally introduced into Lake Julian (Buncombe County, French Broad basin) in ca. 1997 (Tracy, Rohde, and Hogue, 2020). Persisted due to high temperatures in the lake from the thermal discharge of the coal-fired

and gas-fired power generating plant. Thermal discharge ceased in February 2020, and this species was not detected in surveys in September 2021 (Kyle Hussey, Duke Energy, personal communication).

• Colloquial names: armored catfish, pleco.

FIGURE 50. Amazon Sailfin Catfish, *Pterygoplichthys pardalis*. Peru.

MAP 11. Distribution of Amazon Sailfin Catfish, *Pterygoplichthys pardalis*.

Umbridae—Eastern Mudminnow, *Umbra pygmaea* (DeKay, 1842)

• Unique characteristics: Stout, cylindrical body with broad head and short snout. Mouth terminal and not protractile, upper jaw broadly bound to snout with a wide frenum. Caudal fin rounded. Lateral line absent. Body color brown with ten to fourteen thin dark-brown stripes on the sides and a black bar just before the caudal fin base (figure 51).

• Most likely to be confused with (in fresh water): Juvenile Bowfin. The Bowfin has a large gular plate found on the throat between the jaws.

• Colloquial names: mud-fish.

FIGURE 51. Eastern Mudminnow, *Umbra pygmaea*. A—male; B—female.

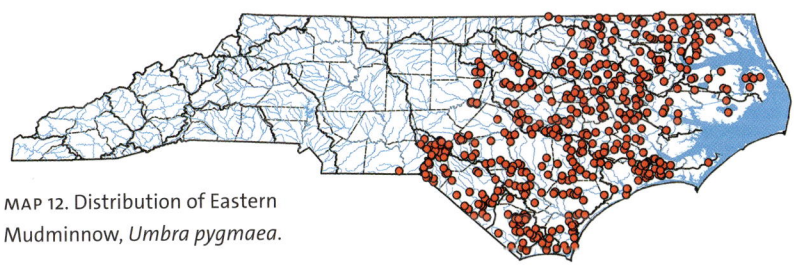

MAP 12. Distribution of Eastern Mudminnow, *Umbra pygmaea*.

Aphredoderidae—Pirate Perch, *Aphredoderus sayanus* (Gilliams, 1824)

- Unique characteristics: Stout-bodied with a large head, terminal mouth, and single dorsal fin. Dark bar below the eye and a larger one at caudal fin base. The anus migrates forward to the throat region between the gills by the late juvenile–early adult stage (figure 52).
- Most likely to be confused with (in fresh water): Small sunfishes (family Centrarchidae). In sunfishes, the position of the anus is immediately anterior to the anal fin in all stages.
- Colloquial names: None.

FIGURE 52. Pirate Perch, *Aphredoderus sayanus*.

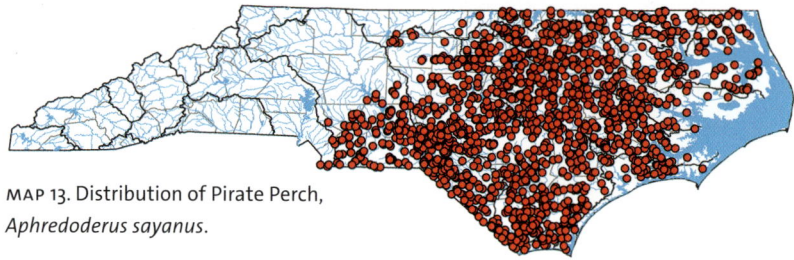

MAP 13. Distribution of Pirate Perch, *Aphredoderus sayanus*.

Amblyopsidae—Swampfish, *Chologaster cornuta* Agassiz, 1853

- **Unique characteristics:** Three dark stripes on the sides and a brown back that contrasts sharply with creamy white to yellow belly. Lacks pelvic fins, and mouth is supraterminal. Eyes are covered with translucent skin. The anus migrates forward to the throat region between the gills by the early juvenile stage. Scales embedded, not visible (figure 53).
- **Most likely to be confused with (in fresh water):** No other species.
- **Colloquial names:** fish of the Dismal Swamp.

FIGURE 53. Swampfish, *Chologaster cornuta*.

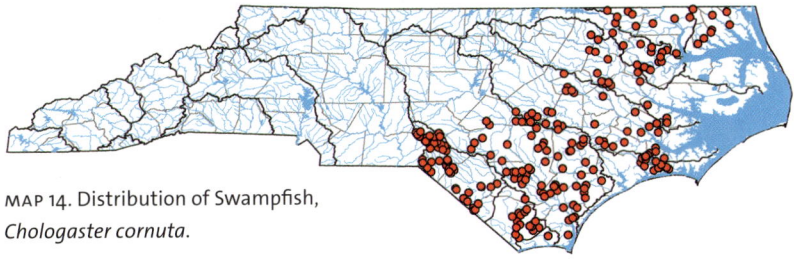

MAP 14. Distribution of Swampfish,
Chologaster cornuta.

Achiridae—Hogchoker, *Trinectes maculatus* (Bloch and Schneider, 1801)

- **Unique characteristics:** Flat, right-facing fishes (that is, the left side of the body lies on the substrate) with small, minute eyes. Lacking pectoral fins and with a small, inconspicuous mouth (figure 54).
- **Most likely to be confused with (in fresh water):** No other species.
- **Colloquial names:** flounder, sole.

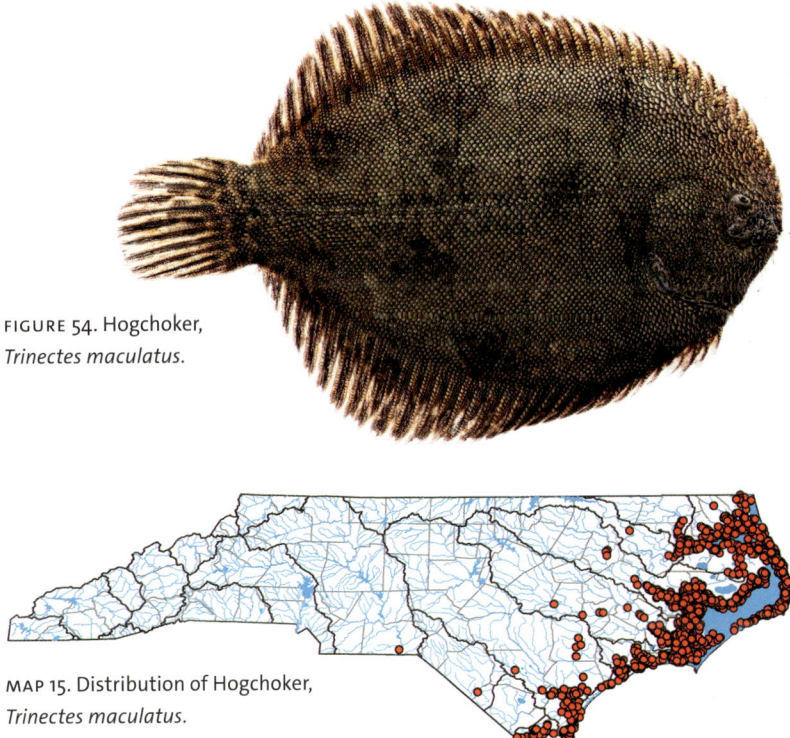

FIGURE 54. Hogchoker,
Trinectes maculatus.

MAP 15. Distribution of Hogchoker,
Trinectes maculatus.

Cyprinodontidae—Sheepshead Minnow, *Cyprinodon variegatus* Lacepède, 1803

- **Unique characteristics:** Deep-bodied with an arched back and five to eight triangular-shaped dark bars on the side. Humeral scales at least twice as large as surrounding scales. One row of tricuspid teeth on each jaw (figure 55).
- **Most likely to be confused with (in fresh water):** No other species.
- **Colloquial names:** variegated minnow, short minnow.

A

B

FIGURE 55. Sheepshead Minnow, *Cyprinodon variegatus*.
A—male; B—female.

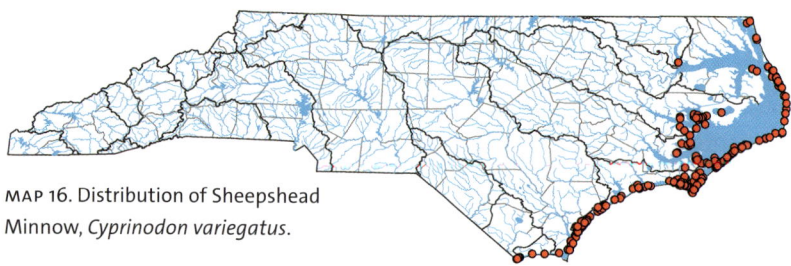

MAP 16. Distribution of Sheepshead
Minnow, *Cyprinodon variegatus*.

Belonidae—Atlantic Needlefish, *Strongylura marina* (Walbaum, 1792)

- **Unique characteristics:** Very small cycloid scales, lateral line scales ca. 300. The caudal fin is homocercal, and the body is slender (figure 56).
- **Most likely to be confused with (in fresh water):** Longnose Gar. The Longnose Gar has large ganoid scales, lateral line scales ca. 57–63. The caudal fin is abbreviate heterocercal, and the body is stout.
- **Colloquial names:** bill-fish, gar-fish, green-gar, doctor-fish.

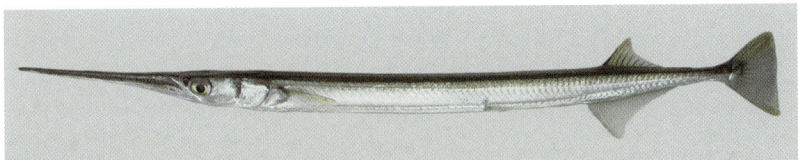

FIGURE 56. Atlantic Needlefish, *Strongylura marina*.

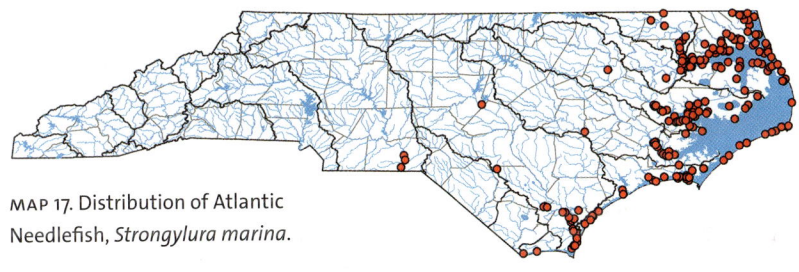

MAP 17. Distribution of Atlantic Needlefish, *Strongylura marina*.

Gasterosteidae—Fourspine Stickleback, *Apeltes quadracus* (Mitchill, 1815)

- **Unique characteristics:** Scaleless body with four dorsal spines angled alternately left to right; pelvic fins below pectoral fins. Long, slender caudal peduncle; no bony plates on sides (figure 57).
- **Most likely to be confused with (in fresh water):** The Threespine Stickleback, *Gasterosteus aculeatus* (figure 58), was documented on two separate occasions, 1974 and 1979, in estuarine waters along North Carolina's coast. The Threespine Stickleback has three dorsal spines, a short caudal peduncle, pelvic fins behind pectoral fins, and bony plates on the sides. While it

is unlikely that one will encounter the Threespine Stickleback in North Carolina waters, these two species may be separated using the following identification key:

1a. Four (three to five) dorsal fin spines angled alternately left to right with the first two longer than the last two and with a wide gap present before the last spine. Pelvic fins below pectoral fins. Long, slender caudal peduncle; lateral keel absent. No bony plates on sides. Tail rounded. Back olive brown; side mottled deep brown; and venter silver-white. . . . . Fourspine Stickleback, *Apeltes quadracus* (Mitchill, 1815)

1b. Three dorsal fin spines. Pelvic fins posterior to pectoral fins. Caudal peduncle short and with a lateral keel. Conspicuous bony plates on sides. Tail straight. Body profusely covered with black specks. . . . . Threespine Stickleback, *Gasterosteus aculeatus* Linnaeus, 1758

• **Colloquial names:** None.

FIGURE 57. Fourspine Stickleback, *Apeltes quadracus*.

FIGURE 58. Threespine Stickleback, *Gasterosteus aculeatus*. Washington.

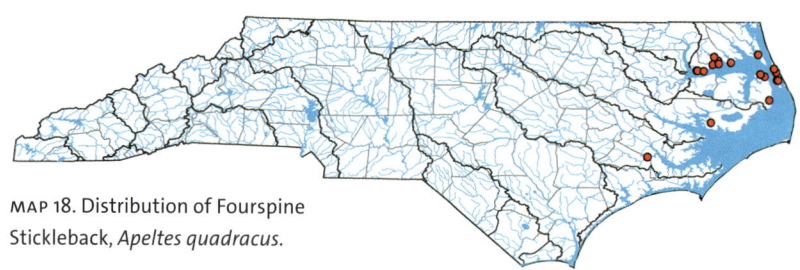

MAP 18. Distribution of Fourspine Stickleback, *Apeltes quadracus*.

Lampreys
Family Petromyzontidae

Lampreys, along with hagfishes, are the most primitive of all fishes, having been around for more than 300 million years. In North Carolina, they constitute a small family with only five species (table 7). Most people are not aware of their existence, unless they are fortunate enough to observe a spawning aggregation in the riffles of a clear Mountain or Coastal Plain stream during the late winter or early spring or hook a fish that has an eel-like "critter" attached. Lampreys are usually mistaken for eels because they are slender and slippery, but lampreys do not have jaws or scales, unlike eels. Lampreys range in size from ca. 100 to 1,200 mm TL. The smallest is the Least Brook Lamprey and the largest the Sea Lamprey.

In North Carolina, they are broadly distributed in many Mountain basins but are mostly absent from streams in the Piedmont portion of the Yadkin, Cape Fear, Neuse, Tar, and Roanoke basins. The only species found in three Coastal basins is the Sea Lamprey (Tracy, Rohde, and Hogue, 2020; table 2). Lampreys are not known to occur in the Savannah, Pigeon, Watauga, New, Broad, Catawba, White Oak, Waccamaw, or Lumber basins; all other basins are known to have at least one species (Tracy, Rohde, and Hogue, 2020). Due to endemism in specific basins, three species are considered imperiled in North Carolina (table 7; NCAC, 2023; NCNHP, 2022; NCWRC, 2021).

The Sea Lamprey is our most widely distributed species, being found in eight basins and occupying rivers and streams from the Fall Zone throughout the Coastal Plain. The probability exists that historically Sea Lampreys occurred in the Broad basin and could currently occur in the Waccamaw, White Oak, and Lumber basins. Cope (1870b) reported Sea Lampreys from the Catawba basin, possibly as far upstream as at Morganton (Burke County), but there are

TABLE 7. Species of lampreys in North Carolina

Scientific Name	Common Name	Imperilment Status
Ichthyomyzon bdellium	Ohio Lamprey	state special concern
Ichthyomyzon greeleyi	Mountain Brook Lamprey	
Lampetra aepyptera	Least Brook Lamprey	state threatened
Lethenteron appendix	American Brook Lamprey	state special concern
Petromyzon marinus	Sea Lamprey	

no verifiable specimens to substantiate his claim (Tracy, Rohde, and Hogue, 2020). In contrast to the Sea Lamprey, the American Brook Lamprey is found only in the French Broad and Roanoke basins. In the French Broad basin, it is extremely rare, being found only in the lower reaches of Spring Creek in Madison County. In the Roanoke basin, it has been found in two small tributaries to the Roanoke River in Halifax County. In many Mountain streams, the Ohio Lamprey inhabits medium-sized to large rivers such as the Tuckasegee, French Broad, and Nolichucky Rivers. The Mountain Brook Lamprey, which sometimes co-occurs with the Ohio Lamprey, usually inhabits much smaller streams than the Ohio Lamprey and is more commonly found. The Least Brook Lamprey resides in sandy-bottom, slow-moving, slightly acidic Coastal Plain streams or small streams along the Fall Zone between the Piedmont and the Coastal Plain in the Neuse and Tar basins. As compared to other families of fishes—for example, sunfishes, catfishes, and minnows—it is our understanding that no species of lamprey has been introduced outside of its native range in North Carolina.

The Sea Lamprey is an anadromous species. It spawns in fresh water, migrates downstream as a juvenile, and matures to adulthood in our estuarine coastal waters or the Atlantic Ocean. Then, when sexually mature, it migrates back upstream, often over long distances, to spawn in small streams (Rohde et al., 1994). All other lamprey species in North Carolina are strictly freshwater inhabitants.

A lamprey in its immature stage is called an ammocoete and only vaguely resembles an adult. Ammocoetes can be found in shallow, sandy, silty areas along the margins of streams. Transforming juveniles and spawning adults are found in riffles, clinging to and moving small stones and pebbles as they construct a spawning nest. Ammocoetes of all species are filter feeders on bacteria, algae, protozoans, and fine particulate organic matter.

Our two parasitic species, the Sea Lamprey and Ohio Lamprey, attach themselves to larger fishes when not spawning. Juvenile and adult Sea Lampreys have
been found attached to (or evidence of previous attachment has been noted)
and feeding upon the American Shad, Hickory Shad, Robust Redhorse, catfishes (family Ictaluridae), black basses (family Centrarchidae), and other large-
bodied fishes. The Ohio Lamprey has been observed attached to (or evidence of
previous attachment has been noted) redhorse suckers (family Catostomidae),
the Northern Hog Sucker, the Common Carp, black basses, catfishes, darters
(family Percidae) (Jenkins and Burkhead, 1994), and River Chub. The American Brook, Least Brook, and Mountain Brook Lampreys are nonparasitic and
do not feed as adults.

Key characteristics for the proper identification of lampreys, visible only
when examined with the aid of a dissecting microscope, include the development and presence or absence of teeth on the oral disk and the number of trunk
myomeres. Identification can be difficult, and knowing the geographical distributions of the species can help, but some do have overlapping distributions.

Identification Key to the Species of Lampreys (Family Petromyzontidae)

Oral area surrounded on top and sides by an oral hood that lacks a marginal
papillary fringe. Mouth cavity having numerous fimbriate oral cirri (appearing sievelike). Teeth and tongue absent.......**Ammocoetes (larvae)**
Oral area lacking a hood, being round or oval, disklike (disk puckered partly
closed in some specimens), the disk having a papillary fringe. Oral cirri
absent. Teeth and tongue present...................................**Adults**

ADULT LAMPREYS

1a. A single indented dorsal fin (figure 59)..........*Ichthyomyzon*, Couplet 2
1b. Two dorsal fins, either widely separate in immature individuals or con-
tiguous in mature individuals (figure 59)..................................3

A

Single indented dorsal fin

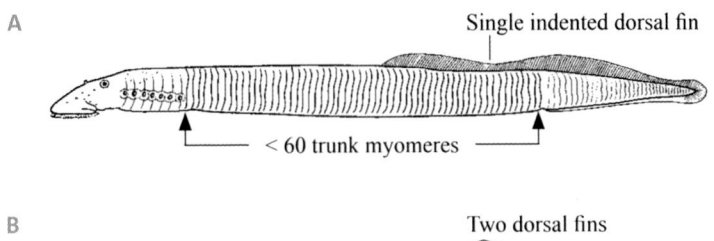

< 60 trunk myomeres

B

Two dorsal fins

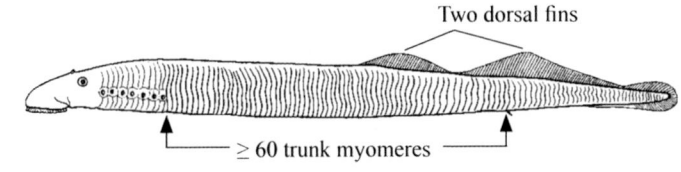

≥ 60 trunk myomeres

FIGURE 59. Side view of an adult lamprey showing the trunk myomeres. A—species with one dorsal fin; B—species with two dorsal fins. Trunk myomeres are counted from the last gill opening to the anus in both adults and ammocoetes. Source: Food and Agriculture Organization of the United Nations, 2011, C. B. Renaud, *Lampreys of the World: An Annotated and Illustrated Catalogue of Lamprey Species Known to Date*, https://www .fao.org/publications/card/en/c/b872b257-e293-566c-aee1-fb1a4aab8bdb. Reproduced with permission.

2a. Posterior field teeth bluntly rounded (figures 60–62). Lateral line neuromasts on the ventral surface of the branchial region unpigmented. Mature adults nonparasitic with a nonfunctional intestine that is thin, granular, and fragmentary (discernible only upon dissection) (map 19)..............
............Mountain Brook Lamprey, *Ichthyomyzon greeleyi* Hubbs and Trautman, 1937

2b. All teeth sharp and strongly pointed (figures 60–62). Lateral line neuromasts on the ventral surface of the branchial region darkly pigmented. Mature adults parasitic with a large, smooth, and round functional intestine (discernible only upon dissection) (map 20).........Ohio Lamprey, *Ichthyomyzon bdellium* (Jordan, 1885)

FIGURE 60. Oral disks. A—Mountain Brook Lamprey, *Ichthyomyzon greeleyi*; B—Ohio Lamprey, *Ichthyomyzon bdellium*.

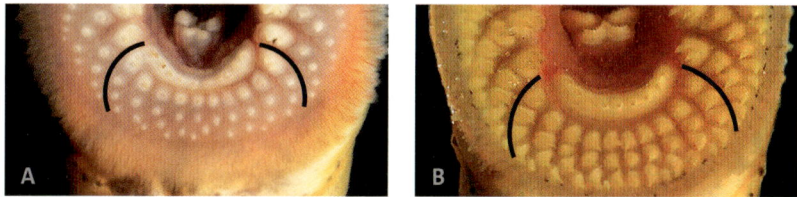

FIGURE 61. Close-up of posterior field of teeth on oral disk, indicated between curved lines. A—teeth bluntly rounded in Mountain Brook Lamprey, *Ichthyomyzon greeleyi*; B—teeth strongly pointed in Ohio Lamprey, *Ichthyomyzon bdellium*.

FIGURE 62. A—Mountain Brook Lamprey, *Ichthyomyzon greeleyi*; B—Ohio Lamprey, *Ichthyomyzon bdellium*.

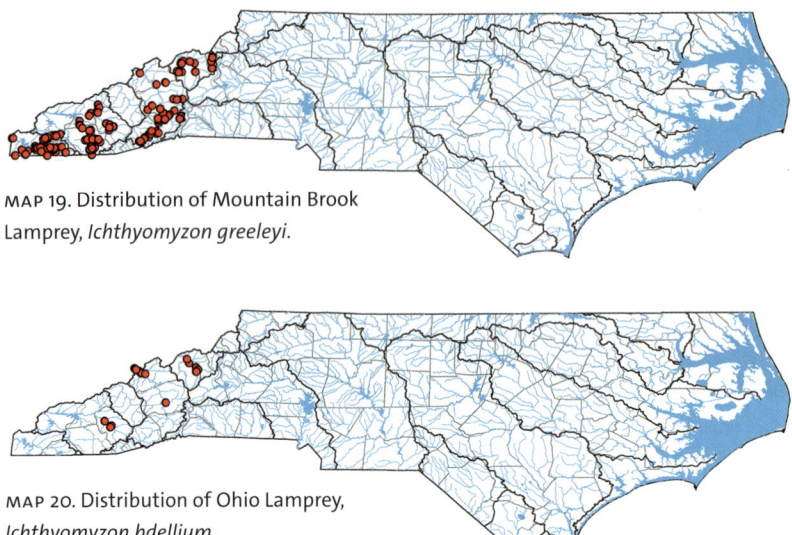

MAP 19. Distribution of Mountain Brook Lamprey, *Ichthyomyzon greeleyi.*

MAP 20. Distribution of Ohio Lamprey, *Ichthyomyzon bdellium.*

3a. Teeth well developed; supraoral lamina a single pointed bicuspid tooth (figure 63). Parasitic (figure 64) (map 21)..................................
.....................Sea Lamprey, *Petromyzon marinus* (Linnaeus, 1758)

3b. Teeth not well developed; supraoral lamina with two teeth, either unicuspid or bicuspid, separated by a wide bridge, which may or may not bear cusps. Nonparasitic..4

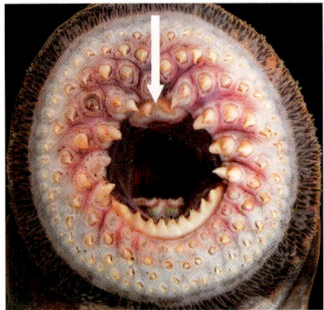

FIGURE 63. Oral disk of Sea Lamprey, *Petromyzon marinus*, with arrow indicating supraoral lamina with a single pointed bicuspid tooth. New Hampshire. Photograph courtesy of Brian Zimmerman.

FIGURE 64. Sea Lamprey, *Petromyzon marinus*, adult. New Hampshire. Photograph courtesy of Brian Zimmerman.

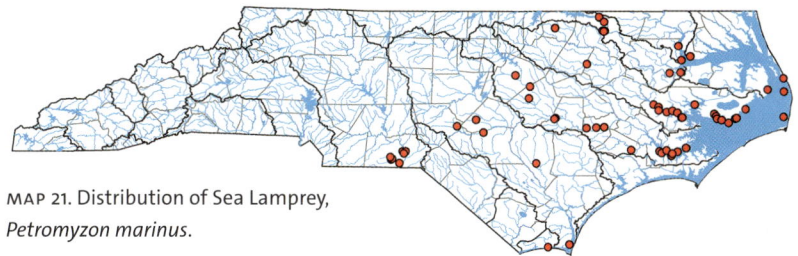

MAP 21. Distribution of Sea Lamprey, *Petromyzon marinus*.

4a. Supraoral tooth cusps widely separated; lateral teeth weakly developed (figures 65 and 66). Trunk myomeres 53–63 (figure 59). Range restricted to Neuse and Tar basins (map 22)...
................Least Brook Lamprey, *Lampetra aepyptera* (Abbott, 1860)

4b. Supraoral tooth cusps not widely separated; lateral teeth moderately developed (figures 65 and 67). Trunk myomeres 59–74 (figure 59). Range restricted to Spring Creek watershed in Madison County, French Broad basin, and to two small tributaries to the Roanoke River in Halifax County, Roanoke basin (map 23)...
..........American Brook Lamprey, *Lethenteron appendix* (DeKay, 1842)

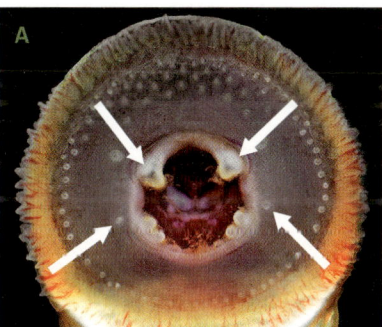 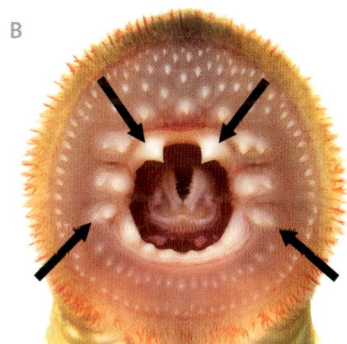

FIGURE 65. Oral disks. A—Least Brook Lamprey, *Lampetra aepyptera*, with arrows indicating well-separated supraoral tooth cusps and weakly developed lateral teeth; B—American Brook Lamprey, *Lethenteron appendix*, with arrows indicating not as widely separated supraoral tooth cusps and moderately developed lateral teeth.

FIGURE 66. Least Brook Lamprey, *Lampetra aepyptera*. A—male; B—female.

FIGURE 67. American Brook Lamprey, *Lethenteron appendix*.

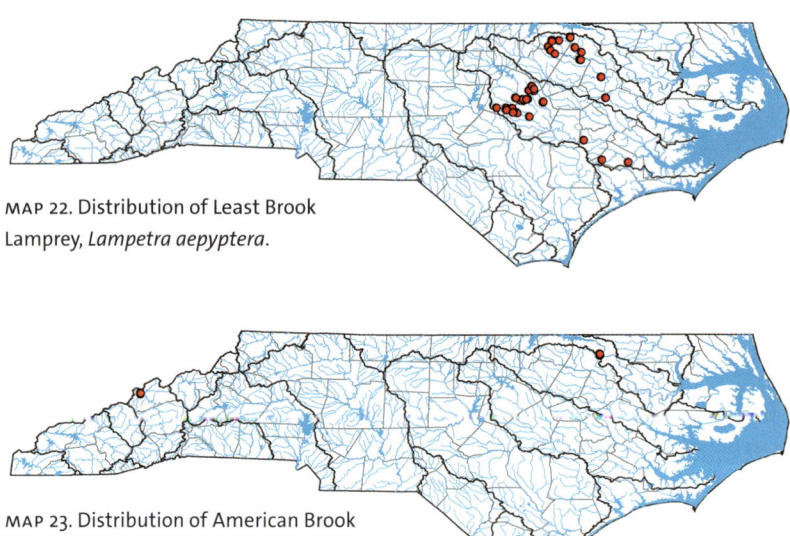

MAP 22. Distribution of Least Brook Lamprey, *Lampetra aepyptera*.

MAP 23. Distribution of American Brook Lamprey, *Lethenteron appendix*.

AMMOCOETE (LARVAL) LAMPREYS

1a. Single slightly indented dorsal fin (figures 59 and 68)......................
....*Ichthyomyzon*, Couplet 2

1b. Two dorsal fins (a low-lying membrane not supported by fin rays may unite the two fins, especially in smaller specimens) (figure 59)..........3

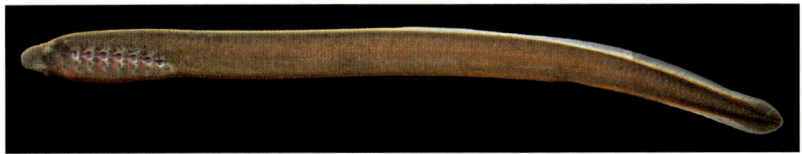

FIGURE 68. Mountain Brook Lamprey, *Ichthyomyzon greeleyi*, ammocoete.

2a. Subocular area unpigmented to lightly pigmented........................
.............Mountain Brook Lamprey, *Ichthyomyzon greeleyi* Hubbs and Trautman, 1937[1]

2b. Subocular area moderately pigmented....................................
...................Ohio Lamprey, *Ichthyomyzon bdellium* (Jordan, 1885)[1]

3a. Trunk myomeres 51–60 (figure 59)...
...............Least Brook Lamprey, *Lampetra aepyptera* (Abbott, 1860)

3b. Trunk myomeres 61–74 (figure 59).......................................4

1. Where their distributions are sympatric, the adult Ohio Lamprey, *Ichthyomyzon bdellium*, may be easily separated from the adult Mountain Brook Lamprey, *I. greeleyi* (Couplets 2a and 2b). However, the ammocoetes (larvae) cannot be easily separated by pigmentation patterns, contrary to Lanteigne (1988) and Renaud (2011) (Jenkins and Burkhead, 1994; B. H. Tracy, personal observation; Tracy, Rohde, and Hogue, 2020).

4a. Specimens 100 mm TL or greater—Tail rounded. Pigmentation in tail extending nearly to fin margin. Pigmented in upper branchial region with a narrow pale area above gill openings. Suborbital area and lower half of lateral portion of oral hood largely pigmented (figure 69)................
........................Sea Lamprey, *Petromyzon marinus* Linnaeus, 1758

4b. Specimens 100 mm TL or greater—Tail slightly pointed, somewhat wedge-like. Pigmentation in tail not extending to fin margin. Pigmentation in upper branchial region leaving wide pale area above gill openings. Suborbital area and lower half of lateral portion of oral hood largely unpigmented (figure 69)..
.........American Brook Lamprey, *Lethenteron appendix* (DeKay, 1842)

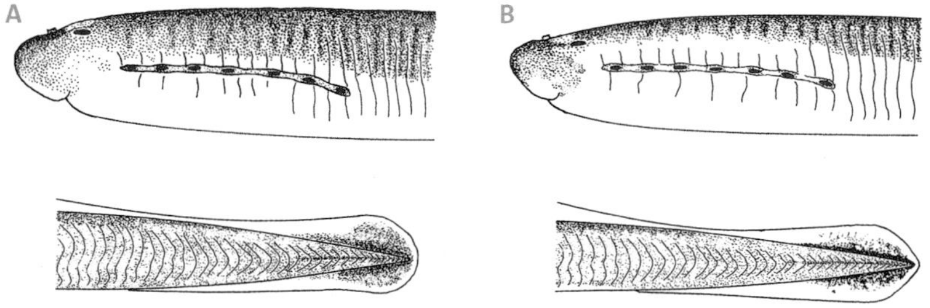

FIGURE 69. Head and caudal pigmentation and caudal fin shape in lamprey ammocoetes. A—Sea Lamprey, *Petromyzon marinus*; B—American Brook Lamprey, *Lethenteron appendix*. Illustration adapted from Jenkins and Burkhead (1994).

Sturgeons
Family Acipenseridae

Three of our largest freshwater fishes, sturgeons, are found in North Carolina's coastal waters, rivers, and estuaries and even in our largest mountain river (table 2). The Atlantic Sturgeon is the largest freshwater species and ranges in size from ca. 880 to 4,300 mm TL; in comparison, the Least Killifish is our smallest freshwater fish at just 36 mm TL. The Lake Sturgeon reaches a maximum length of 2,700 mm TL, and the Shortnose Sturgeon ranges from ca. 430 to 1,090 mm TL (Page and Burr, 2011; Rohde et al., 2009). All three species are listed as imperiled in North Carolina (table 8; NCAC, 2023; NCNHP, 2022; NCWRC, 2021).

The Shortnose Sturgeon is an anadromous species, meaning it migrates from the ocean into fresh water to spawn. However, unlike the Atlantic Sturgeon, the Shortnose Sturgeon does not venture much beyond the high salinity estuaries into the ocean (Rohde et al., 2009; Tracy, Rohde, and Hogue, 2020). Until the late 1980s, the only valid historical record of a Shortnose Sturgeon was from Salmon Creek in Bertie County (Chowan basin) in 1881. In 1985 a gravid female was caught in the Pee Dee River in Anson and Richmond Counties, downstream from the US 74 bridge (Yadkin basin). None have been detected since in the North Carolina portion of the Pee Dee River, but the South Carolina Department of Natural Resources tracked several in 2002–2003 to within 5.6 kilometers of the state line. The first verifiable record from the Cape Fear basin was captured in a gill net in the lower Cape Fear River in 1978. Other more recent records include an adult captured in Albemarle Sound in 1998 and another near the mouth of the Chowan River, downstream from Salmon Creek in 2016 (Tracy, Rohde, and Hogue, 2020).

The Atlantic Sturgeon, also an anadromous species, is found in all the major rivers from the Yadkin to the Lumber basin, except in the Shallotte and

TABLE 8. Species of sturgeons in North Carolina

Scientific Name	Common Name	Imperilment Status
Acipenser brevirostrum	Shortnose Sturgeon	federally endangered
Acipenser fulvescens[a]	Lake Sturgeon	state special concern
Acipenser oxyrinchus	Atlantic Sturgeon	federally endangered

[a] Species found in only one river basin.

Waccamaw basins. It is also found in the sounds and in the Atlantic Ocean (Tracy, Rohde, and Hogue, 2020) and migrates out into the Atlantic Ocean and along the coast (Rohde et al., 2009). There is an anecdotal record from 1882 of the capture of a 201-kilogram specimen from the Haw River, Chatham County (Cape Fear basin). It is possible that the Atlantic Sturgeon historically may have migrated and spawned up into the Fall Zone in the Broad, Catawba, and Cape Fear basins. In the Neuse basin there are recent records from the Neuse River at Goldsboro and Smithfield (Wayne and Johnston Counties, respectively), in the Tar basin from the Tar River near Tarboro (Edgecombe County), in the Roanoke basin from the Roanoke River as far upstream as Weldon (Halifax County), and in the Chowan basin beyond NC 11 in Potecasi Creek (Hertford County). In mid-September 2018 a large adult, perhaps a fall spawning migrant, was detected at Blewett Falls Dam near Rockingham (Richmond County, Yadkin basin) (Tracy, Rohde, and Hogue, 2020) and another in 2019 migrating from the upper Pee Dee River in South Carolina to spawn near Rockingham.

The Lake Sturgeon is potamodromous, meaning it migrates up and down rivers to and from its spawning grounds, always remaining in fresh water. Until recently, the Lake Sturgeon was considered extirpated from the state (NCNHP, 2022). In 2015, a project began to reintroduce the Lake Sturgeon into its native waters in the French Broad basin, the only basin in which it previously occurred. Currently, more than 24,000 young juvenile Lake Sturgeon have been released into the French Broad River between the towns of Hot Springs and Marshall (Madison County) (Tracy, Rohde, and Hogue, 2020).

The recreational and commercial harvesting (take) of any species of sturgeon is prohibited, and any caught must be immediately released (NCDMF 2020; NCWRC, 2022). Please report stranded, injured, or dead sturgeons to the National Oceanic and Atmospheric Administration (NOAA) at (978) 281-9328 or in the Southeast at (844) STURG-911 or (844) 788-7491, or send NOAA an email at Sturg911@noaa.gov. Instructions for resuscitating a sturgeon can be found at https://media.fisheries.noaa.gov/dam-migration-miss/Resuscitation-Cards-120513.pdf.

Key characteristics for the proper identification of sturgeons include the shape and length of the snouts; presence or absence of bony plates between the anal fin and the midlateral scutes; and relative sizes of mouth gape widths. The identification of sturgeons can also be based on geographical distribution (for example, the Lake Sturgeon is the only sturgeon species found west of the Appalachian Mountains).

Identification Key to the Species of Sturgeons (Family Acipenseridae)

1a. Range restricted to French Broad River downstream from the town of Marshall in Madison County (French Broad basin) (figure 70) (map 24) ....Lake Sturgeon, *Acipenser fulvescens* Rafinesque, 1817

1b. Range restricted to Atlantic slope basin rivers and coastal waters........2

FIGURE 70. Lake Sturgeon, *Acipenser fulvescens*.

MAP 24. Distribution of Lake Sturgeon, *Acipenser fulvescens*.

2a. Snout short and blunt (figure 71). Mouth large; the inner gape width is usually greater than 62 percent of the interorbital width (figure 72). One large scute between anal fin and caudal fin (map 25)......................
................Shortnose Sturgeon, *Acipenser brevirostrum* Lesueur, 1818

2b. Snout long and pointed (figure 73). Mouth small; the inner gape width is usually less than 60 percent of the interorbital width (figure 72). Typically, four small scutes, often in two pairs between anal fin and caudal fin (map 26)...........Atlantic Sturgeon, *Acipenser oxyrinchus* Mitchill, 1815

FIGURE 71. Shortnose Sturgeon, *Acipenser brevirostrum*. South Carolina. National Marine Fisheries Service Endangered Species Act Permit 20528–03.

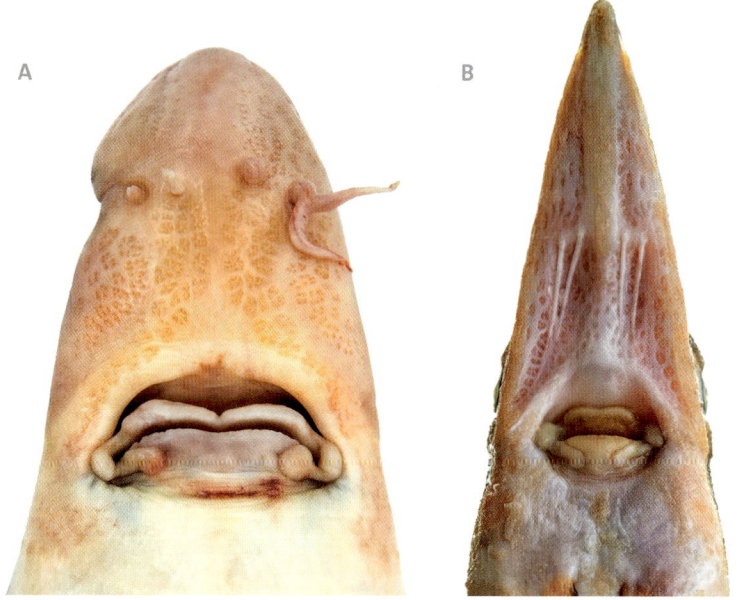

FIGURE 72. Sturgeon mouths showing relative sizes of gape widths. A—Shortnose Sturgeon, *Acipenser brevirostrum*; B—Atlantic Sturgeon, *Acipenser oxyrinchus*. South Carolina. National Marine Fisheries Service Endangered Species Act Permits 20528-03 and 21198.

FIGURE 73. Atlantic Sturgeon, *Acipenser oxyrinchus*. South Carolina. National Marine Fisheries Service Endangered Species Act Permit 21198.

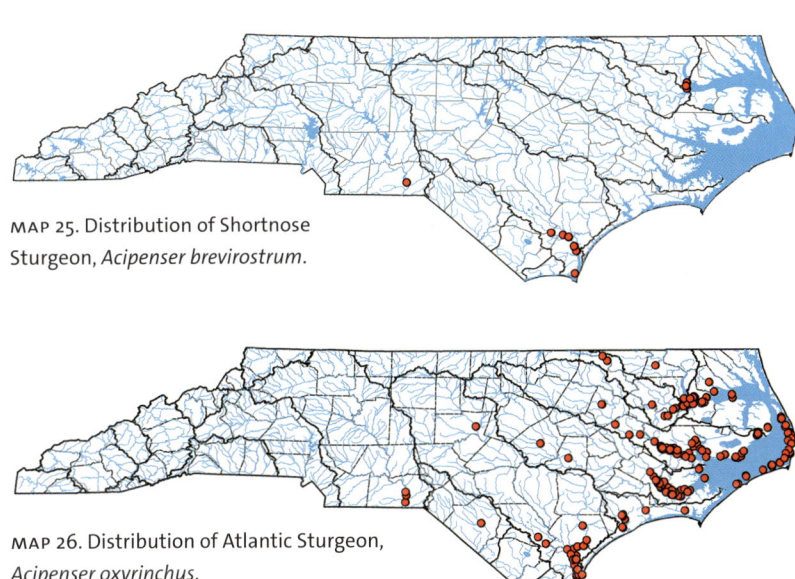

MAP 25. Distribution of Shortnose Sturgeon, *Acipenser brevirostrum*.

MAP 26. Distribution of Atlantic Sturgeon, *Acipenser oxyrinchus*.

6

"Shads"
Families Alosidae and Dorosomatidae

The seven species of Shads and Thread Herrings found in North Carolina were, until recently, classified in the family Clupeidae. These species are now classified in two families: Alosidae, which contains Alewife, American Shad, Atlantic Menhaden, Blueback Herring, and Hickory Shad; and Dorosomatidae, which contains Gizzard Shad and Threadfin Shad (table 9; Wang et al., 2022).

These seven species are widely known by many names, such as bigeye herring, fatback, glut herring, nanny shad, river herring, stink shad, or just plain shad. They occur across the state in freshwater and saltwater environments, but especially in many of our reservoirs, coastal rivers, and estuaries, and offshore (Tracy, Rohde, and Hogue, 2020). Most species are found along the coast, but the Gizzard Shad is our most widely distributed species, found in all basins except for the Savannah, Watauga, and New. None of the species are endemic to a specific river basin in North Carolina (table 2).

TABLE 9. "Shad" species (families Alosidae and Dorosomatidae) found in or along the coast of North Carolina

Scientific Name	Common Name	Management Status
Alosidae		
Alosa aestivalis	Blueback Herring	Managed by NCDMF and NCWRC
Alosa mediocris	Hickory Shad	Managed by NCDMF and NCWRC
Alosa pseudoharengus	Alewife	Managed by NCDMF and NCWRC
Alosa sapidissima	American Shad	Managed by NCDMF and NCWRC
Brevoortia tyrannus	Atlantic Menhaden	Managed by NCDMF and NCWRC
Dorosomatidae		
Dorosoma cepedianum	Gizzard Shad	
Dorosoma petenense	Threadfin Shad	

The Gizzard Shad and Threadfin Shad have been stocked in many reservoirs across the state as forage fish. The Threadfin Shad, a nonindigenous (nonnative) species, was stocked, at times illegally, as a forage fish in many Mountain and Piedmont reservoirs in the Hiwassee, Little Tennessee, Broad, Catawba, Yadkin, and Roanoke basins. Over time, from the Piedmont reservoirs, it has found its way downstream to the coast and is now widely distributed from the Albemarle Sound to the Shallotte basins. The Alewife, indigenous to our Atlantic slope streams, was illegally introduced into Lake Norman (Catawba basin) as a potential new forage fish for the Striped Bass and black basses fisheries. Similar to most poorly thought-out and illegal introductions, the Alewife, along with the introduced Blueback Herring, has caused more harm than good to the fisheries in reservoirs such as lakes James and Norman (Catawba basin), and Hiwassee (Hiwassee basin).

"Shads" vary greatly in size, from a length of ca. 180 mm TL for the Threadfin Shad to ca. 760 mm TL for the American Shad. Because of their size and abundance, "shads" were and continue to be a commercially and recreationally important group of fishes with seasonal and river-basin specific creel and landing limits (see, for example, NCDMF, 2020; NCWRC, 2019b, 2022).

Species of "shads," except for land-locked reservoir populations, make late winter–early spring migratory spawning runs up the coastal rivers. Historically, the runs occurred far upstream into the Piedmont until dam construction along the Fall Zone and overfishing halted their migrations. Implementation of strict harvesting quotas has helped the Alewife, Blueback Herring, and American Shad on their road to recovery, but full recovery may take a long time. Although some populations are severely depleted, no species is listed as imperiled in North Carolina (table 9; NCNHP, 2022; NCWRC, 2021).

Key characteristics for the proper identification of "shads" include the shape and position of the upper jaw; presence or absence of a long filament in the dorsal fin; and body, caudal fin, and peritoneum pigmentation. Identification can be difficult, especially in the field.

Identification Key to the Species of Shads (Family Alosidae) and Thread Herrings (Family Dorosomatidae)

1a. Last dorsal fin ray filamentous (figure 74b)....................................
....Family Dorosomatidae, Couplet 2
1b. Last dorsal fin ray not filamentous............Family Alosidae, Couplet 3

2a. Mouth inferior and subterminal; snout bulbous and fleshy, projecting past upper jaw. Upper jaw with a notch in the posterior ventral margin. Anal fin rays 25–37. More than 50 (52–70) scales in lateral series. Caudal fin dusky (figure 74) (map 27)...
........................Gizzard Shad, *Dorosoma cepedianum* (Lesueur, 1818)

2b. Mouth terminal; snout not bulbous and not extending anterior to upper jaw. Upper jaw without a notch in the posterior ventral margin. Anal fin rays 17–27. Fewer than 50 (41–48) scales in lateral series. Caudal fin yellow (figure 74) (map 28)...
........................Threadfin Shad, *Dorosoma petenense* (Günther, 1867)

FIGURE 74. A—Gizzard Shad, *Dorosoma cepedianum*; B—Threadfin Shad, *Dorosoma petenense*.

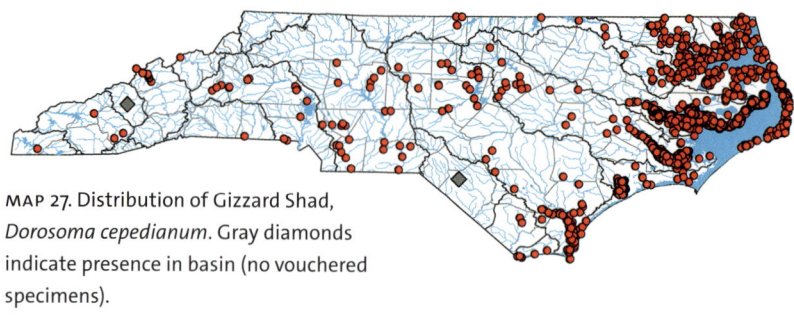

MAP 27. Distribution of Gizzard Shad, *Dorosoma cepedianum*. Gray diamonds indicate presence in basin (no vouchered specimens).

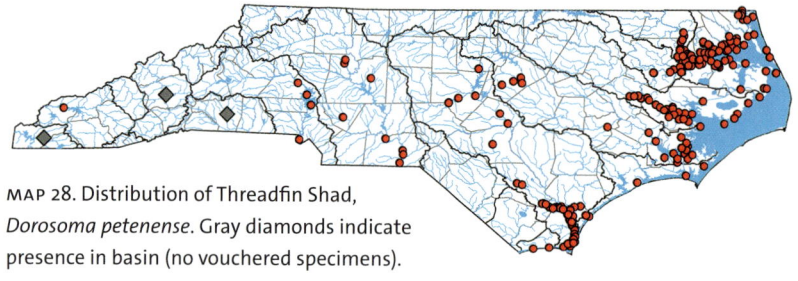

MAP 28. Distribution of Threadfin Shad, *Dorosoma petenense*. Gray diamonds indicate presence in basin (no vouchered specimens).

3a. Predorsal area with a fatty ridge that bears greatly enlarged scales on either side of the dorsal midline (figure 75). Scales crowded anterodorsally; other scales deeply overlapping with exposed margins strongly serrated. Pectoral fin with an axillary scale. Six branched pelvic fin rays (figure 76) (map 29)........Atlantic Menhaden, *Brevoortia tyrannus* (Latrobe, 1802)

3b. Predorsal scales along dorsal midline normal, neither enlarged nor serrated. Scales with smooth posterior margins. Pectoral fin without an axillary scale. Eight branched pelvic fin rays.................*Alosa*, Couplet 4

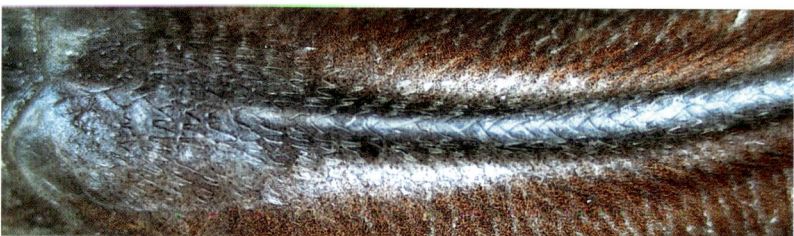

FIGURE 75. Predorsal scales on Atlantic Menhaden, *Brevoortia tyrannus*.

FIGURE 76. Atlantic Menhaden, *Brevoortia tyrannus*.

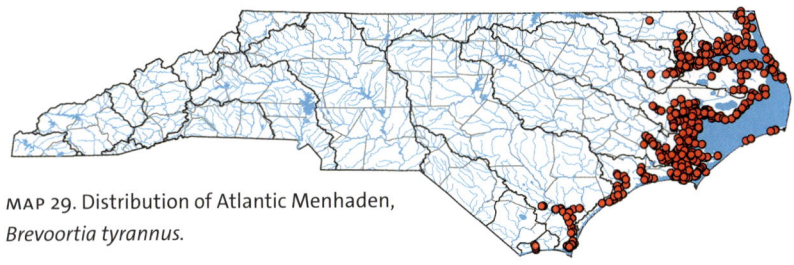

MAP 29. Distribution of Atlantic Menhaden, *Brevoortia tyrannus*.

4a. Upper margin of lower jaw rising steeply within mouth (figure 77)......5
4b. Upper margin of lower jaw rising gradually within mouth (figure 77)...6

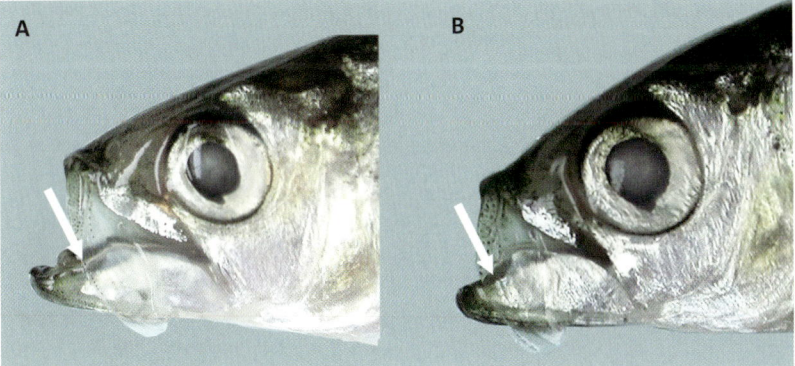

FIGURE 77. Upper margin of lower jaw with arrows indicating gradual (A) or steep (B) rise. A—American Shad, *Alosa sapidissima*; B—Blueback Herring, *Alosa aestivalis*.

5a. Eye diameter greater than length of snout. Peritoneum silvery to pale gray (visible upon dissection). Dorsum grayish green in life (figure 78) (map 30) ....Alewife, *Alosa pseudoharengus* (Wilson, 1811)

5b. Eye diameter less than length of snout. Peritoneum sooty or black (visible upon dissection). Dorsum distinctly blue in life (figure 78) (map 31)...... ....Blueback Herring, *Alosa aestivalis* (Mitchill, 1814)

FIGURE 78. A—Alewife, *Alosa pseudoharengus*; B—Blueback Herring, *Alosa aestivalis*.

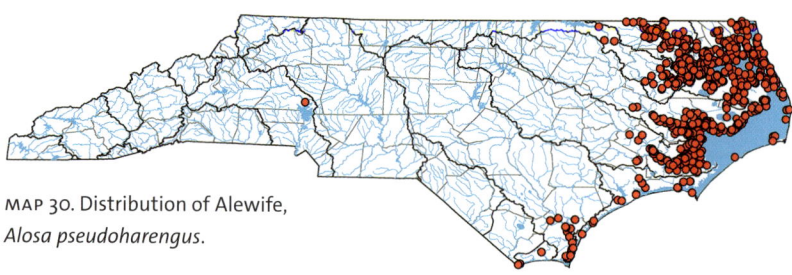

MAP 30. Distribution of Alewife, *Alosa pseudoharengus*.

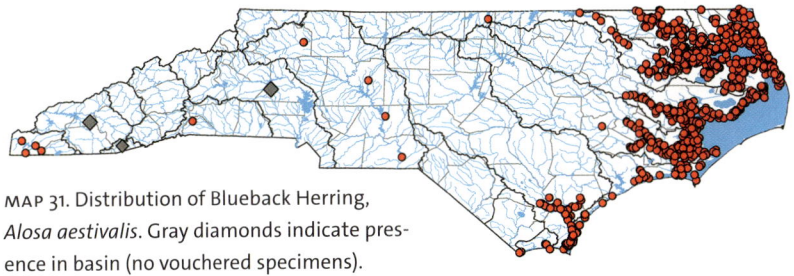

MAP 31. Distribution of Blueback Herring,
Alosa aestivalis. Gray diamonds indicate presence in basin (no vouchered specimens).

6a. Mouth superior; lower jaw part of the dorsal profile with tip projecting forward of the snout when mouth is closed (figures 79 and 80). Jaw teeth present. Gill rakers on lower limb of anterior arch 18–23 and widely spaced (map 32)..................Hickory Shad, *Alosa mediocris* (Mitchill, 1814)

6b. Mouth terminal; lower jaw not part of the dorsal profile with tip even with or projecting slightly forward of the upper jaw when mouth is closed (figures 79 and 80). Jaw teeth minute or absent in adults. Cheek deeper than wide. Gill rakers on lower limb of the first gill arch 59–76 and crowded, 26–43 in specimens less than 125 mm TL (map 33)..............
........................American Shad, *Alosa sapidissima* (Wilson, 1811)

FIGURE 79. Positioning of the lower jaw in relation to the snout. A—Hickory Shad, *Alosa mediocris*; B—American Shad, *Alosa sapidissima*.

A

B

FIGURE 80. A—Hickory Shad, *Alosa mediocris*; B—American Shad, *Alosa sapidissima*.

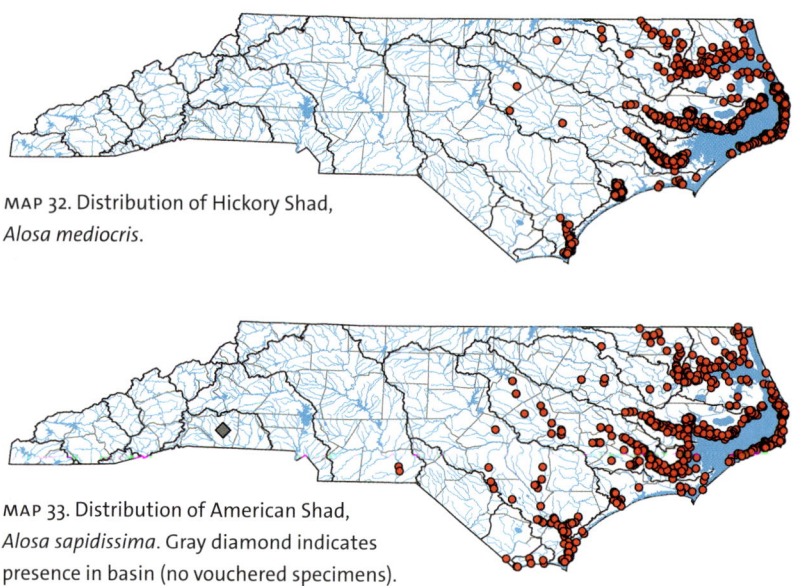

MAP 32. Distribution of Hickory Shad,
Alosa mediocris.

MAP 33. Distribution of American Shad,
Alosa sapidissima. Gray diamond indicates
presence in basin (no vouchered specimens).

Suckers
Family Catostomidae

There are twenty-nine species of suckers including five undescribed species inhabiting North Carolina waters (tables 2 and 10; Tracy, Rohde, and Hogue, 2020; Tracy, Smith, and Rohde, 2021). Colloquial names for these species include creek trout, hoovers, mullets, razor backs, redhorses, and many others. Twelve species are considered imperiled in North Carolina (table 10; NCAC, 2023; NCNHP, 2022; NCWRC, 2021).

Suckers are found throughout North Carolina in streams, big rivers, and reservoirs from Cherokee County in the mountains to Dare County along the Albemarle Sound. Every county has at least one species, but Madison and Stokes

TABLE 10. Species of suckers in North Carolina

Scientific Name	Common Name	Imperilment Status
Carpiodes carpio	River Carpsucker	state special concern
Carpiodes cyprinus	Quillback	state significantly rare
Carpiodes sp. "Atlantic" Highfin Carpsucker		state special concern
Carpiodes sp. "Carolina" Quillback		state significantly rare
Catostomus commersonii	White Sucker	
Erimyzon oblongus	Eastern Creek Chubsucker	
Erimyzon sucetta	Lake Chubsucker	
Hypentelium nigricans	Northern Hog Sucker	
Hypentelium roanokense	Roanoke Hog Sucker	
Ictiobus bubalus	Smallmouth Buffalo	state significantly rare

TABLE 10. (*continued*)

Scientific Name	Common Name	Imperilment Status
Ictiobus cyprinellus	Bigmouth Buffalo	
Ictiobus niger[a]	Black Buffalo	state significantly rare
Minytrema melanops	Spotted Sucker	
Moxostoma anisurum	Silver Redhorse	
Moxostoma ariommum[a]	Bigeye Jumprock	state threatened
Moxostoma breviceps	Smallmouth Redhorse	state significantly rare
Moxostoma carinatum	River Redhorse	
Moxostoma cervinum	Blacktip Jumprock	
Moxostoma collapsum	Notchlip Redhorse	
Moxostoma duquesnei	Black Redhorse	
Moxostoma erythrurum	Golden Redhorse	
Moxostoma macrolepidotum	Shorthead Redhorse	
Moxostoma pappillosum	V-lip Redhorse	
Moxostoma robustum	Robust Redhorse	state endangered
Moxostoma rupiscartes	Striped Jumprock	
Moxostoma sp. "Brassy" Jumprock		
Moxostoma sp. "Carolina" Redhorse		state threatened
Moxostoma sp. "Sicklefin" Redhorse		state threatened
Thoburnia hamiltoni[a]	Rustyside Sucker	state endangered

Note: Names in quotation marks denote scientifically undescribed species.
[a] Species found in only one river basin.

Counties take the prize for having the most—eleven species each. In Madison County they are the Black Buffalo, Black Redhorse, Golden Redhorse, Northern Hog Sucker, Quillback, River Carpsucker, River Redhorse, Silver Redhorse, Smallmouth Buffalo, Smallmouth Redhorse, and White Sucker (Tracy, Rohde, and Hogue, 2020). In Stokes County they are the Bigeye Jumprock, Blacktip Jumprock, "Brassy" Jumprock, Golden Redhorse, Northern Hog Sucker, Notchlip Redhorse, Quillback, Roanoke Hog Sucker, Rustyside Sucker, V-lip

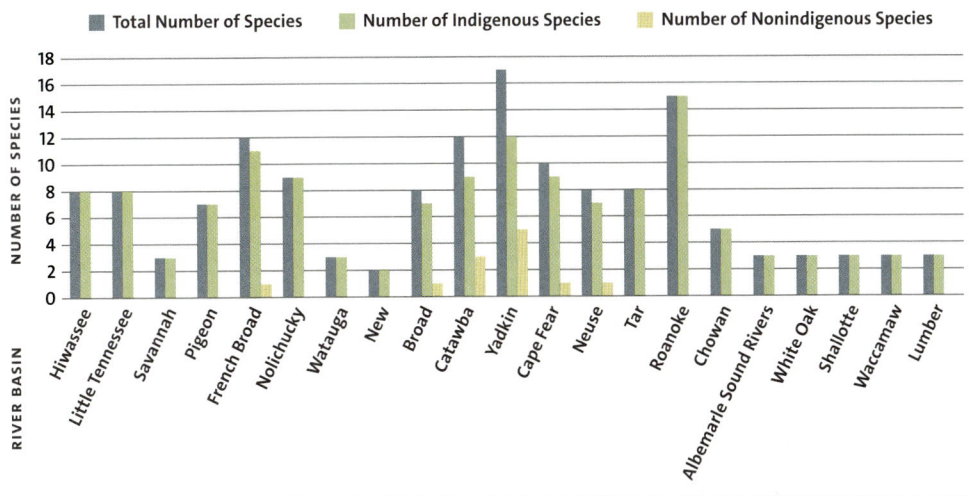

GRAPH 1. Diversity of suckers across North Carolina's river basins (extant species).

Redhorse, and White Sucker (Beane, 2017; Hogue and Tracy, 2014; Tracy, Rohde, and Hogue, 2020).

Three species are found in only one river basin: the Bigeye Jumprock and Rustyside Sucker in the upper Roanoke basin and the Black Buffalo in the lower French Broad basin (tables 2 and 10). The White Sucker is our most widely distributed species and is found in fifteen of our twenty-one river basins. It is absent from waters east of Interstate 95 in the Chowan, Albemarle Sound, White Oak, Shallotte, Waccamaw, and Lumber basins (table 2; Tracy, Rohde, and Hogue, 2020). The population of Robust Redhorse in the Pee Dee River (Yadkin basin) is being augmented with North Carolina Wildlife Resources Commission hatchery–reared fish, and NCWRC staff are also reintroducing the Smallmouth Buffalo into the upper French Broad River near Hendersonville (Henderson County, French Broad basin) (Brena Jones and Luke Etchison, NCWRC, personal communication).

There are more species of suckers, seventeen, found in the Yadkin basin than in any other basin in North Carolina (graph 1; table 2; Tracy, Rohde, and Hogue, 2020). Four of those species have been introduced from other basins in North Carolina—the Northern Hog Sucker, Roanoke Hog Sucker, Smallmouth Buffalo, and Striped Jumprock—and one species has been introduced from outside the state, the Bigmouth Buffalo. Our least speciose basin is the New basin where only the Northern Hog Sucker and White Sucker are found (graph 1).

Key characteristics for the proper identification of suckers include the shape and texture of the lips (figures 81 and 82; Hogue and Tracy, 2014); lateral line scale counts; dorsal fin ray counts; and pharyngeal teeth structure.

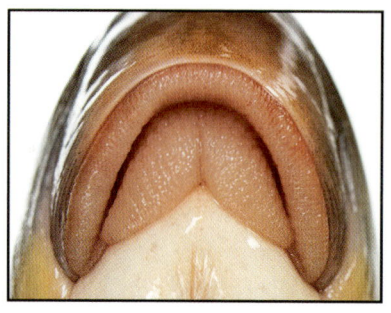

Silver Redhorse
Moxostoma anisurum

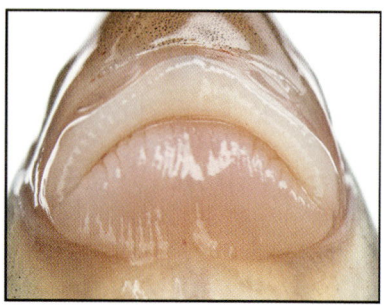

Smallmouth Redhorse
Moxostoma breviceps

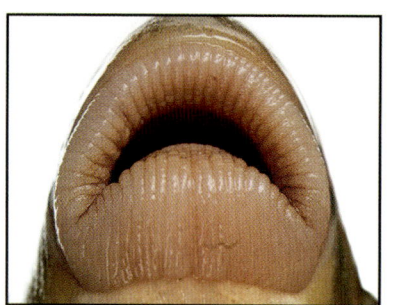

River Redhorse
Moxostoma carinatum

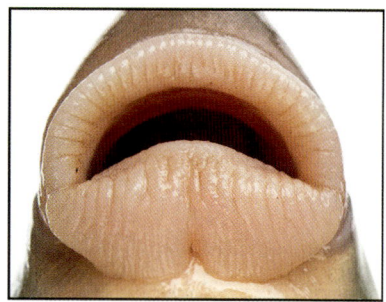

Black Redhorse
Moxostoma duquesnei

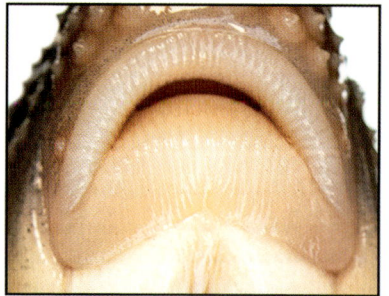

Golden Redhorse
Moxostoma erythrurum

"Sicklefin" Redhorse
Moxostoma sp.

FIGURE 81. Lips of various species of Mississippi River drainage *Moxostoma*.

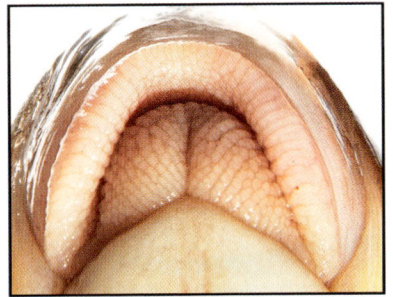

Notchlip Redhorse
Moxostoma collapsum

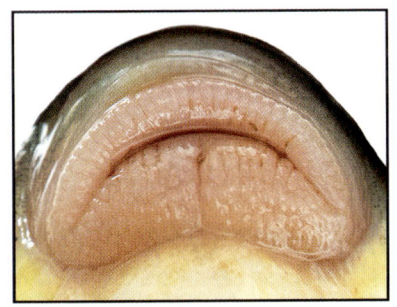

Shorthead Redhorse
Moxostoma macrolepidotum

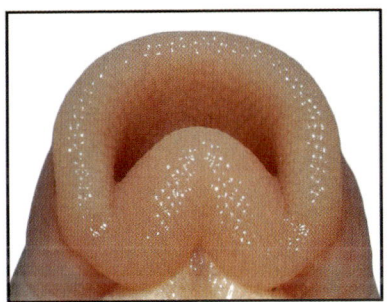

V-lip Redhorse
Moxostoma pappillosum

Robust Redhorse
Moxostoma robustum

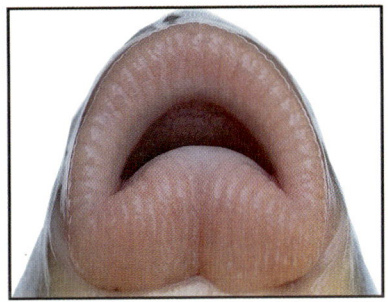

"Brassy" Jumprock
Moxostoma sp.

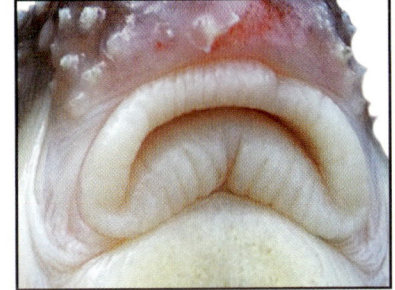

"Carolina" Redhorse
Moxostoma sp.

FIGURE 82. Lips of various species of Atlantic slope *Moxostoma*.

Identification Key to the Species of Suckers
(Family Catostomidae)

1a. Dorsal fin base short (figure 83); ten to eighteen dorsal fin rays; dorsal fin margin not strongly falcate; anterior dorsal rays not greatly elongated. .2

1b. Dorsal fin base long (figure 83); twenty-two to thirty dorsal fin rays; dorsal fin margin strongly falcate (figure 83); anterior dorsal rays greatly elongated
....23

FIGURE 83. Length of dorsal fin base. A—short in White Sucker, *Catostomus commersonii*; B—long and margin strongly falcate in "Atlantic" Highfin Carpsucker, *Carpiodes* sp.

2a. Lateral line absent or virtually so...3

2b. Lateral line complete and visible...5

3a. In adults each scale with a dark spot at the base, forming longitudinal stripes (figure 84) (map 34)..
................Spotted Sucker, *Minytrema melanops* (Rafinesque, 1820)
3b. Side with single wide dark stripe in young, occasionally blotches in adults..*Erimyzon*, Couplet 4

FIGURE 84. Spotted Sucker, *Minytrema melanops*.

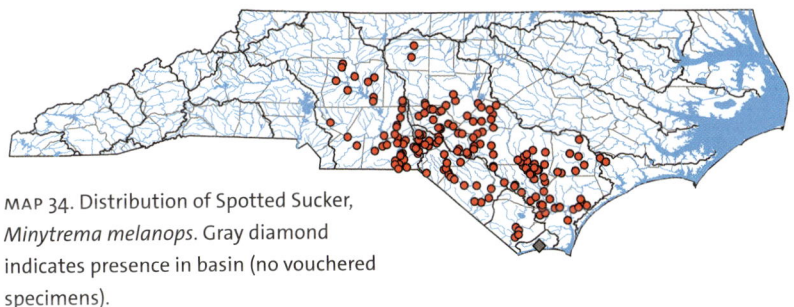

MAP 34. Distribution of Spotted Sucker, *Minytrema melanops*. Gray diamond indicates presence in basin (no vouchered specimens).

4a. Lateral scales usually 34–38. Anterior tip of upper lip generally near the level of the lower rim of the eye (figures 85 and 86). In mature specimens, eye diameter ca. 1.5–1.8 times snout length. Configuration of body in cross section—more laterally compressed, deeper bodied; body width ca. 1.8–2.2 times depth (map 35)..
......................Lake Chubsucker, *Erimyzon sucetta* (Lacepède, 1803)

4b. Lateral scales usually 40–46. Anterior tip of upper lip generally well below level of lower rim of the eye (figures 85 and 86). In mature specimens, eye diameter ca. 2.1–2.6 times snout length. Configuration of body in cross section—less compressed, somewhat more cylindrical; body width ca. 1.4–1.7 times depth. Introduced into French Broad basin (map 36).......
..........Eastern Creek Chubsucker, *Erimyzon oblongus* (Mitchell, 1814)

FIGURE 85. Position of upper lip relative to eye and eye size relative to snout length. A—Lake Chubsucker, *Erimyzon sucetta*; B—Eastern Creek Chubsucker, *Erimyzon oblongus*.

FIGURE 86. A—Lake Chubsucker, *Erimyzon sucetta*, South Carolina; B—Eastern Creek Chubsucker, *Erimyzon oblongus*, tuberculate male.

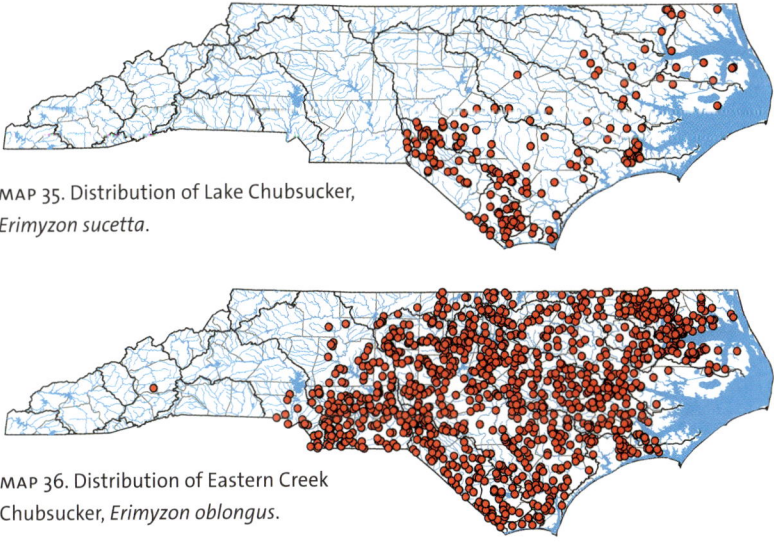

MAP 35. Distribution of Lake Chubsucker, *Erimyzon sucetta*.

MAP 36. Distribution of Eastern Creek Chubsucker, *Erimyzon oblongus*.

5a. Body scales becoming progressively smaller from the caudal peduncle to the head (figure 87) (map 37)...
......................White Sucker, *Catostomus commersonii* (Lacepède, 1803)

5b. Body scales not becoming progressively smaller from the caudal peduncle to the head...6

FIGURE 87. White Sucker, *Catostomus commersonii.*

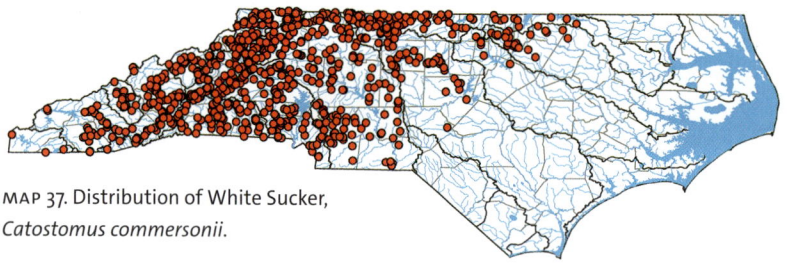

MAP 37. Distribution of White Sucker, *Catostomus commersonii.*

6a. Head concave between the eyes..................*Hypentelium*, Couplet 7

6b. Head not concave between the eyes......................................8

7a. Upper lip with smaller papillae and no plicae (figure 88). Lateral line scales (44) 45–48 (50). Body form elongate. Dark and light horizontal stripes usually absent or faint (figure 89) (map 38).........................
..............Northern Hog Sucker, *Hypentelium nigricans* (Lesueur, 1817)

7b. Upper lip coarsely papillose on the outer border but plicate or subplicate on the inner edge (figure 88). Lateral line scales (38) 40–43 (44). Body form stocky anteriorly. Dark and light horizontal stripes usually moderately developed (figure 89). Range restricted to Roanoke basin and upper Ararat River watershed (Yadkin basin) (map 39).........................
Roanoke Hog Sucker, *Hypentelium roanokense* Raney and Lachner, 1947

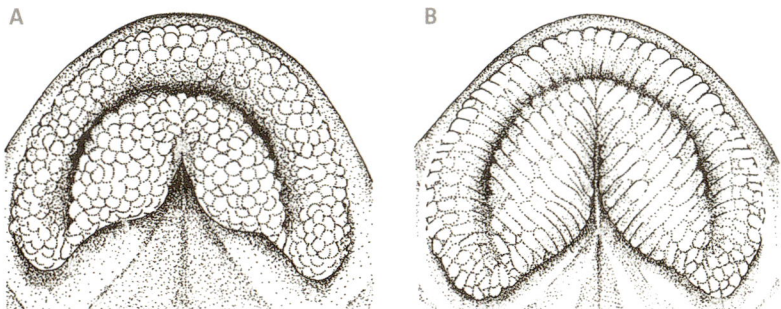

FIGURE 88. Lips. A—Northern Hog Sucker, *Hypentelium nigricans*; B—Roanoke Hog Sucker, *Hypentelium roanokense*. Illustration adapted from Jenkins and Burkhead (1994).

FIGURE 89. A—Northern Hog Sucker, *Hypentelium nigricans*; B—Roanoke Hog Sucker, *Hypentelium roanokense*.

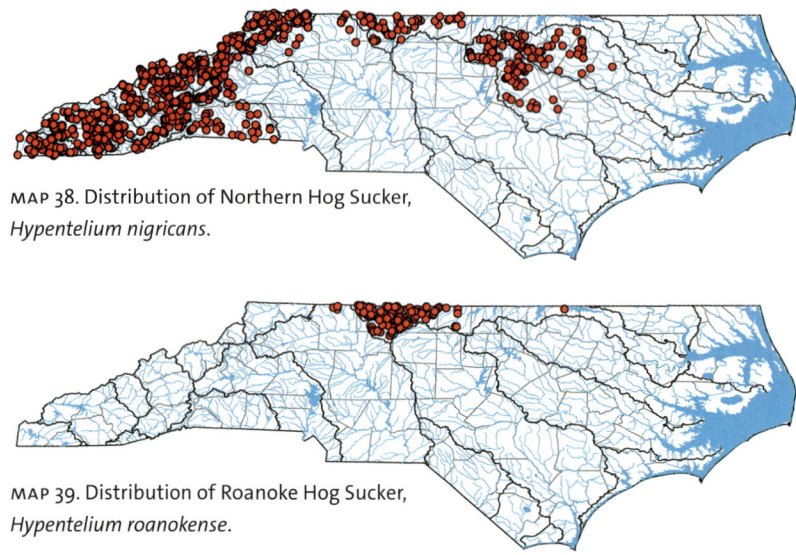

MAP 38. Distribution of Northern Hog Sucker, *Hypentelium nigricans*.

MAP 39. Distribution of Roanoke Hog Sucker, *Hypentelium roanokense*.

8a. Circumpeduncle scales 16 (14–16) (figure 90). Dorsal fin rays 10–12 (13)..9
8b. Circumpeduncle scales 12 or 13 (figure 90). Dorsal fin rays almost always
     13 or more..13

FIGURE 90.
Circumpeduncle
scales. Count using
zigzag pattern.
Golden Redhorse,
*Moxostoma
erythrurum*.

9a. Posterior part of lower lip papillose, flared posteriorly to form a free flap (figure 91). Inner surface of lips with firm, smooth rim, often separated from outer part of lips by a narrow groove. Range restricted to upper Roanoke basin..10

9b. Posterior part of lower lip plicate or semi-plicate, not flared posteriorly to form a free flap (figure 91). Inner surface of lips without a firm, smooth rim. Range not restricted to upper Roanoke basin.......................11

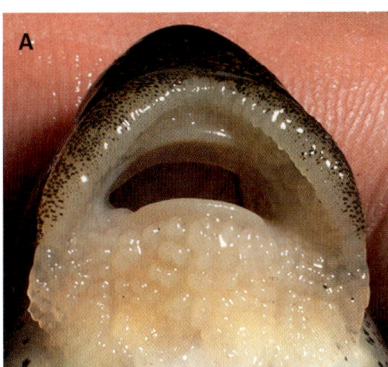 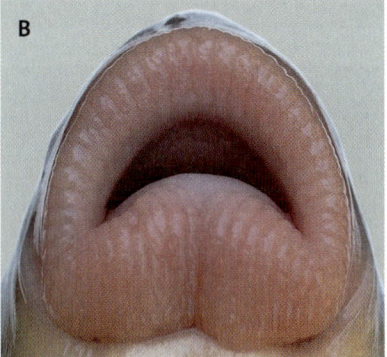

FIGURE 91. Lower lips. A—Papillose lips, Bigeye Jumprock, *Moxostoma ariommum*; B—plicate lips, "Brassy" Jumprock, *Moxostoma* sp.

10a. Upper lip papillose (figure 92). Eye and head large. Caudal base lacking two large pale areas (figure 93) (map 40).....................................
........Bigeye Jumprock, *Moxostoma ariommum* Robins and Raney, 1956

10b. Upper lip plicate (figure 92). Eye and head small. Caudal base with two large pale areas (figure 93) (map 41).......................................
.......Rustyside Sucker, *Thoburnia hamiltoni* Raney and Lachner, 1946

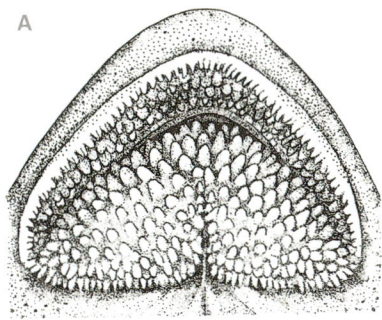 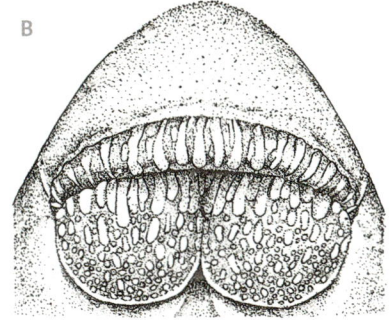

FIGURE 92. Lips. A—Bigeye Jumprock, *Moxostoma ariommum*; B—Rustyside Sucker, *Thoburnia hamiltoni*. Illustration adapted from Jenkins and Burkhead (1994).

FIGURE 93. A—Bigeye Jumprock, *Moxostoma ariommum*; B—Rustyside Sucker, *Thoburnia hamiltoni*. Virginia.

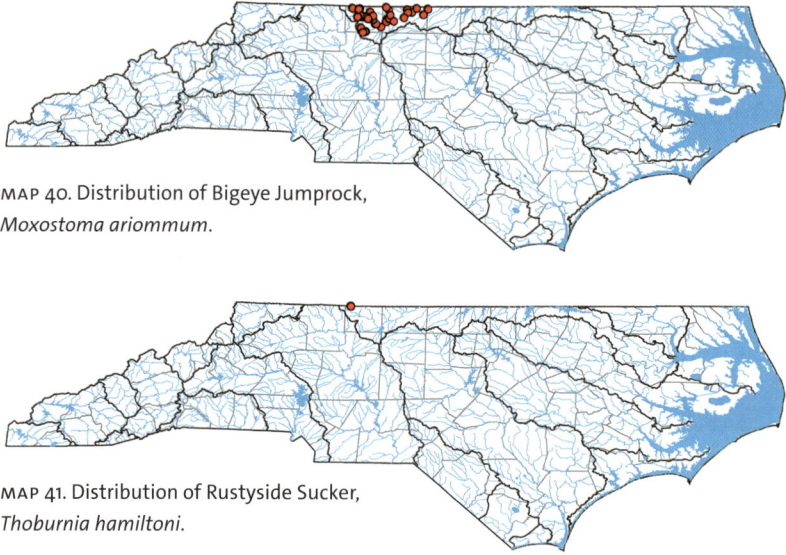

MAP 40. Distribution of Bigeye Jumprock, *Moxostoma ariommum*.

MAP 41. Distribution of Rustyside Sucker, *Thoburnia hamiltoni*.

11a. Dorsal and caudal fins with black tips (figure 94). Range restricted to Neuse, Tar, and Roanoke basins (map 42)....................................
...................Blacktip Jumprock, *Moxostoma cervinum* (Cope, 1868)

11b. Dorsal and caudal fins without black tips. Range does not include Neuse, Tar or Roanoke basins..12

FIGURE 94. Blacktip Jumprock, *Moxostoma cervinum*.

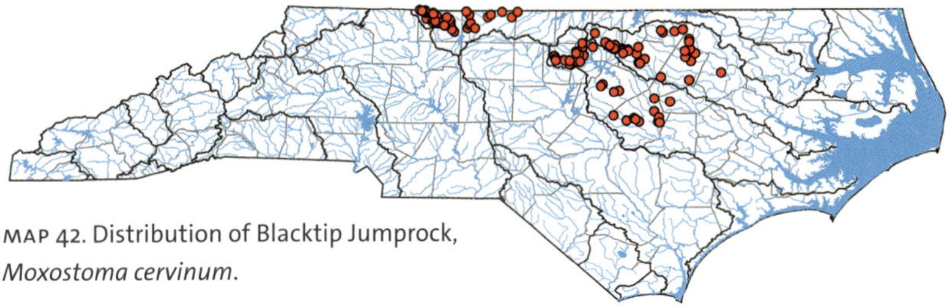

MAP 42. Distribution of Blacktip Jumprock, *Moxostoma cervinum*.

12a. Dorsal rays (11) 12 or 13. Gill rakers 28–34. Body form stout. Head between eyes strongly convex, well elevated above the orbit. In juveniles and small adults, lateral body stripes below the lateral line—pale stripes wider than or equal in width to dark stripes (figure 95). Lateral body blotches absent in medium-sized juveniles and adults. Lower lip plicae with few or no deep transverse grooves (figure 82) (map 43)...........................
.... "Brassy" Jumprock, *Moxostoma* sp.

12b. Dorsal rays 10 or 11 (12). Gill rakers 23–26. Body form elongate in adults. Head between eyes flat or slightly convex, slightly or not at all elevated above orbit. In juveniles and small adults, lateral body stripes below the lateral line—pale stripes narrower in width than dark stripes (figure 95). Lateral body blotches retained in juveniles and adults, often very blotchy and mottled. Lower lip plicae with numerous transverse grooves (subplicate) (map 44)...
......Striped Jumprock, *Moxostoma rupiscartes* Jordan and Jenkins, 1889

FIGURE 95. A—"Brassy" Jumprock, *Moxostoma* sp.; B—Striped Jumprock, *Moxostoma rupiscartes*.

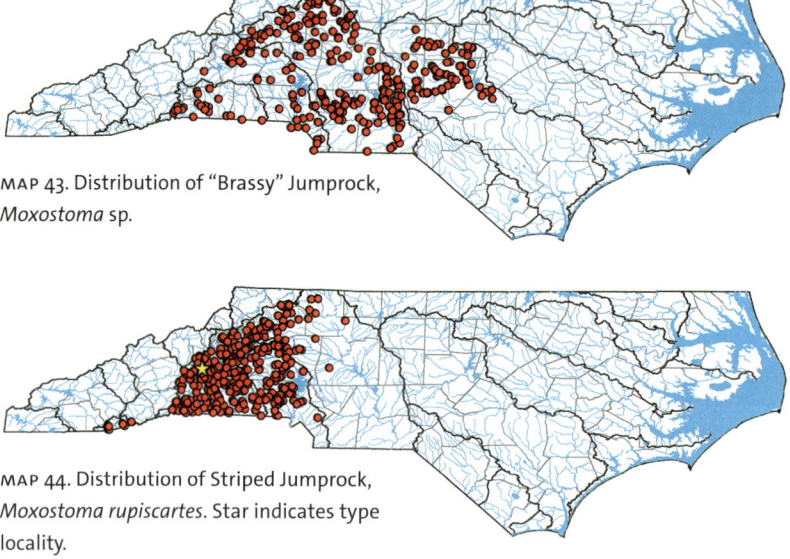

MAP 43. Distribution of "Brassy" Jumprock, *Moxostoma* sp.

MAP 44. Distribution of Striped Jumprock, *Moxostoma rupiscartes*. Star indicates type locality.

13a. Lips fully or nearly fully papillose or semi-papillose (figure 96). V-lipped; halves of lower lip mostly unconnected medially. Posterior margin of lower lip medially forming a moderately or very acute angle............14

13b. Lips plicate or sometimes appearing corrugate or upper lip plicate and lower lip subplicate. Full-lipped; halves of lower lip mostly fully connected medially (figure 96). Posterior margin of lower lip forming slightly acute to very obtuse angle, or margin straight..........................16

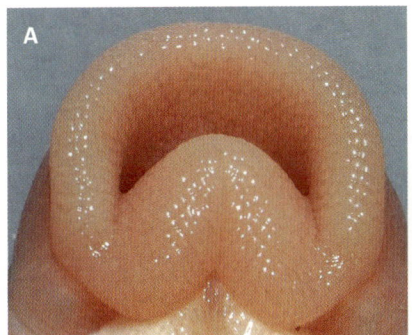 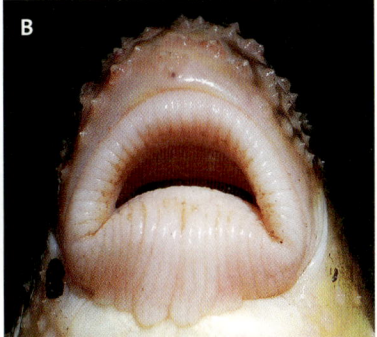

FIGURE 96. Lower lips. A—papillose lips, V-lip Redhorse, *Moxostoma pappillosum*; B—plicate lips, Robust Redhorse, *Moxostoma robustum*.

14a. Lower lip papillose, papillae rounded at edges, regularly arranged, small, and mostly subequal or equal in size (figures 82 and 97). Front of upper lip papillose in medium-sized juveniles to adults. Lower lip with smoothly curved posterior margin, not abruptly thinned at a point distinctly anterior to its juncture with upper lip. Dorsal fin margin of large juveniles and adults almost always slightly to moderately falcate (figure 98). Profile elongate, slightly or not at all elevated toward dorsal fin. Scale bases dark (map 45)..........V-lip Redhorse, *Moxostoma pappillosum* (Cope, 1870)

14b. Lower lip semi-papillose. Plicae deeply, transversely, and somewhat irregularly dissected, which result in papillae-like subdivisions that are irregularly arranged and unequal in size (figures 82 and 97). Front of upper lip smooth, lacking papillae. Lower lip abruptly thinned at a point distinctly anterior to its juncture with upper lip. Dorsal fin margin of large juveniles and adults convex, straight, or slightly concave. Profile moderate or highly elevated toward dorsal fin. Scale bases pale...............................15

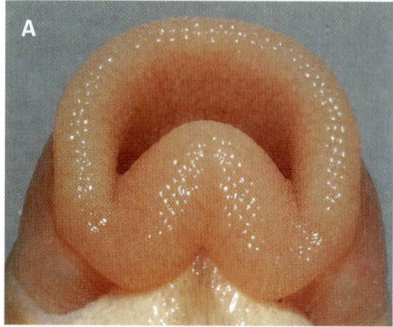 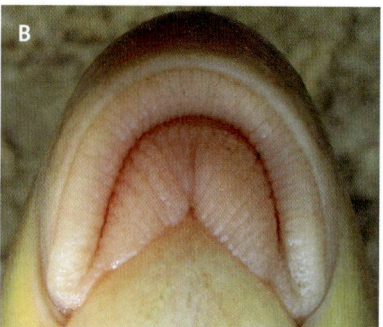

FIGURE 97. Lower lips. A—V-lip Redhorse, *Moxostoma pappillosum*; B—Notchlip Redhorse, *Moxostoma collapsum*.

FIGURE 98. V-lip Redhorse, *Moxostoma pappillosum*.

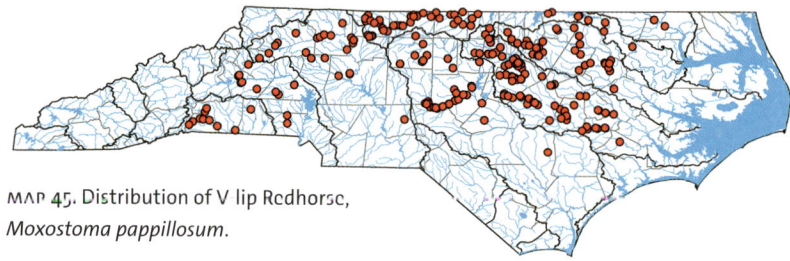

MAP 45. Distribution of V-lip Redhorse, *Moxostoma pappillosum*.

15a. Profile elevated; body depth at dorsal fin origin (27) 28–32 (34) percent SL. Dorsal fin margin usually slightly convex in adults (figure 99). Range restricted to Hiwassee, Little Tennessee, French Broad, and Nolichucky basins (map 46)..
...................Silver Redhorse, *Moxostoma anisurum* (Rafinesque, 1820)

15b. Body form moderate; body depth at dorsal fin origin (23) 24–28 (30) percent SL. Dorsal fin margin slightly concave or straight in adults (figure 99). Range restricted to all of the basins in the Piedmont and the Chowan basin (map 47)..
....................Notchlip Redhorse, *Moxostoma collapsum* (Cope, 1870)

FIGURE 99. A—Silver Redhorse, *Moxostoma anisurum*; B—Notchlip Redhorse, *Moxostoma collapsum*. South Carolina.

MAP 46. Distribution of Silver Redhorse, *Moxostoma anisurum*.

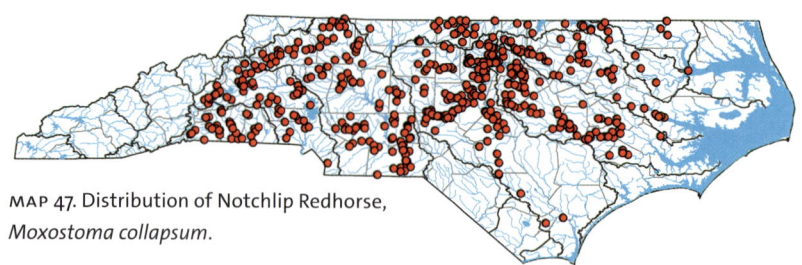

MAP 47. Distribution of Notchlip Redhorse, *Moxostoma collapsum*.

16a. Dorsal fin very falcate (figure 100), the anterior tip when appressed exceeding the posterior tip, usually markedly. Lower lip usually subplicate, posterior margin truncate or nearly so (figure 81). Range restricted to Hiwassee and Little Tennessee basins (map 48).............................
...."Sicklefin" Redhorse, *Moxostoma* sp.

16b. Dorsal fin margin straight to moderately falcate (rarely very falcate in adult Smallmouth Redhorse, *Moxostoma breviceps*) (figure 100)........17

FIGURE 100. Dorsal fin margins. A—"Sicklefin" Redhorse, *Moxostoma* sp., strongly falcate; B—"Carolina" Redhorse, *Moxostoma* sp., moderately straight.

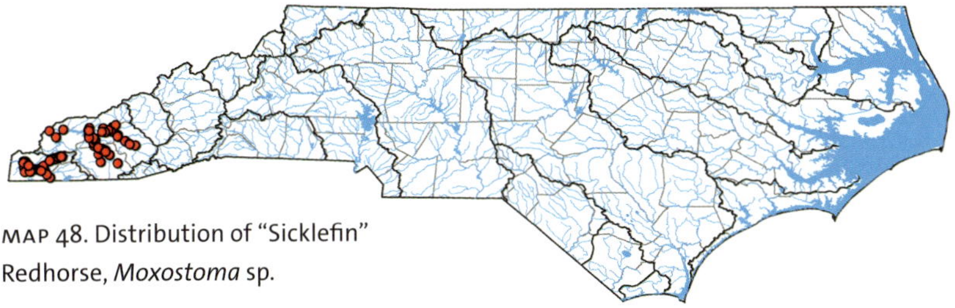

MAP 48. Distribution of "Sicklefin" Redhorse, *Moxostoma* sp.

17a. Pharyngeal arch stout; lower teeth large, molariform; six to nine teeth on lower half of tooth row. Caudal and dorsal fins bright red in life. (Note: The only other redhorse species with bright red dorsal and caudal fins are the Shorthead Redhorse, *Moxostoma macrolepidotum*, and Smallmouth Redhorse, *M. breviceps*, which both have a tiny mouth with the posterior border of the lower lip straight, a moderately falcate dorsal fin, and a very slab-sided rather than cylindrical body cross section.) Snout of breeding males with medium-sized nuptial tubercles.............................18

17b. Pharyngeal arch slender; lower teeth thin, comblike; twelve to thirty teeth on lower half of tooth row. Caudal and dorsal fins not bright red in life (except occasionally in young and in the Shorthead Redhorse, *Moxostoma macrolepidotum*, and Smallmouth Redhorse, *M. breviceps*). Snout of breeding males with or without medium-sized breeding tubercles…19

18a. Body uniformly shaded with dark and pale scale pattern gradually changing shade from anterior to posterior. Stripes and irregular dusky patches absent. Upper body dominantly brassy, coppery, or olive (figure 101). Pelvic fin rays usually 9. Males with few or no nuptial tubercles on opercle. Range restricted to Hiwassee, Little Tennessee, Pigeon, and French Broad basins (map 49)....River Redhorse, *Moxostoma carinatum* (Cope, 1870)

18b. Body with dark and pale scale pattern not uniformly shaded from anterior to posterior; with stripes and/or irregular dusky patches (figure 101). Upper body dominantly golden brown. Pelvic fin rays usually 10. Males with many nuptial tubercles on opercle. Range restricted to Catawba (extirpated) and lower Yadkin basins (map 50)..............................
....................Robust Redhorse, *Moxostoma robustum* (Cope, 1870)

FIGURE 101. A—River Redhorse, *Moxostoma carinatum*; B—Robust Redhorse, *Moxostoma robustum*. South Carolina.

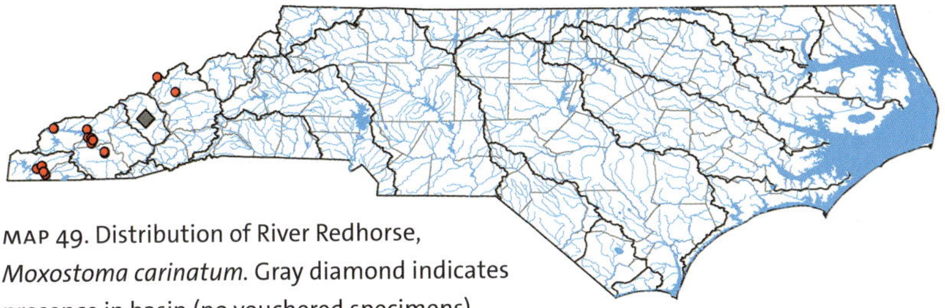

MAP 49. Distribution of River Redhorse, *Moxostoma carinatum*. Gray diamond indicates presence in basin (no vouchered specimens).

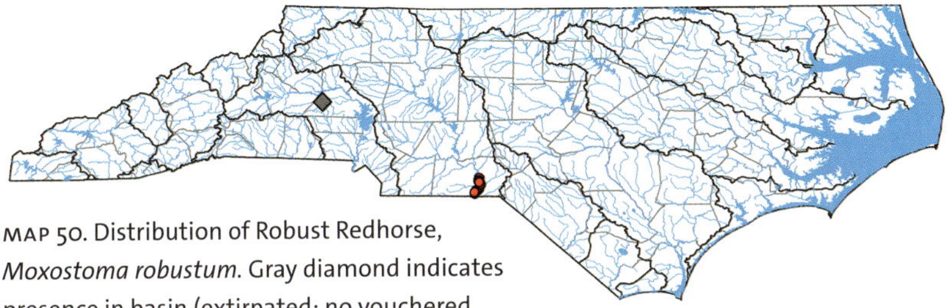

MAP 50. Distribution of Robust Redhorse, *Moxostoma robustum*. Gray diamond indicates presence in basin (extirpated; no vouchered specimens).

19a. Lower lip subplicate; plicae transected into small to large oval elements (figure 102). Head small and short, length ca. 20 percent of SL in large juveniles and adults..20

19b. Lower lip plicate; plicae undissected or corrugated (figure 102). Head not small or short, length greater than ca. 20 percent of SL in large juveniles and adults..21

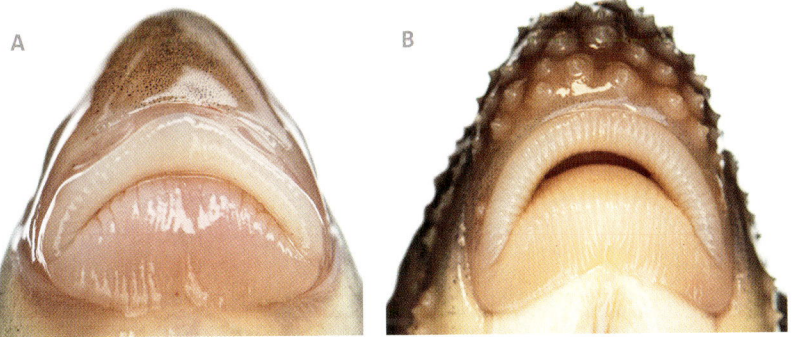

FIGURE 102. Lower lips. A—subplicate, Smallmouth Redhorse, *Moxostoma breviceps*; B—undissected plicae, Golden Redhorse, *Moxostoma erythrurum*.

20a. Pelvic fin rays modally 10. Dorsal fin rays modally 12 (figure 103). Range restricted to Hiwassee, Little Tennessee, Pigeon, French Broad, and Nolichucky basins (map 51)...
................Smallmouth Redhorse, *Moxostoma breviceps* (Cope, 1870)
20b. Pelvic fin rays modally 9. Dorsal fin rays modally 13 (figure 103). Range restricted to Catawba, Yadkin, Cape Fear, Neuse, Tar, Roanoke, Chowan, and Albemarle Sound basins (map 52)....................................
........Shorthead Redhorse, *Moxostoma macrolepidotum* (Lesueur, 1817)

FIGURE 103. A—Smallmouth Redhorse, *Moxostoma breviceps*; B—Shorthead Redhorse, *Moxostoma macrolepidotum*.

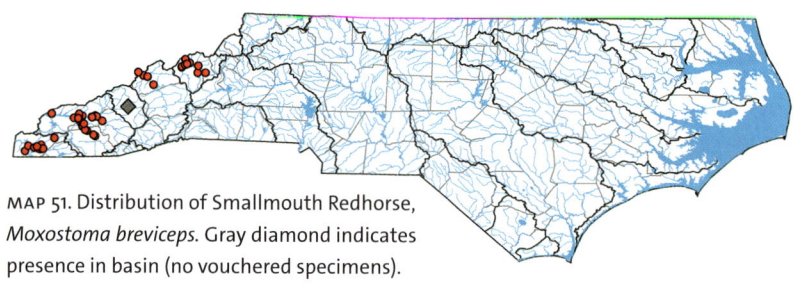

MAP 51. Distribution of Smallmouth Redhorse, *Moxostoma breviceps*. Gray diamond indicates presence in basin (no vouchered specimens).

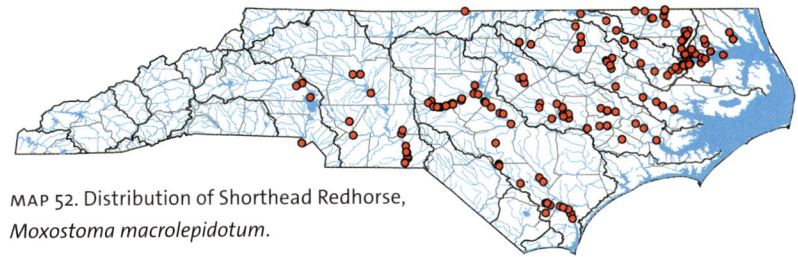

MAP 52. Distribution of Shorthead Redhorse, *Moxostoma macrolepidotum*.

21a. Lateral line scales (43) 44–48 (51). Breast with embedded scales, usually with a small scaleless area anteromedially. Snout of breeding males with minute or no nuptial tubercles. Pelvic fin rays modally 10 (but modally 9 in the Blue Ridge region of the Tennessee River drainage). Angle of posterior edge of lower lip (95°) 120°–160° (175°) (figures 81 and 104) (map 53) ....Black Redhorse, *Moxostoma duquesnei* (Lesueur, 1817)

21b. Lateral line scales (37) 39–43 (45). Breast fully scaled; scales not embedded. Snout of breeding males with medium to large-sized nuptial tubercles. Pelvic fin rays modally 9. Angle of posterior edge of lower lip 90°–130° (155°)...22

FIGURE 104. Black Redhorse, *Moxostoma duquesnei*.

MAP 53. Distribution of Black Redhorse, *Moxostoma duquesnei*.

22a. Supratemporal canal usually interrupted medially. Dorsal fin rays 14 or 15. Pectoral fin rays 10. Lateral line scales 43 or 44 (figure 105). Range restricted to middle Cape Fear and lower Yadkin basins (map 54)........ ...."Carolina" Redhorse, *Moxostoma* sp.

22b. Supratemporal canal not usually interrupted medially. Dorsal fin rays 12 or 13. Pectoral fin rays usually 9. Lateral line scales usually less than 42 (figure 105). Range restricted to Hiwassee, Little Tennessee, Pigeon, French Broad, Nolichucky, and Roanoke basins (map 55)................. ....Golden Redhorse, *Moxostoma erythrurum* (Rafinesque, 1818)

FIGURE 105. A—"Carolina" Redhorse, *Moxostoma* sp.; B—Golden Redhorse, *Moxostoma erythrurum*.

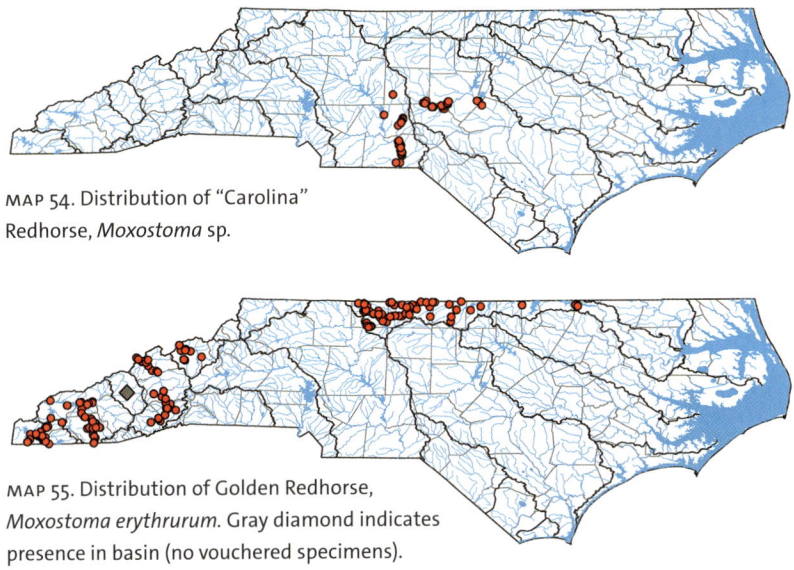

MAP 54. Distribution of "Carolina" Redhorse, *Moxostoma* sp.

MAP 55. Distribution of Golden Redhorse, *Moxostoma erythrurum*. Gray diamond indicates presence in basin (no vouchered specimens).

23a. Subopercle broadest below middle, its lower posterior margin somewhat angular (triangular) (figure 106). Pelvic and anal fins scarcely or not at all speckled with melanophores. Body silvery.........*Carpiodes*, Couplet 24
23b. Subopercle broadest at middle, its lower posterior margin evenly rounded (figure 106). Pelvic and anal fins densely speckled with melanophores. Body brownish or blackish............................*Ictiobus*, Couplet 27

FIGURE 106. Arrows indicating location of subopercle. A—triangular subopercle, *Carpiodes* spp.; B—rounded subopercle, *Ictiobus* spp. Photograph courtesy of Brian Zimmerman (*Ictiobus* spp.).

24a. Nipple-like projection present on tip of lower lip (figure 107). Junction of upper and lower lips extending posteriorly to or beyond the anterior portion of the eye (figure 108). Lateral line scales (33) 36 (37). Dorsal fin rays usually 27 or fewer...25

24b. Nipple-like projection absent from tip of lower lip (figure 107). Junction of upper and lower lips not extending posteriorly to or beyond the anterior portion of the eye (figure 108). Lateral line scales (36) 37 (40). Dorsal fin rays usually 26 or more...26

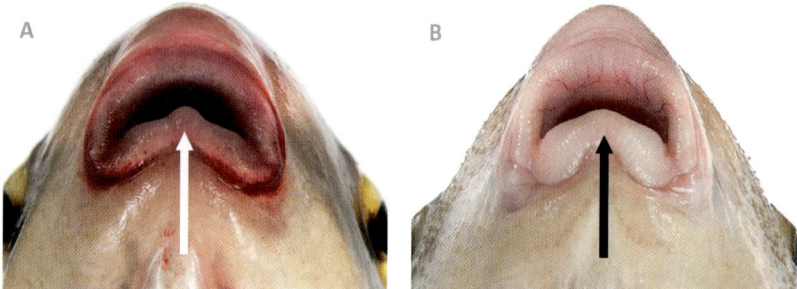

FIGURE 107. Arrows indicating presence/absence of nipple-like projection in lower lip. A—with projection, "Atlantic" Highfin Carpsucker, *Carpiodes* sp. South Carolina; B—without projection, "Carolina" Quillback, *Carpiodes* sp. South Carolina.

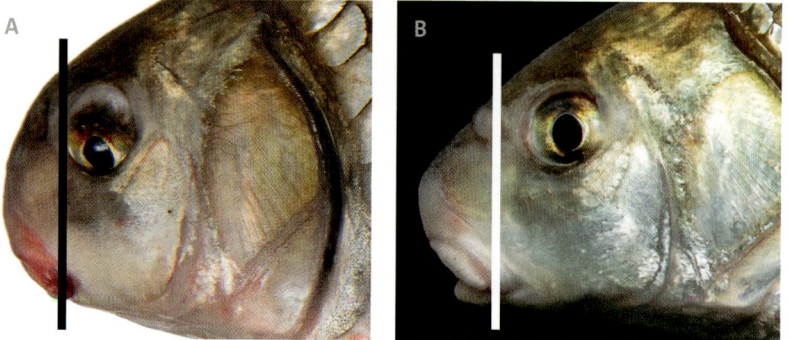

FIGURE 108. A—bar indicating junction of lips extending posteriorly to or beyond the anterior portion of the eye, "Atlantic" Highfin Carpsucker, *Carpiodes* sp. South Carolina. B—bar indicating junction of lips not extending posteriorly to or beyond the anterior portion of the eye, Quillback, *Carpiodes cyprinus*. Photograph courtesy of Luke Etchison (Quillback).

25a. Snout of adult blunt and rounded (figure 109). Breeding tubercles large, larger on snout and top of head than on operculum. Tubercles on most of body scales. Range restricted to Catawba, Yadkin, and Cape Fear basins (map 56)....................."Atlantic" Highfin Carpsucker, *Carpiodes* sp.

25b. Snout of adult angular (figure 109). Breeding tubercles small, smaller on snout and top of head than on operculum. Tubercles absent from most of body. Range restricted to French Broad and Nolichucky basins (map 57) ....River Carpsucker, *Carpiodes carpio* (Rafinesque, 1820)

FIGURE 109. A—"Atlantic" Highfin Carpsucker, *Carpiodes* sp. South Carolina. B—River Carpsucker, *Carpiodes carpio*. Ohio. Photograph courtesy of Brian Zimmerman (River Carpsucker).

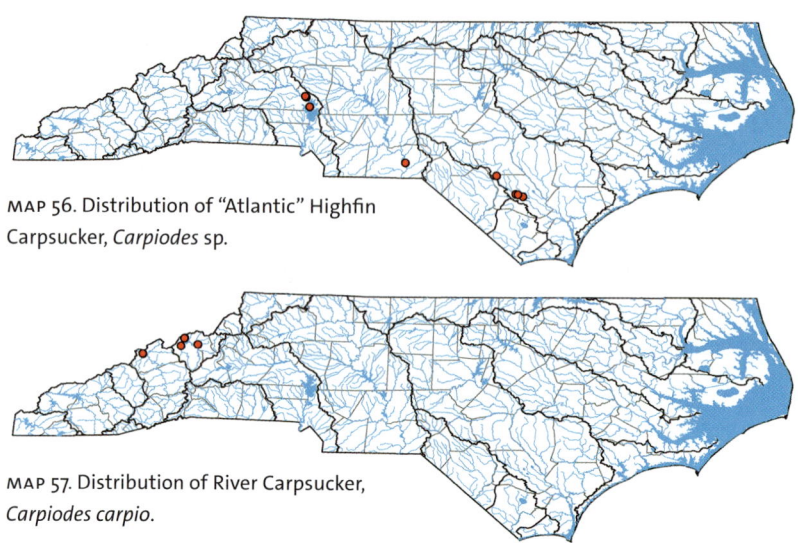

MAP 56. Distribution of "Atlantic" Highfin Carpsucker, *Carpiodes* sp.

MAP 57. Distribution of River Carpsucker, *Carpiodes carpio*.

26a. Range restricted to Pigeon, French Broad, Nolichucky, and Roanoke basins (figure 110) (map 58).......Quillback, *Carpiodes cyprinus* (Lesueur, 1817)
26b. Range restricted to Broad, Catawba, and Yadkin basins (figure 110) (map 59) ...."Carolina" Quillback, *Carpiodes* sp.

FIGURE 110. A—Quillback, *Carpiodes cyprinus*; B—"Carolina" Quillback, *Carpiodes* sp. South Carolina. Photograph courtesy of Luke Etchison (Quillback).

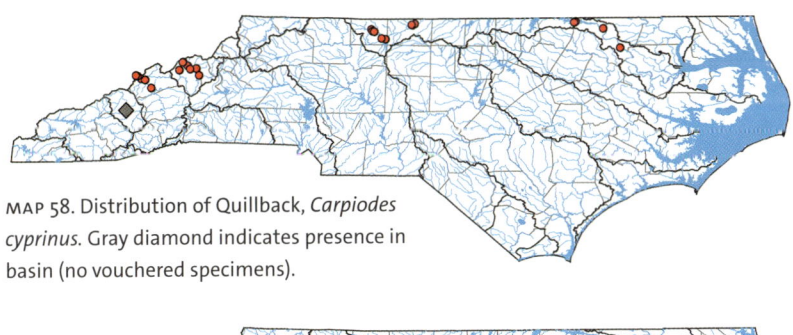

MAP 58. Distribution of Quillback, *Carpiodes cyprinus*. Gray diamond indicates presence in basin (no vouchered specimens).

MAP 59. Distribution of "Carolina" Quillback, *Carpiodes* sp.

27a. Tip of upper lip nearly level with lower margin of eye (figures 111 and 112). Mouth oblique; large. Length of upper jaw nearly equal to snout length. Lips nearly smooth; upper lip thin, only shallowly grooved. Introduced into Catawba and Yadkin basins (map 60)...............................
............Bigmouth Buffalo, *Ictiobus cyprinellus* (Valenciennes, 1844)

27b. Tip of upper lip far below lower margin of eye (figure 111). Mouth nearly horizontal; small. Length of upper jaw much less than snout length. Upper lip thick and deeply plicate.......................................28

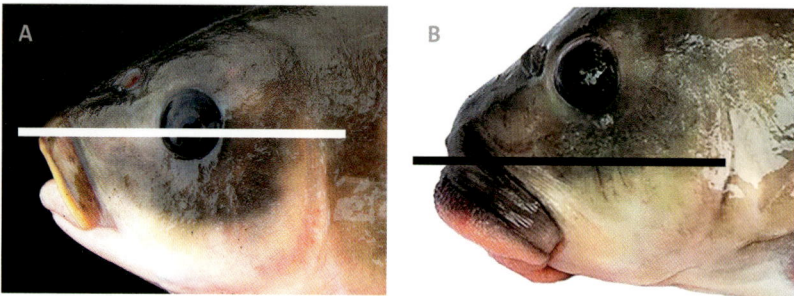

FIGURE 111. Tip of upper lip. A—nearly level with lower margin of eye, Bigmouth Buffalo, *Ictiobus cyprinellus*. Ohio. B—far below eye, Black Buffalo, *Ictiobus niger*. Photographs courtesy of Brian Zimmerman (Bigmouth Buffalo) and Luke Etchison (Black Buffalo).

FIGURE 112. Bigmouth Buffalo, *Ictiobus cyprinellus*. Ohio. Photograph courtesy of Brian Zimmerman.

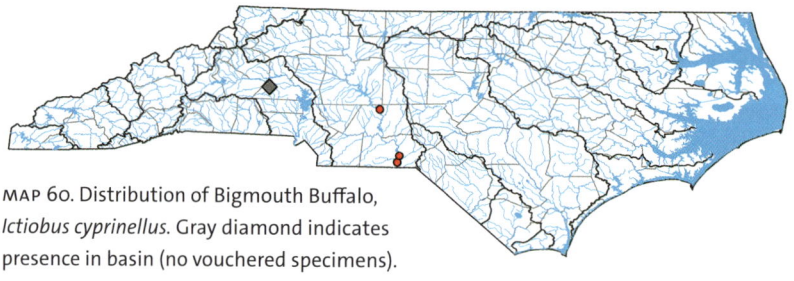

MAP 60. Distribution of Bigmouth Buffalo, *Ictiobus cyprinellus*. Gray diamond indicates presence in basin (no vouchered specimens).

28a. Body deep and compressed; body depth, even in juveniles, usually less than 2.8 times in SL (figure 113). Back strongly keeled or ridged from nape to dorsal fin. Mouth almost horizontal and inferior. Range restricted to French Broad and Nolichucky basins; introduced into Catawba and Yadkin basins (map 61)..
................Smallmouth Buffalo, *Ictiobus bubalus* (Rafinesque, 1818)

28b. Body depth usually greater than 2.8 times in SL (figure 113). Back rounded or weakly keeled from nape to dorsal fin. Mouth slightly to almost oblique and terminal. Range restricted to lower French Broad basin (map 62)....
.............................Black Buffalo, *Ictiobus niger* (Rafinesque, 1819)

A

B

FIGURE 113. A—Smallmouth Buffalo, *Ictiobus bubalus*. South Carolina. B—Black Buffalo, *Ictiobus niger*. Photograph courtesy of Luke Etchison (Black Buffalo).

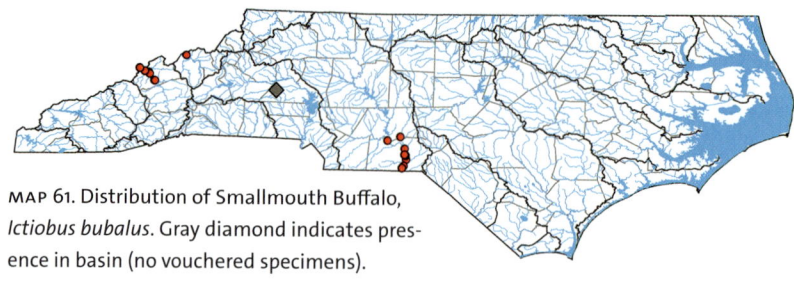

MAP 61. Distribution of Smallmouth Buffalo, *Ictiobus bubalus*. Gray diamond indicates presence in basin (no vouchered specimens).

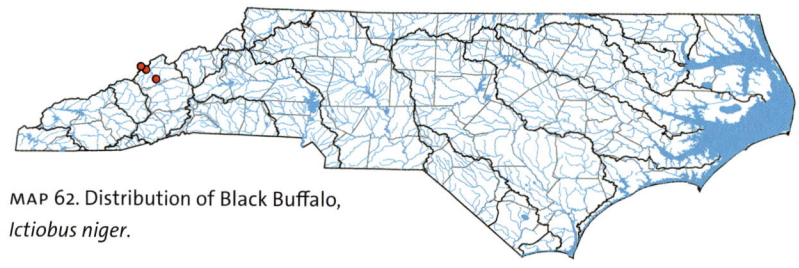

MAP 62. Distribution of Black Buffalo, *Ictiobus niger*.

8

"Minnows"
Families Cyprinidae, Leuciscidae, and Xenocyprididae

"Minnows", until recently, were classified in the family Cyprinidae, along with the Common Carp, Koi, Goldfish, and Grass Carp. Currently, our sixty-eight indigenous (native) species are classified in the family Leuciscidae, a former subfamily of cyprinid fishes (Tan and Armbruster, 2018). The nonindigenous (nonnative or introduced) Grass Carp is now classified in the family Xenocyprididae (Tan and Armbruster, 2018), and the nonindigenous Common Carp, Goldfish, and Koi remain in the family Cyprinidae.

There are seventy-two species of "minnows" in North Carolina, including thirteen species found in only one river basin, three waiting to be scientifically described, and a few that may be renamed or split into additional species (tables 2 and 11) (Tracy, Rohde, and Hogue, 2020). The family Leuciscidae, with sixty-eight species, is our most diverse family of North Carolina's freshwater fish assemblage (Tracy, Rohde, and Hogue, 2020). No species has been extirpated from our state, but three species (Eastern Silvery Minnow, Spotfin Chub, and Spotfin Shiner) have been extirpated from a portion of their native ranges (Tracy, Rohde, and Hogue, 2020), and eighteen species are considered imperiled, many due to their endemism in specific basins in North Carolina (table 11; NCAC, 2023; NCNHP, 2022; NCWRC, 2021). Introductions of species into other basins are often the result of bait bucket dumping by fishermen, or for aquatic plant management (Grass Carp), or historically by the aquaculture trade (Common Carp, Goldfish, and Koi) (table 12).

"Minnows" are called by many names, including Baltimore minnows, crappie minnows, horned daces, horny-heads, knotty heads, minners, shad roaches, spawn eaters, and many other colloquial names.

TABLE 11. "Minnow" species (families Cyprinidae, Leuciscidae, and Xenocyprididae) in North Carolina

Family, Scientific Name	Common Name	Imperilment Status
Cyprinidae		
Carassius auratus	Goldfish	
Cyprinus carpio	Common Carp	
Cyprinus rubrofuscus	Koi	
Xenocyprididae		
Ctenopharyngodon idella	Grass Carp	
Leuciscidae		
Alburnops chalybaeus	Ironcolor Shiner	state threatened
Alburnops petersoni	Coastal Shiner	
Campostoma anomalum	Central Stoneroller	
Chrosomus oreas	Mountain Redbelly Dace	
Clinostomus funduloides	Rosyside Dace	
Clinostomus sp. "Hiwassee" Dace[a]		state special concern
Clinostomus sp. "Smoky" Dace[a]		state special concern
Coccotis coccogenis	Warpaint Shiner	
Cyprinella analostana	Satinfin Shiner	
Cyprinella chloristia	Greenfin Shiner	
Cyprinella galactura	Whitetail Shiner	
Cyprinella labrosa	Thicklip Chub	state threatened
Cyprinella lutrensis	Red Shiner	
Cyprinella nivea	Whitefin Shiner	
Cyprinella pyrrhomelas	Fieryblack Shiner	
Cyprinella spiloptera	Spotfin Shiner	
Cyprinella zanema	Santee Chub	state threatened
Cyprinella sp. "Thinlip" Chub		state special concern
Erimonax monachus	Spotfin Chub	federally threatened
Erimystax insignis	Blotched Chub	state threatened
Exoglossum laurae[a]	Tonguetied Minnow	state significantly rare
Exoglossum maxillingua[a]	Cutlip Minnow	state special concern

TABLE 11. (*continued*)

Family, Scientific Name	Common Name	Imperilment Status
Hudsonius altipinnis	Highfin Shiner	
Hudsonius hudsonius	Spottail Shiner	
Hybognathus regius	Eastern Silvery Minnow	
Hybopsis amblops	Bigeye Chub	
Hybopsis hypsinotus	Highback Chub	
Hybopsis rubrifrons[a]	Rosyface Chub	state threatened
Hydrophlox chiliticus	Redlip Shiner	
Hydrophlox chlorocephalus[a]	Greenhead Shiner	
Hydrophlox lutipinnis	Yellowfin Shiner	state special concern
Hydrophlox rubricroceus	Saffron Shiner	
Hydrophlox sp. "Piedmont" Shiner[a]		
Luxilus albeolus	White Shiner	
Luxilus cerasinus	Crescent Shiner	
Luxilus chrysocephalus	Striped Shiner	state special concern
Lythrurus ardens	Rosefin Shiner	
Lythrurus matutinus	Pinewoods Shiner	
Miniellus alborus	Whitemouth Shiner	
Miniellus mekistocholas[a]	Cape Fear Shiner	federally endangered
Miniellus procne	Swallowtail Shiner	
Miniellus scabriceps[a]	New River Shiner	
Nocomis leptocephalus	Bluehead Chub	
Nocomis micropogon	River Chub	
Nocomis platyrhynchus[a]	Bigmouth Chub	
Nocomis raneyi	Bull Chub	
Notemigonus crysoleucas	Golden Shiner	
Notropis amoenus	Comely Shiner	
Notropis bifrenatus	Bridle Shiner	state endangered
Notropis maculatus	Taillight Shiner	

TABLE 11. (*continued*)

Family, Scientific Name	Common Name	Imperilment Status
Notropis micropteryx	Highland Shiner	
Notropis photogenis	Silver Shiner	
Notropis scepticus	Sandbar Shiner	
Notropis telescopus	Telescope Shiner	
Notropis sp. "Kanawha" Rosyface Shiner[a]		state significantly rare
Paranotropis leuciodus	Tennessee Shiner	
Paranotropis spectrunculus	Mirror Shiner	
Paranotropis volucellus	Mimic Shiner	
Phenacobius crassilabrum	Fatlips Minnow	
Phenacobius teretulus[a]	Kanawha Minnow	state special concern
Pimephales notatus	Bluntnose Minnow	
Pimephales promelas	Fathead Minnow	
Pteronotropis cummingsae	Dusky Shiner	
Rhinichthys atratulus[a]	Blacknose Dace	
Rhinichthys cataractae	Longnose Dace	
Rhinichthys obtusus	Western Blacknose Dace	
Semotilus atromaculatus	Creek Chub	
Semotilus lumbee	Sandhills Chub	state special concern

Note: Names in quotation marks denote scientifically undescribed species.
[a] Species found in only one river basin.

TABLE 12. "Minnow" species introduced into or from within North Carolina

Family, Scientific Name	Common Name
Cyprinidae	
Carassius auratus[a]	Goldfish
Cyprinus carpio[a]	Common Carp
Cyprinus rubrofuscus[a]	Koi
Xenocyprididae	
Ctenopharyngodon idella[a]	Grass Carp
Leuciscidae	
Campostoma anomalum	Central Stoneroller
Chrosomus oreas	Mountain Redbelly Dace
Clinostomus funduloides	Rosyside Dace
Coccotis coccogenis	Warpaint Shiner
Cyprinella chloristia	Greenfin Shiner
Cyprinella galactura	Whitetail Shiner
Cyprinella lutrensis[b]	Red Shiner
Hybopsis hypsinotus	Highback Chub
Hydrophlox chiliticus	Redlip Shiner
Hydrophlox lutipinnis	Yellowfin Shiner
Hydrophlox rubricroceus	Saffron Shiner
Luxilus cerasinus	Crescent Shiner
Lythrurus ardens	Rosefin Shiner
Miniellus alborus	Whitemouth Shiner
Miniellus procne	Swallowtail Shiner
Nocomis leptocephalus	Bluehead Chub
Notemigonus crysoleucas	Golden Shiner
Notropis amoenus	Comely Shiner
Paranotropis leuciodus	Tennessee Shiner
Paranotropis spectrunculus	Mirror Shiner
Paranotropis volucellus	Mimic Shiner
Pimephales promelas[b]	Fathead Minnow

[a] Introduced from outside the United States.
[b] Introduced from outside North Carolina.

A few superlatives associated with our "minnow" fauna (tables 2 and 11; Tracy, Rohde, and Hogue, 2020):

1. Found in all twenty-one river basins—the Common Carp and Golden Shiner
2. Found nowhere else in the world—the Cape Fear Shiner and Pinewoods Shiner
3. Smallest (native)—the Bridle Shiner (ca. 50 mm SL) and Ironcolor Shiner (ca. 55 mm SL)
4. Largest (native)—the Bull Chub (ca. 270 mm SL)
5. Rarest—the Bridle Shiner
6. Most geographically restricted—the Bigmouth Chub, Bridle Shiner, Cape Fear Shiner, Cutlip Minnow, Kanawha Minnow, "Kanawha" Rosyface Shiner, New River Shiner, Rosyface Chub, Sandhills Chub, Spotfin Chub, Striped Shiner, Tonguetied Minnow, and Yellowfin Shiner
7. Most commonly encountered and abundant—the Central Stoneroller (Mountains), Bluehead Chub (Piedmont), and Dusky Shiner, Highfin Shiner, and Swallowtail Shiner (Coastal Plain)
8. Introduced from outside the United States—the Common Carp, Goldfish, Grass Carp, and Koi
9. Introduced from other states—the Fathead Minnow and Red Shiner
10. Indigenous to North Carolina but have been introduced into other basins within North Carolina—twenty species.

In terms of species diversity, the Catawba basin has the greatest number of species—thirty-five—with twenty-eight of them indigenous and seven nonindigenous (graph 2). In the Linville River watershed of the Catawba basin, the Warpaint Shiner is considered indigenous through stream capture, but in other parts of the upper Catawba basin it is considered nonindigenous through bait bucket introductions (table 2; Tracy, Rohde, and Hogue, 2020). The basin with the fewest number of species is the Shallotte, with four indigenous (Coastal Shiner, Dusky Shiner, Golden Shiner and Ironcolor Shiner) and two nonindigenous (Common Carp and Grass Carp) species. The basin with the greatest number of nonindigenous (introduced) species is the Yadkin, with twelve nonindigenous species—the Central Stoneroller, Comely Shiner, Common Carp, Fathead Minnow, Goldfish, Grass Carp, Greenfin Shiner, Mountain Redbelly Dace, Red Shiner, Rosefin Shiner, Swallowtail Shiner, and Warpaint

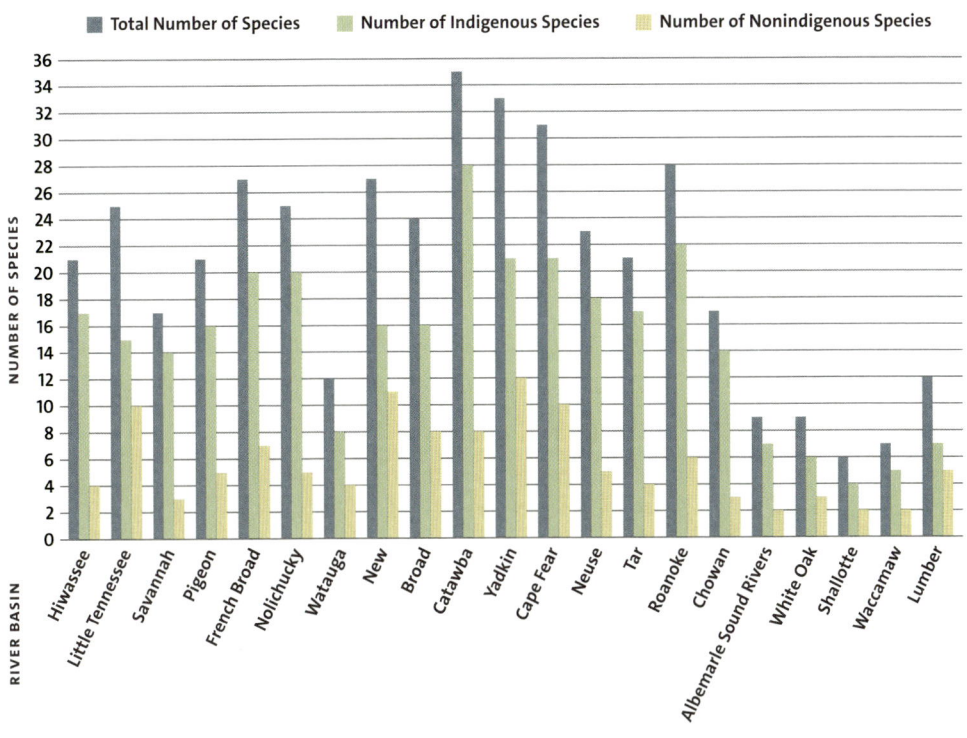

GRAPH 2. Diversity of "minnows" across North Carolina's river basins (extant species).

Shiner. The basins with fewest number of introduced species are the Albemarle Sound, Shallotte, and Waccamaw basins, each with two species (Common Carp and Grass Carp).

Key characteristics for the proper identification of "minnows" include the presence or absence of a frenum; lateral stripe width and length; the number of lateral line scales; the positioning and pigmentation of the dorsal fin; the anal ray count; presence of spines versus rays; the position of the mouth; the pharyngeal teeth count; the presence/absence and length and shape of maxillary barbels; the overall color pattern; and the geographical distributions of the species. The identification of minnows with seven or eight anal rays and immature and female *Nocomis* spp., where species co-occur, can be very challenging.

Identification Key to the Carps, Minnows, and Sharpbellies (Families Cyprinidae, Leuciscidae, and Xenocyprididae)

1a. Dorsal fin long with a stout, saw-toothed spine, followed by thirteen or more branched rays. Anal fin with a stout spine............................ ....Family Cyprinidae, Couplet 2

1b. Dorsal fin without a stout, saw-toothed spine, followed by twelve or fewer branched rays. Anal fin without a stout spine...........................4

2a. One pair of fleshy barbels on each side located near the corner of the mouth on the upper jaw (anterior barbels may sometimes be missing or poorly developed) (figure 114). Lateral line scales 29–40.................3

2b. Barbels absent (figure 114). Lateral line scales 28–32 (map 63)............. ....Goldfish, *Carassius auratus* (Linnaeus, 1758)

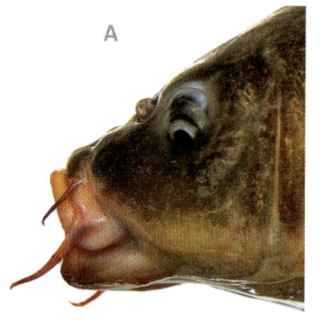 

FIGURE 114. A—Common Carp, *Cyprinus carpio*; B—Goldfish, *Carassius auratus*, wild color.

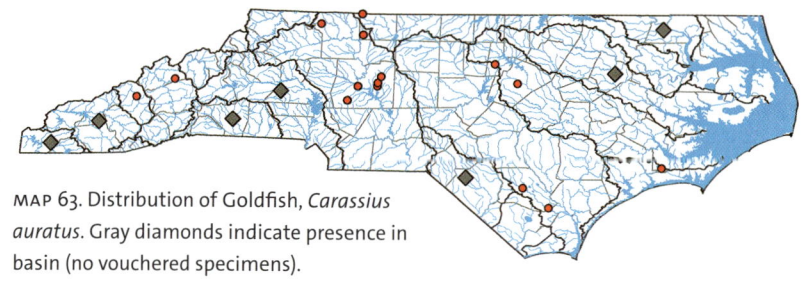

MAP 63. Distribution of Goldfish, *Carassius auratus*. Gray diamonds indicate presence in basin (no vouchered specimens).

3a. Lateral line scales 29–33 (36). Eighteen to twenty-three branched dorsal rays. Silvery body with red pelvic, anal, and lower caudal lobes (wild populations) (figure 115) (map 64) . . . . Koi, *Cyprinus rubrofuscus* Lacepède, 1803

3b. Lateral line scales 33–37 (40). Some genetic strains may have only a few large lateral line scales ("mirror" carp) or be lacking in lateral lines scales entirely ("leather" carp). Seventeen to twenty-one branched dorsal rays. Gray to bronze body (figure 115) (map 65) . . . . Common Carp, *Cyprinus carpio* Linnaeus, 1758

FIGURE 115. A—Koi, *Cyprinus rubrofuscus*; B—Common Carp, *Cyprinus carpio*.

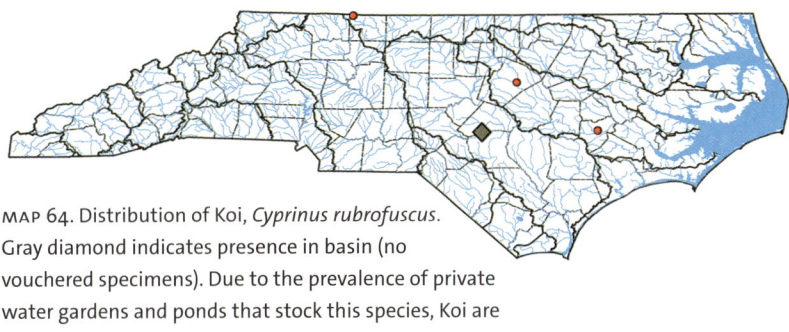

MAP 64. Distribution of Koi, *Cyprinus rubrofuscus*. Gray diamond indicates presence in basin (no vouchered specimens). Due to the prevalence of private water gardens and ponds that stock this species, Koi are more widely distributed than shown.

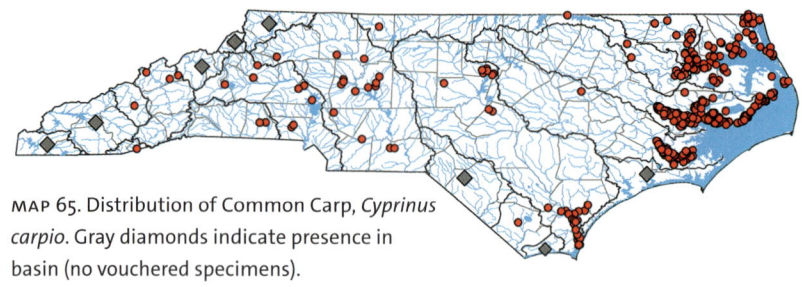

MAP 65. Distribution of Common Carp, *Cyprinus carpio*. Gray diamonds indicate presence in basin (no vouchered specimens).

    4a. Distance from anal fin origin to tip of snout three or more times as long as the distance from the anal fin origin to the caudal fin base (figure 116) (map 66)............................Family Xenocyprididae, Grass Carp, *Ctenopharyngodon idella* (Valenciennes, 1844)

    4b. Distance from anal fin origin to tip of snout less than three times as long as the distance from the anal fin origin to the caudal fin base..............
....Family Leuciscidae, Couplet 5

FIGURE 116. Grass Carp, *Ctenopharyngodon idella*. South Carolina.

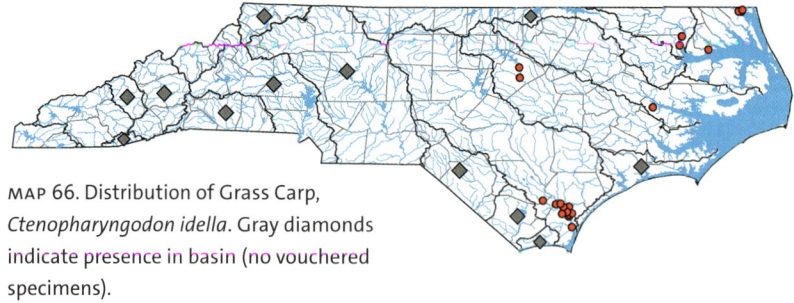

MAP 66. Distribution of Grass Carp, *Ctenopharyngodon idella*. Gray diamonds indicate presence in basin (no vouchered specimens).

5a. Frenum present; premaxillae nonprotractile, attached to the snout with skin (figure 117)..6

5b. Frenum absent; premaxillae protractile, separated from the snout by a groove (figure 117)...10

A   B

FIGURE 117. Frenum. A—present in Tonguetied Minnow, *Exoglossum laurae*; B—absent in Creek Chub, *Semotilus atromaculatus*.

6a. Lower lip appears deformed and split; central part cartilaginous and side lobes fleshy...*Exoglossum*, Couplet 7

6b. Lower lip normal; central part not cartilaginous and stiff.................. ....*Rhinichthys*, Couplet 8

7a. Barbel present. Lower jaw tapers gradually anteriorly, not lobed, not producing a three-lobed outline (figures 118 and 119). Range restricted to New basin (map 67)... ....Tonguetied Minnow, *Exoglossum laurae* (Hubbs, 1931)

7b. Barbel absent. Lower jaw lobes produce a three-lobed outline anteriorly (figures 118 and 119). Range restricted to upper Roanoke basin (map 68) ....Cutlip Minnow, *Exoglossum maxillingua* (Lesueur, 1817)

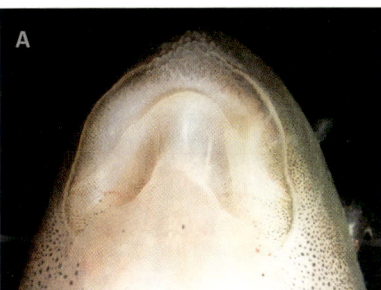 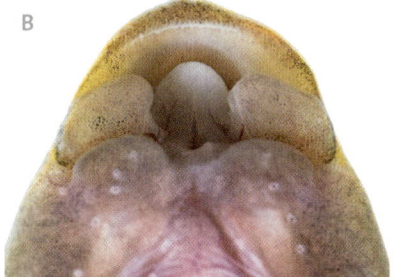

A   B

FIGURE 118. Ventral view of mouths of *Exoglossum*. A—Tonguetied Minnow, *E. laurae*; B—Cutlip Minnow, *E. maxillingua*.

FIGURE 119. A—Tonguetied Minnow, *Exoglossum laurae*; B—Cutlip Minnow, *Exoglossum maxillingua*.

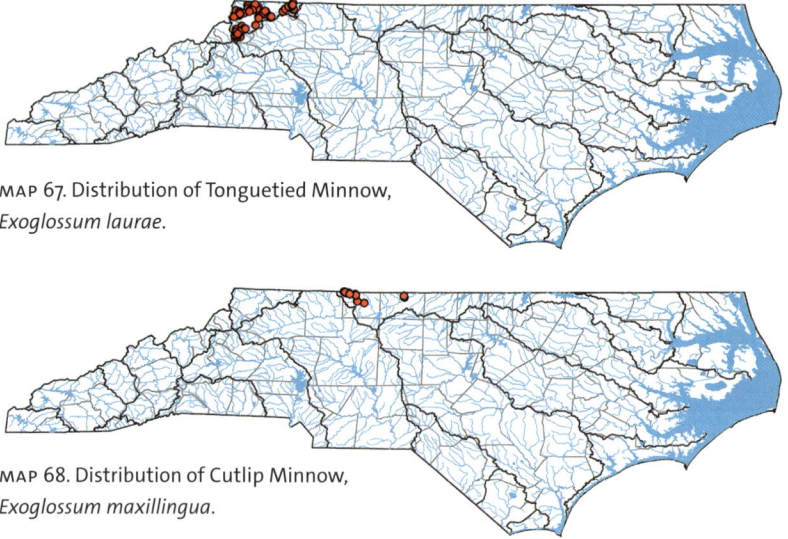

MAP 67. Distribution of Tonguetied Minnow, *Exoglossum laurae*.

MAP 68. Distribution of Cutlip Minnow, *Exoglossum maxillingua*.

8a. Mouth very inferior, snout projecting well forward of the mouth. Barbel absent. Distance from snout tip to anterior point of lower jaw greater than or equal to the diameter of the eye (figure 120) (map 69)...................
................Longnose Dace, *Rhinichthys cataractae* (Valenciennes, 1842)

8b. Mouth terminal or subterminal, snout projecting barely forward of the mouth. Barbel present. Distance from snout tip to anterior point of the lower jaw less than the diameter of the eye................................9

FIGURE 120. Longnose Dace, *Rhinichthys cataractae*.

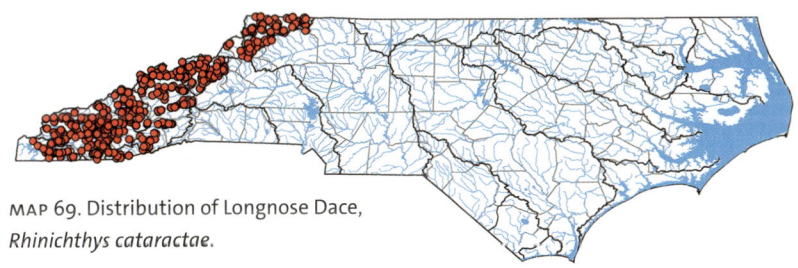

MAP 69. Distribution of Longnose Dace, *Rhinichthys cataractae*.

9a. Lateral line scales (46) 51–58 (63). Breeding males with a red-orange band that occurs adjacent to and above black lateral stripes (and in some slightly below) (figure 121). Range restricted to upper Dan River (Roanoke basin) (map 70).......Blacknose Dace, *Rhinichthys atratulus* (Hermann, 1804)

9b. Lateral line scales (56) 57–64 (70). In breeding males, the black lateral stripe is replaced by a red-orange band (figure 122). Range not restricted to upper Roanoke basin (map 71)...
...............Western Blacknose Dace, *Rhinichthys obtusus* Agassiz, 1854

FIGURE 121. Blacknose Dace, *Rhinichthys atratulus*. Pennsylvania. A—male; B—female.
Photographs courtesy of Robert Criswell.

FIGURE 122. Western Blacknose Dace, *Rhinichthys obtusus*. A—from New basin; B—from
Little Tennessee basin.

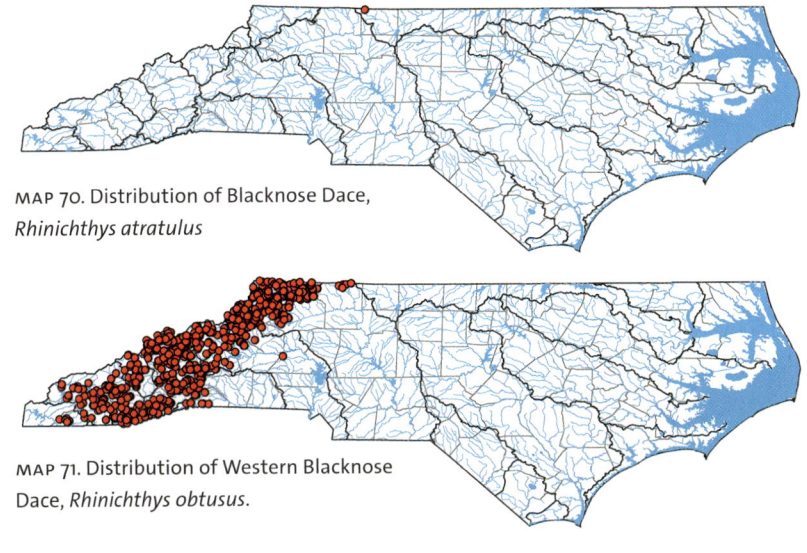

MAP 70. Distribution of Blacknose Dace, *Rhinichthys atratulus*

MAP 71. Distribution of Western Blacknose Dace, *Rhinichthys obtusus*.

10a. Lower jaw with a firm cartilaginous ridge (figure 123) (map 72)..........
..........Central Stoneroller, *Campostoma anomalum* (Rafinesque, 1820)
10b. Lower jaw lacking a firm cartilaginous ridge.............................11

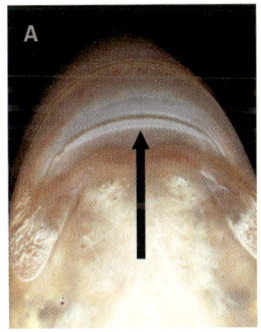

FIGURE 123. Central Stoneroller, *Campostoma anomalum*. A—ventral view of mouth with a cartilaginous ridge; B—full body.

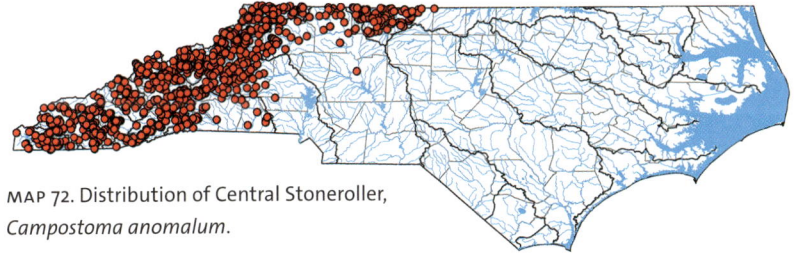

MAP 72. Distribution of Central Stoneroller, *Campostoma anomalum*.

11a. Barbel usually present (occasionally lacking in *Hybopsis* spp.), located in groove above the maxilla at or slightly anterior to the posterior end of the maxilla (figure 124)...12

11b. Barbel absent; if barbel is present, as in the Santee Chub, *Cyprinella zanema*; Spotfin Chub, *Erimonax monachus*; Thicklip Chub, *C. labrosa*; and "Thinlip" Chub, *Cyprinella* sp., then dorsal fin membranes are profusely infused with pigment, either anteriorly, posteriorly, or across the base of the dorsal fin; barbel also present in nuptial males of the Bluntnose Minnow, *Pimephales notatus* (figure 124)..................................21

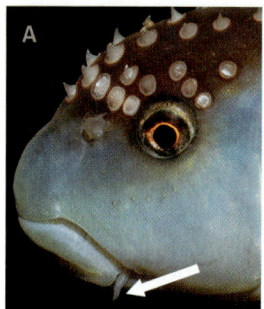

FIGURE 124. A—arrow indicating maxillary barbel in Bluehead Chub, *Nocomis leptocephalus*; B—arrows indicating maxillary barbel and dorsal fin infused with pigment in Thicklip Chub, *Cyprinella labrosa*. South Carolina.

12a. Barbel small, flat, and triangular, located in maxillary groove anterior to end of the jaw.......................................*Semotilus*, Couplet 13

12b. Barbel conical, located at posterior tip of the maxilla...................14

13a. Eight dorsal fin rays. Dark spot near dorsal fin origin (figure 125). Lateral line scales 49–57 (map 73)..
.....................Creek Chub, *Semotilus atromaculatus* (Mitchill, 1818)

13b. Nine dorsal fin rays. No dark spot near dorsal fin origin (figure 125). Lateral line scales 46–48. Range restricted to Sand Hills ecoregion of Yadkin, Cape Fear, and Lumber basins (map 74).................................
.............Sandhills Chub, *Semotilus lumbee* Snelson and Suttkus, 1978

FIGURE 125. A—Creek Chub, *Semotilus atromaculatus*; B—Sandhills Chub, *Semotilus lumbee*. South Carolina.

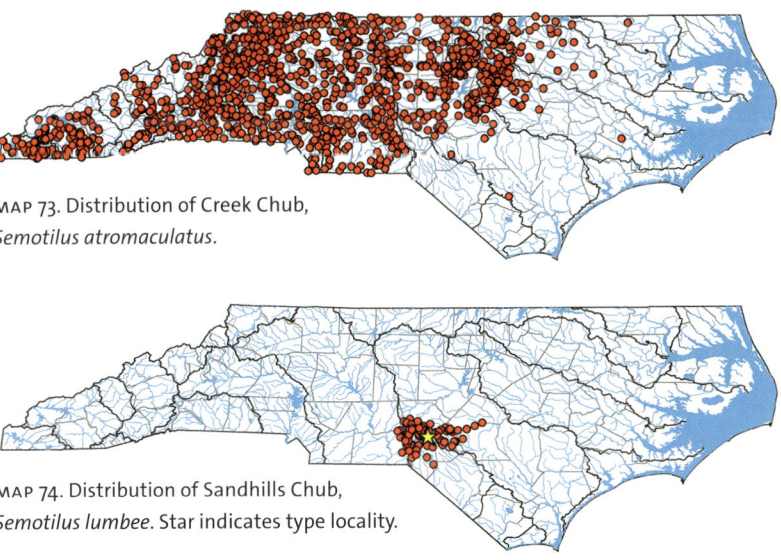

MAP 73. Distribution of Creek Chub, *Semotilus atromaculatus*.

MAP 74. Distribution of Sandhills Chub, *Semotilus lumbee*. Star indicates type locality.

14a. Eye large and mouth small. Eye larger than or equal to upper jaw. Mouth distinctly subterminal to inferior, nearly horizontal....................15

14b. Eye small and mouth large. Eye smaller than upper jaw. Mouth subterminal, often oblique......................................*Nocomis*, Couplet 18

15a. Lateral blotches along the sides (figure 126) (map 75).....................
............Blotched Chub, *Erimystax insignis* (Hubbs and Crowe, 1956)

15b. No lateral blotches along the sides...................*Hybopsis*, Couplet 16

FIGURE 126. Blotched Chub, *Erimystax insignis*.

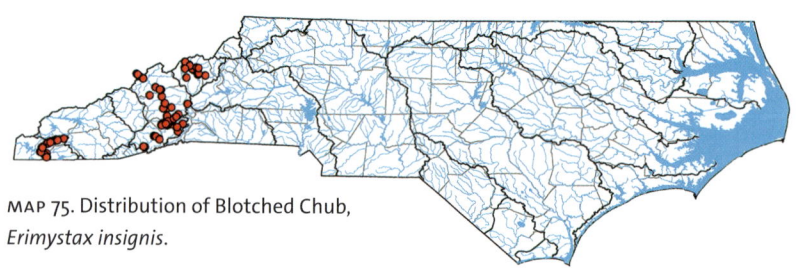

MAP 75. Distribution of Blotched Chub, *Erimystax insignis*.

16a. Breast well scaled. Body thick (figure 127). Barbel sometimes absent. Predorsal scales 15–20. Range restricted to Broad, Catawba, and Yadkin basins; introduced into Little River watershed (New basin) (map 76).......
......................Highback Chub, *Hybopsis hypsinotus* (Cope, 1870)

16b. Breast naked or with embedded scales. Body thin (figure 128). Barbel sometimes absent in Bigeye Chub, *Hybopsis amblops*. Predorsal scales 12–15
....17

FIGURE 127. Highback Chub, *Hybopsis hypsinotus*, with arrow indicating relative thickness of body.

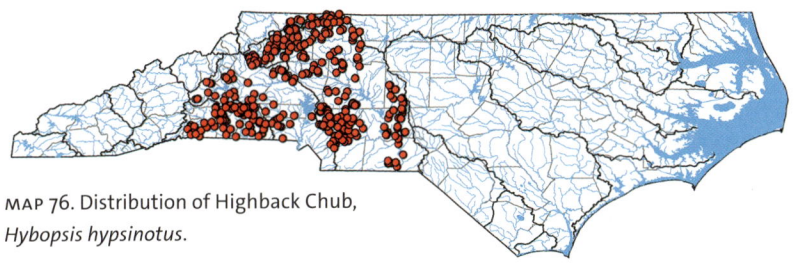

MAP 76. Distribution of Highback Chub, *Hybopsis hypsinotus*.

17a. Eye width ca. 1.2 times snout length (figure 128). Range restricted to Savannah basin (map 77)..
....................Rosyface Chub, *Hybopsis rubrifrons* (Jordan, 1877)

17b. Eye width ca. equal to snout length. Barbel less than half the length of the pupil, often absent (figure 128). Range restricted to Hiwassee, French Broad, Pigeon, and Nolichucky basins (map 78)...........................
....................Bigeye Chub, *Hybopsis amblops* (Rafinesque, 1820)

FIGURE 128. Arrows indicating relative thinness of bodies. A—Rosyface Chub, *Hybopsis rubrifrons*; B—Bigeye Chub, *Hybopsis amblops*.

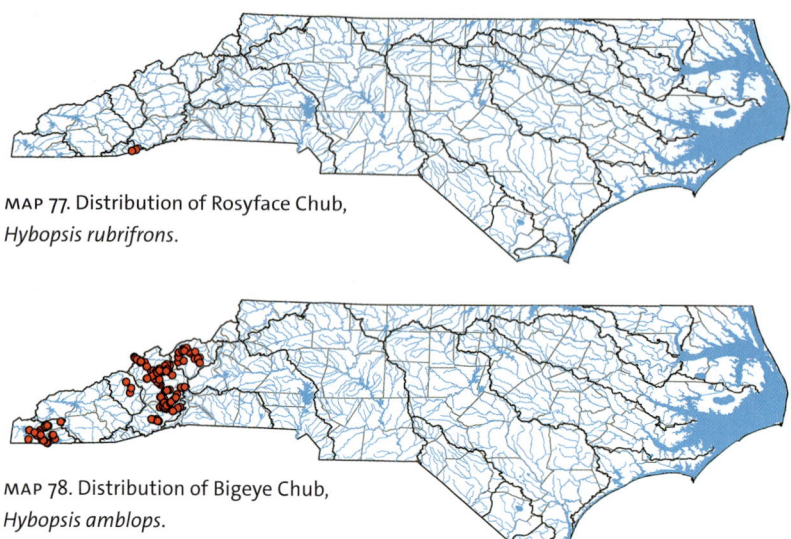

MAP 77. Distribution of Rosyface Chub, *Hybopsis rubrifrons*.

MAP 78. Distribution of Bigeye Chub, *Hybopsis amblops*.

18a. Head tubercles or scars present in prenasal and subnasal areas in adult males (figure 129). For specimens of the Bigmouth Chub, *Nocomis platyrhynchus*, from the New basin, circumbody scales (29) 32–36 (39). For specimens from Atlantic drainages (i.e., Bull Chub, *Nocomis raneyi*), breast less than 40 percent (usually less than 20 percent) scaled. Scales above lateral line 8. Upon dissection, intestine with a simple S-shaped configuration (figure 130) 19

18b. Head tubercles or scars absent from prenasal and subnasal area in adult males (figure 129). For specimens from the New basin, circumbody scales (26) 28–31 (33). For specimens from Atlantic drainages, breast greater than 50 percent scaled. Scales above lateral line 7. Upon dissection, intestine moderately whorled (figures 130 and 131) (map 79) . . . .
. . . . .Bluehead Chub, *Nocomis leptocephalus* (Girard, 1856)

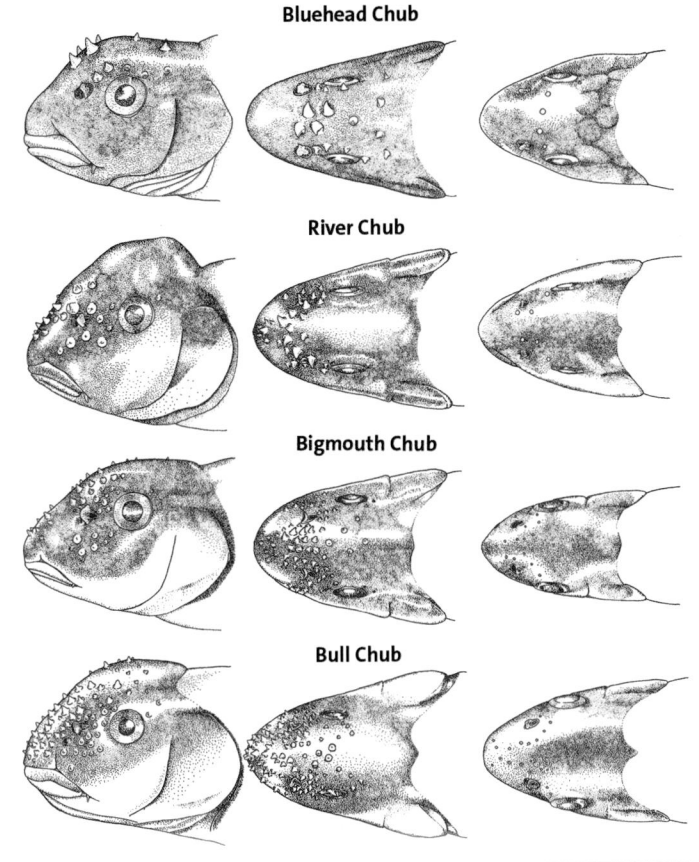

Bluehead Chub

River Chub

Bigmouth Chub

Bull Chub

FIGURE 129. Tuberculation in *Nocomis* spp. *Left and center columns*—nuptial males; *right column*—juveniles. Illustration adapted from Jenkins and Burkhead (1994).

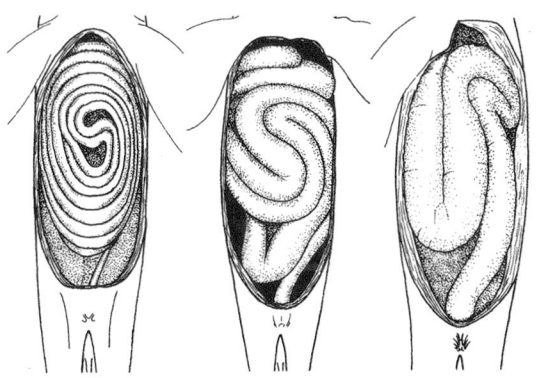

FIGURE 130. Intestinal coiling. *Left*—Eastern Silvery Minnow, *Hybognathus regius*; *center*—Bluehead Chub, *Nocomis leptocephalus*; *right*—Bull Chub, *Nocomis raneyi*. Illustration adapted from Jenkins and Burkhead (1994).

FIGURE 131. Bluehead Chub, *Nocomis leptocephalus*, male.

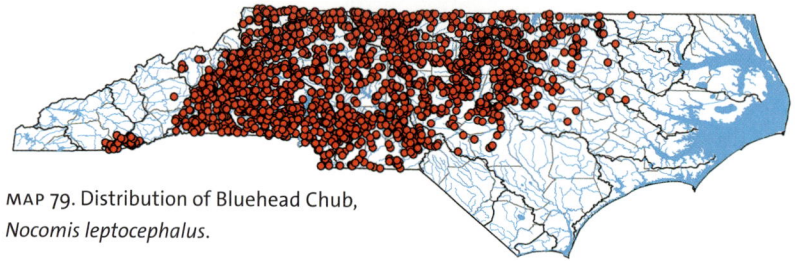

MAP 79. Distribution of Bluehead Chub,
Nocomis leptocephalus.

19a. Tubercles usually fewer than 60 on head, not extending into interorbital area (figure 129). Predorsal circumferential scales usually (28) 30–33 (36) (figure 132) (map 80)......River Chub, *Nocomis micropogon* (Cope, 1865)

19b. Tubercles 60–200 on head in specimens greater than 100 mm, extending into interorbital area (figure 129). Predorsal circumferential scales usually (29) 32–36 (39)..20

FIGURE 132. River Chub, *Nocomis micropogon*, male.

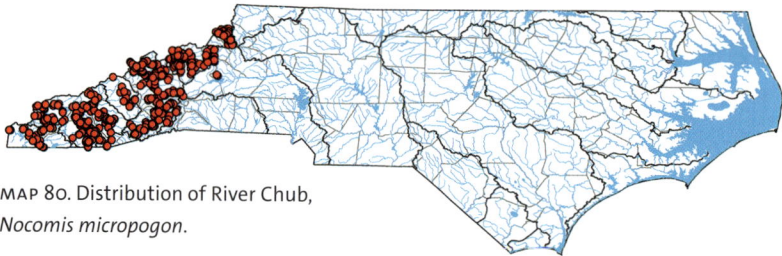

MAP 80. Distribution of River Chub,
Nocomis micropogon.

20a. Range restricted to Neuse, Tar, Roanoke, and Chowan basins (figure 133) (map 81)..........Bull Chub, *Nocomis raneyi* Lachner and Jenkins, 1971
20b. Range restricted to New basin (figure 133) (map 82)........................ ......Bigmouth Chub, *Nocomis platyrhynchus* Lachner and Jenkins, 1971

FIGURE 133. A—Bull Chub, *Nocomis raneyi*; B—Bigmouth Chub, *Nocomis platyrhynchus*.

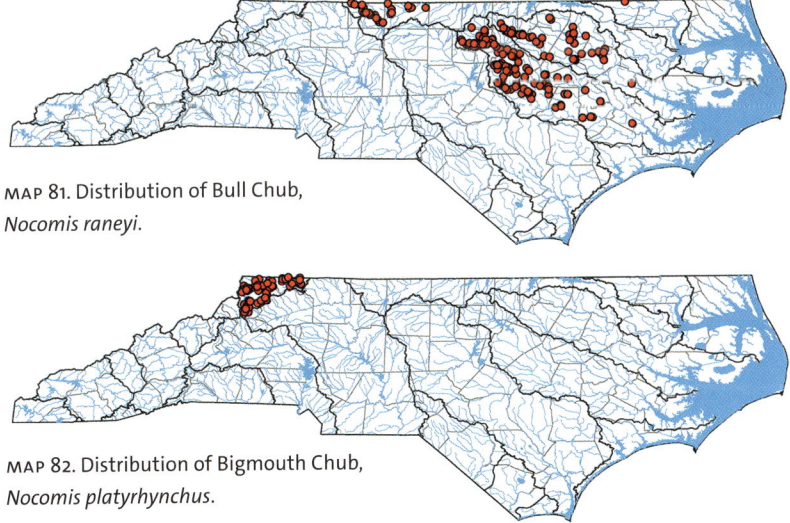

MAP 81. Distribution of Bull Chub, *Nocomis raneyi*.

MAP 82. Distribution of Bigmouth Chub, *Nocomis platyrhynchus*.

21a. Midline of belly between pelvic fins and anal fin with a nonscaled keel. Anal fin falcate with twelve or more rays. Lateral line strongly curved downward (figure 134) (map 83). . . . .
. . . . Golden Shiner, *Notemigonus crysoleucas* (Mitchill, 1814)

21b. Midline of belly between pelvic fins and anal fin without a nonscaled keel. Anal fin not falcate with eleven or fewer rays. Lateral line not strongly curved downward. 22

FIGURE 134. Golden Shiner, *Notemigonus crysoleucas*. A—adult; B—juvenile.

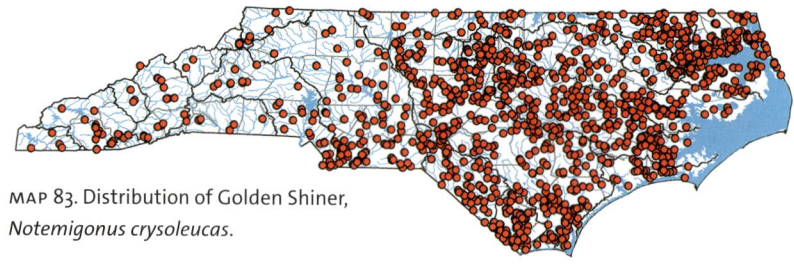

MAP 83. Distribution of Golden Shiner, *Notemigonus crysoleucas*.

22a. Predorsal scales smaller than lateral scales; appear crowded behind the head (figure 135). First dorsal fin ray in adults significantly shorter than other rays..*Pimephales*, Couplet 23

22b. Predorsal scales nearly same size as lateral scales; not crowded (figure 135). First dorsal fin ray in adults not significantly shorter than other rays
....24

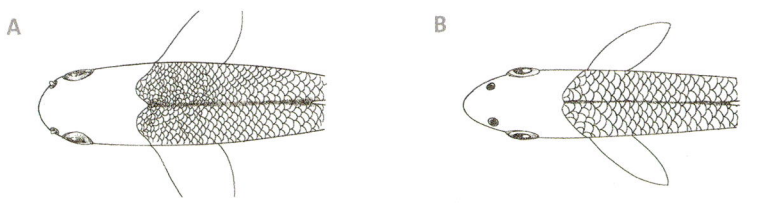

FIGURE 135. Predorsal squamation. A—scales small and crowded anteriorly; B—typical scale pattern. Illustration adapted from Jenkins and Burkhead (1994).

23a. Lateral line complete. Basicaudal spot distinct. Mouth inferior, almost horizontal. Nuptial males with maxillary barbel (figure 136) (map 84)....
.............Bluntnose Minnow, *Pimephales notatus* (Rafinesque, 1820)

23b. Lateral line usually incomplete. Basicaudal spot indistinct or absent. Mouth nearly terminal, oblique. Nuptial males without maxillary barbel (figure 136) (map 85)...
................Fathead Minnow, *Pimephales promelas* Rafinesque, 1820

FIGURE 136. A—Bluntnose Minnow, *Pimephales notatus*; B—Fathead Minnow, *Pimephales promelas*.

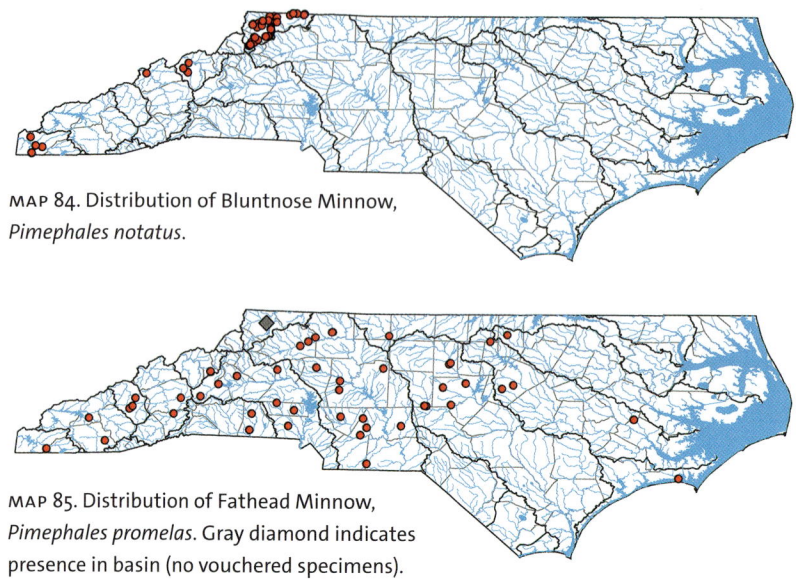

MAP 84. Distribution of Bluntnose Minnow,
Pimephales notatus.

MAP 85. Distribution of Fathead Minnow,
Pimephales promelas. Gray diamond indicates
presence in basin (no vouchered specimens).

24a. Lateral line short or absent. Lateral line scales 70–90. Scales small and
thin, difficult to observe (figure 137) (map 86)............................
....................Mountain Redbelly Dace, *Chrosomus oreas* Cope, 1868
24b. Lateral line usually complete. Lateral line scales less than 70. Scales clearly
visible...25

FIGURE 137. Mountain Redbelly Dace, *Chrosomus oreas*.

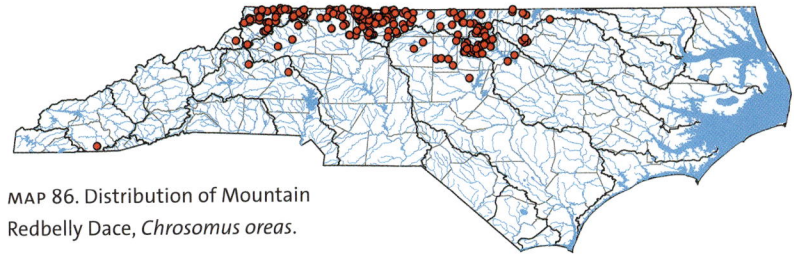

MAP 86. Distribution of Mountain
Redbelly Dace, *Chrosomus oreas*.

25a. Lips papillose, sucker-like; lower lip forms a fleshy lobe (figure 138). Mouth inferior . . . . *Phenacobius*, Couplet 26
25b. Lips not papillose, sucker-like, or forming fleshy lobe. Mouth variable 27

FIGURE 138. Inferior mouth with papillose lips. Fatlips Minnow, *Phenacobius crassilabrum*.

26a. Lateral line scales 43–48 (49). Basicaudal spot usually absent (figure 139). Range restricted to New basin (map 87) . . . . Kanawha Minnow, *Phenacobius teretulus* Cope, 1867
26b. Lateral line scales (54) 56–65 (68). Basicaudal spot distinct to faint (figure 139). Range restricted to Little Tennessee, French Broad, and Nolichucky basins (map 88) . . . . Fatlips Minnow, *Phenacobius crassilabrum* Minckley and Craddock, 1962

FIGURE 139. A—Kanawha Minnow, *Phenacobius teretulus*; B—Fatlips Minnow, *Phenacobius crassilabrum*.

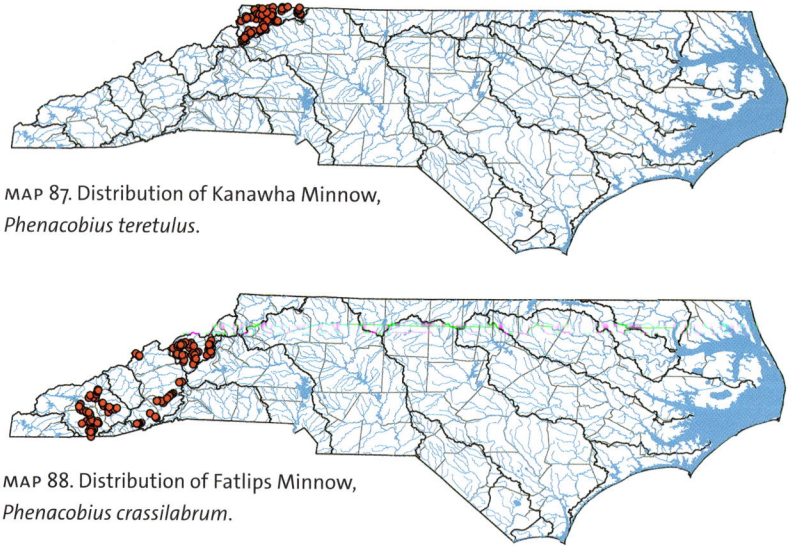

MAP 87. Distribution of Kanawha Minnow, *Phenacobius teretulus*.

MAP 88. Distribution of Fatlips Minnow, *Phenacobius crassilabrum*.

27a. Upper jaw extends beyond front of eye. Chin narrow and elongate; lower jaws converge inwardly when viewed from below. Mouth terminal and oblique. Scales small. Lateral line scales (44) 47–54 (58).................. ....*Clinostomus*, Couplet 28

27b. Upper jaw seldom extends beyond front of eye. Chin and lower jaws variable. Mouth terminal, subterminal, or inferior; angle variable. Scale size variable. Lateral line scales usually less than 43 (except in *Lythrurus* spp., which usually have 40–51 lateral line scales and a black spot at base of dorsal fin)...29

28a. Range restricted to Hiwassee basin (figure 140) (map 89)................. ...."Hiwassee" Dace, *Clinostomus* sp.

28b. Range restricted to Little Tennessee basin (figure 140) (map 90).......... ...."Smoky" Dace, *Clinostomus* sp.

28c. Widely distributed from Piedmont to New basin and seven of the Atlantic slope basins; introduced into Little Tennessee, French Broad, Nolichucky, and Watauga basins (figure 140) (map 91)......................... ....Rosyside Dace, *Clinostomus funduloides* Girard, 1856

FIGURE 140. A—"Hiwassee" Dace, *Clinostomus* sp.; B—"Smoky" Dace, *Clinostomus* sp.; C—Rosyside Dace, *Clinostomus funduloides.*

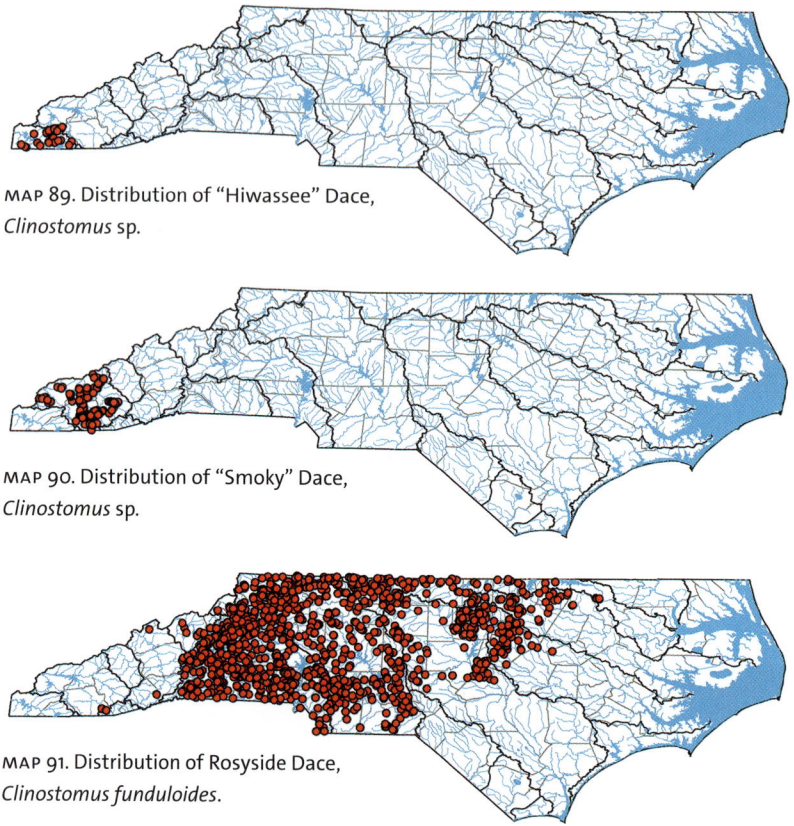

MAP 89. Distribution of "Hiwassee" Dace,
Clinostomus sp.

MAP 90. Distribution of "Smoky" Dace,
Clinostomus sp.

MAP 91. Distribution of Rosyside Dace,
Clinostomus funduloides.

29a. Mouth small; crescent shaped. Small groove along anteroventral edge of
lachrymal bone extends markedly dorsally toward snout from groove that
separates side of snout from upper lip (figure 141). Upon dissection, intestine long and strongly coiled (figure 130) (map 92)........................
...............Eastern Silvery Minnow, *Hybognathus regius* Girard, 1856

29b. Mouth variable; not crescent shaped. Lachrymal groove, if present, not extending markedly dorsally. Upon dissection, intestine short and not long
or strongly coiled (except in the Cape Fear Shiner, *Miniellus mekistocholas*)
....30

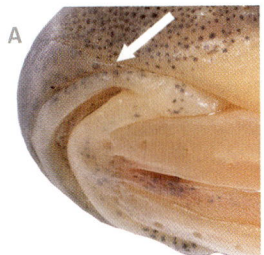

FIGURE 141. Eastern Silvery Minnow, *Hybognathus regius*. A—close-up of mouth with arrow indicating location of lachrymal groove; B—full body.

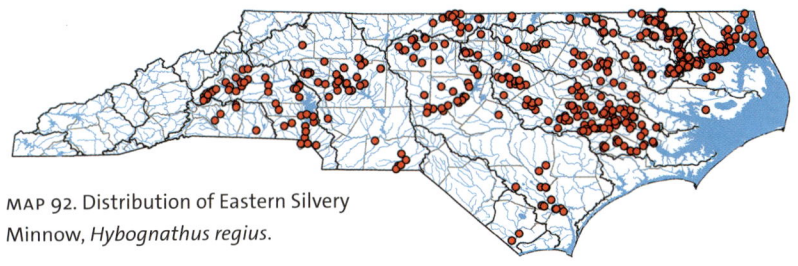

MAP 92. Distribution of Eastern Silvery Minnow, *Hybognathus regius*.

30a. Dorsal fin infused with black pigment on last two to four interradial membranes (figure 142) (except in the Red Shiner, *Cyprinella lutrensis*, in which entire dorsal fin is infused with pigment). Maxillary barbel present or absent..........................*Erimonax* and *Cyprinella*, Couplet 31

30b. Dorsal fin not infused with pigment, or if pigmented, pigment not restricted on last two to four interradial membranes (figure 142). Maxillary barbel absent...41

FIGURE 142.
A—dorsal fin membranes with black pigment and small maxillary barbel in Santee Chub, *Cyprinella zanema*; B—dorsal fin membranes without black pigment or maxillary barbel in Sandbar Shiner, *Notropis scepticus*.

31a. Maxillary barbel present..
.............................*Erimonax* and *Cyprinella* (in part), Couplet 32
31b. Maxillary barbel absent...................*Cyprinella* (in part), Couplet 35

32a. Range restricted to Little Tennessee basin; historically found in French
Broad basin (figure 143) (map 93)...
..........................Spotfin Chub, *Erimonax monachus* (Cope, 1868)
32b. Range not restricted to Little Tennessee basin...........................33

FIGURE 143. Spotfin Chub, *Erimonax monachus*. A—males; B—female. Photographs courtesy of Derek Wheaton (males) and Luke Etchison (female).

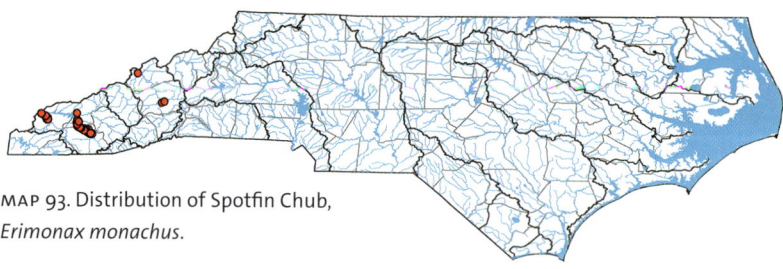

MAP 93. Distribution of Spotfin Chub, *Erimonax monachus*.

33a. Dorsal fin origin approximately over origin of pelvic fin (figure 144). Small dark blotches and cross-hatching on back and side. Body deep; depth less than five times into SL. White spot at anterior base of dorsal fin; minimal pigment below lateral line (map 94)...
.............................Thicklip Chub, *Cyprinella labrosa* (Cope, 1870)

33b. Dorsal fin origin two or three scales behind origin of pelvic fin (figure 144). No dark blotches or cross-hatching on back and side. Body slender; depth greater than five times into SL. White spot absent at anterior base of dorsal fin; well pigmented below lateral line.............................34

FIGURE 144. Dorsal fin origin relative to pelvic fin origin. A—Thicklip Chub, *Cyprinella labrosa*, dorsal fin approximately over pelvic fin origin; B—Santee Chub, *Cyprinella zanema*, dorsal fin approximately two or three scales behind pelvic fin origin.

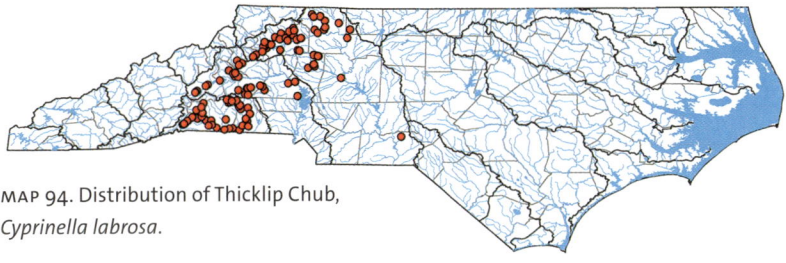

MAP 94. Distribution of Thicklip Chub, *Cyprinella labrosa*.

34a. Barbel 1.2–1.5 times pupil width. Upper lip broad; width (measured ventrally at midpoint) less than two times pupil width (figure 145). Range restricted to Broad and Catawba basins (map 95)...........................
............Santee Chub, *Cyprinella zanema* (Jordan and Brayton, 1878)

34b. Barbel 2.0–2.2 times pupil width. Upper lip thin; width (measured ventrally at midpoint) more than two times pupil width (figure 145). Range restricted to lower Yadkin, Cape Fear and Lumber basins (map 96)......
...."Thinlip" Chub, *Cyprinella* sp.

FIGURE 145. A—Santee Chub, *Cyprinella zanema*; B—"Thinlip" Chub, *Cyprinella* sp.

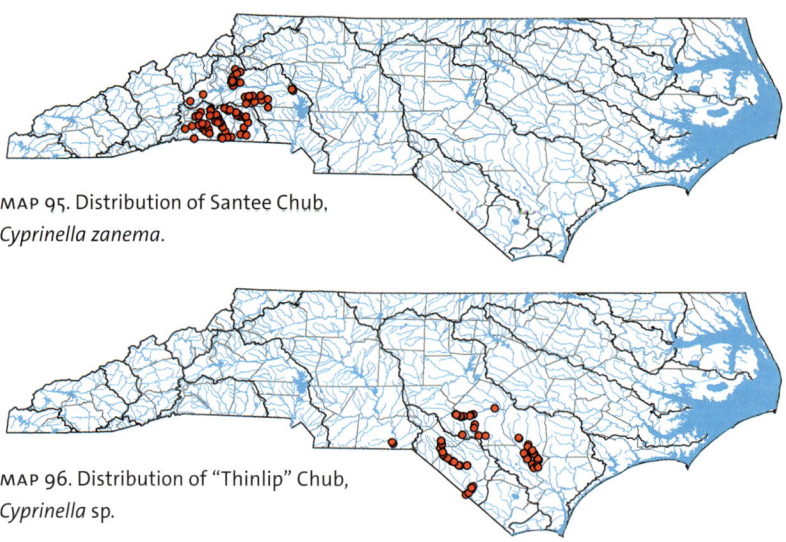

MAP 95. Distribution of Santee Chub, *Cyprinella zanema*.

MAP 96. Distribution of "Thinlip" Chub, *Cyprinella* sp.

35a. Entire dorsal fin evenly pigmented. Body deep in adults; depth less than 3.6 times in SL (figure 146) (map 97)..
................Red Shiner, *Cyprinella lutrensis* (Baird and Girard, 1853)

35b. Dorsal fin pigmentation restricted to or concentrated in last two or three interradial membranes. Body slender; depth usually greater than 3.6 times in SL (figure 146)...36

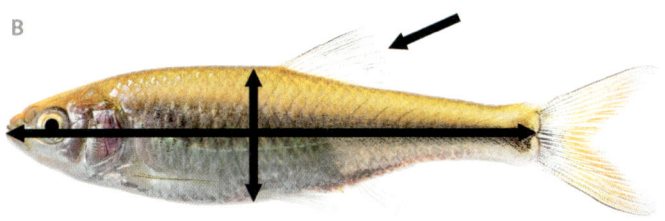

FIGURE 146. Arrows indicating dorsal fin pigmentation and body depth into SL. A (Missouri) and B—Red Shiner, *Cyprinella lutrensis*; C—Satinfin Shiner, *C. analostana*.

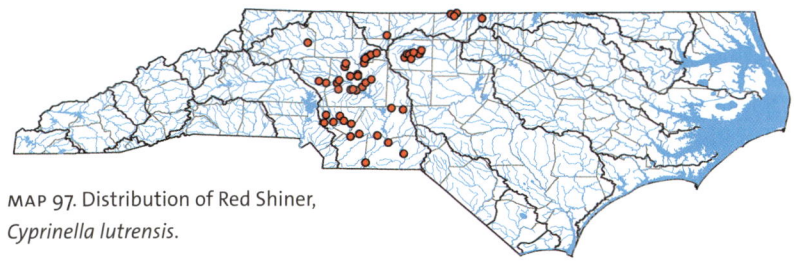

MAP 97. Distribution of Red Shiner, *Cyprinella lutrensis*.

36a. Anal fin rays usually 8, rarely 9 37
36b. Anal fin rays usually 9–11 39

37a. Predorsal circumferential scales above lateral line (13) 15. Lateral line scales 37–40. Lateral stripe black; distinct anterior-to-dorsal-fin origin. Mouth inferior (figure 147) (map 98). . . . .
. . . . Whitefin Shiner, *Cyprinella nivea* (Cope, 1870)
37b. Predorsal circumferential scales above lateral line 11–13. Lateral line scales 32–38. Lateral stripe not distinct anterior-to-dorsal-fin origin. Mouth usually subterminal, slightly oblique. 38

FIGURE 147. Whitefin Shiner, *Cyprinella nivea*.

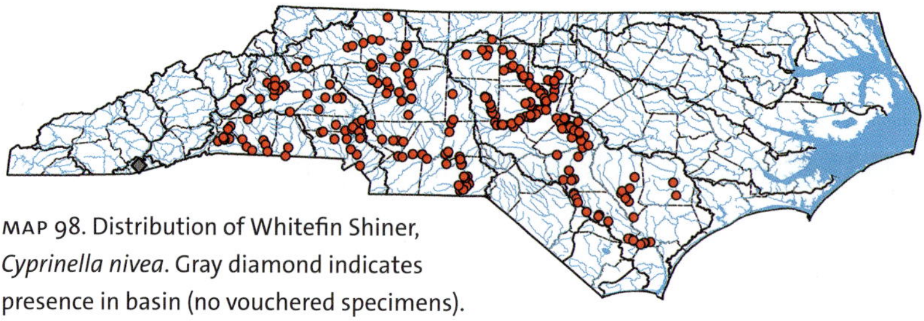

MAP 98. Distribution of Whitefin Shiner, *Cyprinella nivea*. Gray diamond indicates presence in basin (no vouchered specimens).

38a. Predorsal circumferential scales above lateral line usually 11, rarely 13. Lateral line scales 32–35 (figure 148). Range restricted to Broad and Catawba basins (one record from Yadkin basin) (map 99). . . . .
. . . . Greenfin Shiner, *Cyprinella chloristia* (Jordan and Brayton, 1878)
38b. Predorsal circumferential scales above lateral line usually 13. Lateral line scales 35–38 (figure 148). Range restricted to Hiwassee, Pigeon, French Broad, Nolichucky, and New basins (map 100). . . . .
. . . . Spotfin Shiner, *Cyprinella spiloptera* (Cope, 1867)

FIGURE 148. A—Greenfin Shiner, *Cyprinella chloristia*; B—Spotfin Shiner, *Cyprinella spiloptera*.

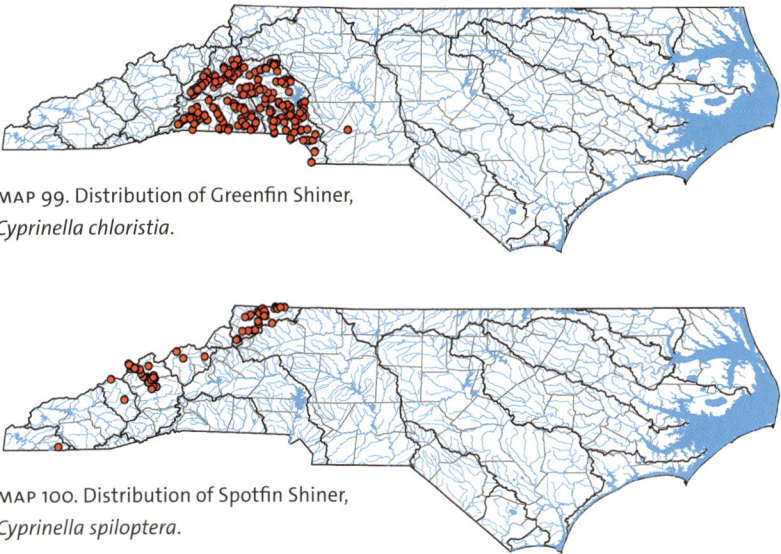

MAP 99. Distribution of Greenfin Shiner, *Cyprinella chloristia*.

MAP 100. Distribution of Spotfin Shiner, *Cyprinella spiloptera*.

39a. Upper and lower portions of caudal fin base with distinct large pale patches (figure 149). Predorsal stripe dark (map 101)......................
......................Whitetail Shiner, *Cyprinella galactura* (Cope, 1868)

39b. Upper and lower portions of caudal fin base without distinct large pale patches. Predorsal stripe variable..40

FIGURE 149. Whitetail Shiner, *Cyprinella galactura*. A—female or juvenile; B—male in breeding colors. Photograph courtesy of Luke Etchison (male).

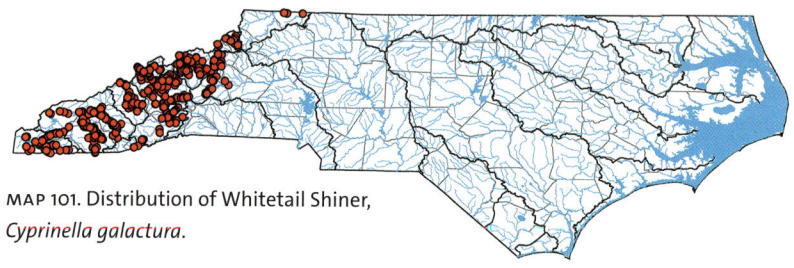

MAP 101. Distribution of Whitetail Shiner, *Cyprinella galactura*.

40a. Anal fin rays usually 10 or 11. Caudal fin with black edge preceded by red in large males. No lightly pigmented area at base of caudal fin. Humeral bar black (figure 150) (map 102)...
..................Fieryblack Shiner, *Cyprinella pyrrhomelas* (Cope, 1870)

40b. Anal fin rays usually 9. No pigment in caudal fin. Light basicaudal bar behind caudal spot. Humeral bar faint (figure 150) (map 103)............
......................Satinfin Shiner, *Cyprinella analostana* Girard, 1859

FIGURE 150. Arrows indicating location of humeral bar. A—Fieryblack Shiner, *Cyprinella pyrrhomelas*; B—Satinfin Shiner, *Cyprinella analostana*.

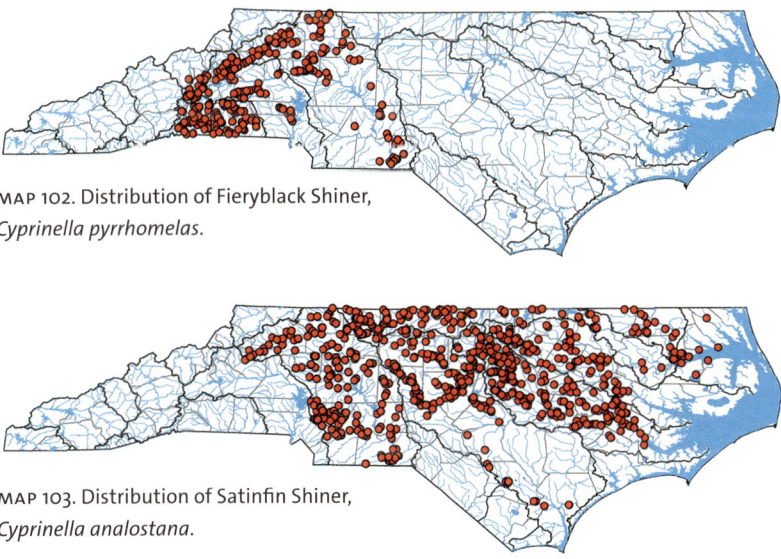

MAP 102. Distribution of Fieryblack Shiner, *Cyprinella pyrrhomelas*.

MAP 103. Distribution of Satinfin Shiner, *Cyprinella analostana*.

41a. Exposed portion (lunula) of anterior lateral line scales greater than twice as high as wide (figure 151a). Anal fin rays modally 9 .42

41b. Exposed portion (lunula) of anterior lateral line scales less than or equal to twice as high as wide (except for the Mimic Shiner, *Paranotropis volucellus*, which has anterior lateral line scale lunula greater than twice as high as wide) (figure 151b and c). Anal fin rays 7–12 .45

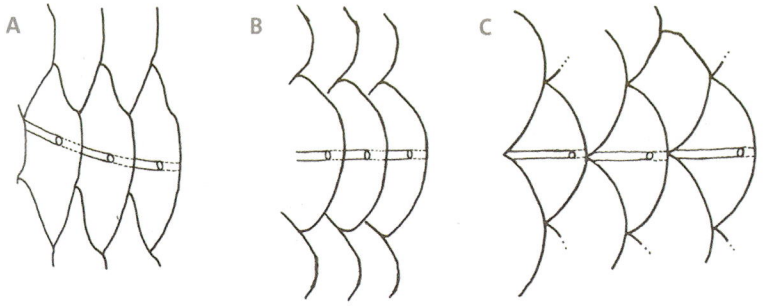

FIGURE 151. Height relative to width of exposed portion (lunula) of anterior lateral line scales. A—*Luxilus* spp.; B—Mimic Shiner, *Paranotropis volucellus*; C—typical of most species. Illustration adapted from Jenkins and Burkhead (1994).

42a. Caudal and dorsal fins with light base and dark submarginal stripe. Humeral bar black; dark in young. Mouth terminal and oblique (figure 152) (map 104) . . . . .Warpaint Shiner, *Coccotis coccogenis* (Cope, 1868)

42b. Caudal and dorsal fins with no dark stripe. Humeral bar seldom black and distinct. Mouth subterminal or terminal . . . .*Luxilus*, Couplet 43

FIGURE 152. Warpaint Shiner, *Coccotis coccogenis*.

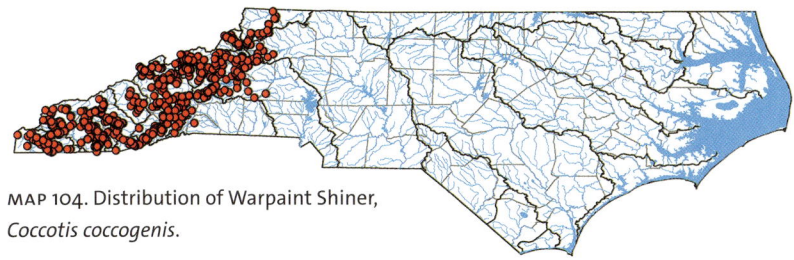

MAP 104. Distribution of Warpaint Shiner, *Coccotis coccogenis*.

43a. Predorsal scales 13–16; mid-row usually distinct and scales easily counted. Two or three wavy, dark horizontal lines running between dorsal fin insertion and lateral stripe (figure 153). Range restricted to Hiwassee, Pigeon, French Broad, and Nolichucky basins (map 105).............................
.....................Striped Shiner, *Luxilus chrysocephalus* Rafinesque, 1820

43b. Predorsal scales 17–25; mid-row seldom straight, scales usually difficult to count. Range restricted to New and Atlantic slope basins..............44

FIGURE 153. Striped Shiner, *Luxilus chrysocephalus*.

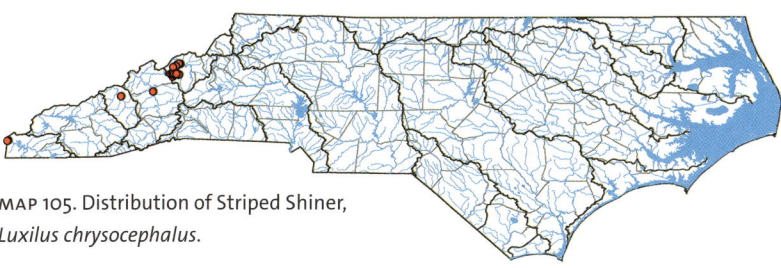

MAP 105. Distribution of Striped Shiner, *Luxilus chrysocephalus*.

44a. Sides with irregularly scattered black scales extending below lateral line (figure 154). Dorsal anterior rays extending two to three scales beyond posterior rays in appressed fin. Range restricted to Roanoke and Chowan basins; introduced into upper Haw River watershed (Cape Fear basin) (map 106)................Crescent Shiner, *Luxilus cerasinus* (Cope, 1868)

44b. Sides with no irregularly scattered black scales (occasionally with irregular dark bars two or three scales deep) (figure 154). Dorsal anterior rays extending zero to one scale beyond posterior rays in appressed fin. Range restricted to New, Cape Fear, Neuse, Tar, Roanoke, and Chowan basins (map 107).................White Shiner, *Luxilus albeolus* (Jordan, 1889)

FIGURE 154. A—Crescent Shiner, *Luxilus cerasinus*; B—White Shiner, *Luxilus albeolus*.

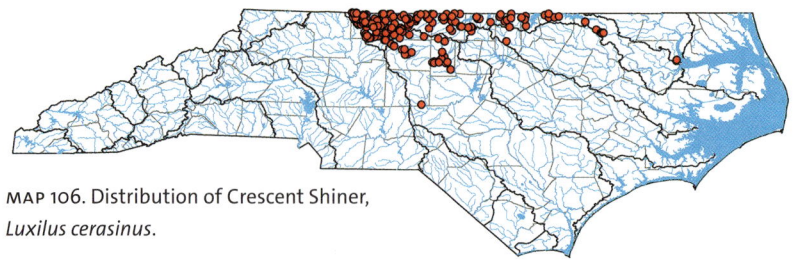

MAP 106. Distribution of Crescent Shiner, *Luxilus cerasinus*.

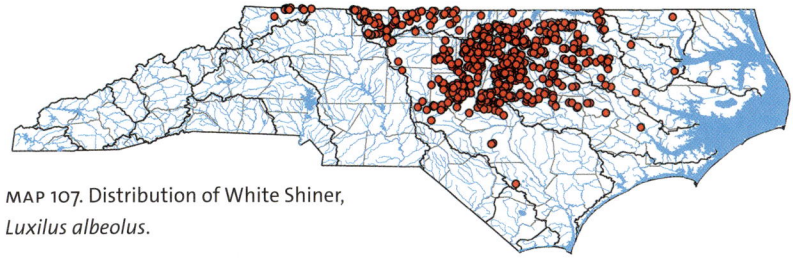

MAP 107. Distribution of White Shiner, *Luxilus albeolus*.

45a. Black spot on dorsal fin near origin. Antero-dorsolateral scales distinctly smaller than postdorsal scales. Mouth terminal; oblique. Body very elongate. Dorsal fin origin moderately or much posterior to pelvic fin base. Anal rays modally 11. Breast usually naked........*Lythrurus*, Couplet 46

45b. No black spot on dorsal fin near origin. Antero-dorsolateral scales ca. the same size as the postdorsal scales. Mouth position and body shape, dorsal fin origin, number of anal fin rays, and breast scalation variable.......47

46a. Vivid, blood-red chromatic colors restricted to upper head and median fins (figure 155). Females in nuptial condition masculinized, with tubercle development on dorsum of head equal to that of males. Range restricted to Neuse and Tar basins (map 108).......................................
....................Pinewoods Shiner, *Lythrurus matutinus* (Cope, 1870)

46b. Less extensive development of red on fins (figure 155). Females in nuptial condition not masculinized, with no or weakly developed tubercles on dorsum of head. Range restricted to Roanoke basin with introduced populations in Yadkin and Cape Fear basins (map 109)..........................
...........................Rosefin Shiner, *Lythrurus ardens* (Cope, 1868)

FIGURE 155. A—Pinewoods Shiner, *Lythrurus matutinus*; B—Rosefin Shiner, *Lythrurus ardens*.

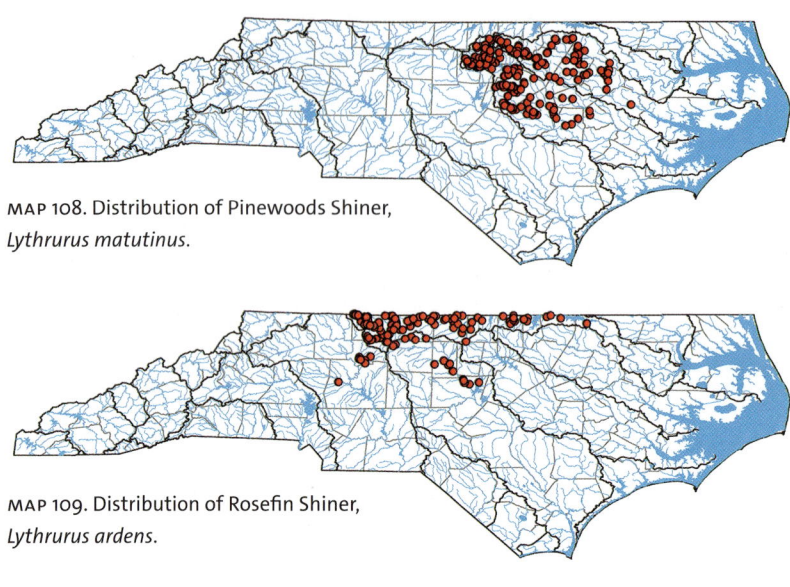

MAP 108. Distribution of Pinewoods Shiner, *Lythrurus matutinus*.

MAP 109. Distribution of Rosefin Shiner, *Lythrurus ardens*.

47a. Anal fin rays usually 10–12...48
47b. Anal fin rays usually 7 or 8...55

48a. Dark lateral stripe extends from dark caudal fin spot to around tip of snout and on the lips. Lateral line often incomplete, not set off by dark dashes. Anal fin base often dark...49
48b. Dark lateral stripe not encircling the snout. Caudal fin spot seldom dark. Lateral line complete, often set off with black spots or dashes. Anal fin base unpigmented or with faint pigment...............................50

49a. Dark lateral stripe extending anteriorly onto opercle, confined to upper 60 percent of opercle with a horizontal ventral margin (figure 156). Black lateral stripe above lateral line in region of pelvic fin. Light zone above dark lateral stripe usually continuous to head. Anal fin base generally not pigmented. Pharyngeal teeth usually 2,4–4,2 (map 110)..................
......................Highfin Shiner, *Hudsonius altipinnis* (Cope, 1870)
49b. Dark lateral stripe extending anteriorly onto opercle, covering most of opercle with ventral margin sloping anteroventrally (figure 156). Black lateral stripe extending one or two scale rows below the lateral line in region of pelvic fin. Light zone above dark lateral stripe usually slightly obscured

anteriorly behind head by dark pigment edging scales. Anal fin base black; pigment extending to under caudal peduncle. Pharyngeal teeth usually 1,4–4,1 (map 111)...
....................Dusky Shiner, *Pteronotropis cummingsae* (Myers, 1925)

FIGURE 156. A—Highfin Shiner, *Hudsonius altipinnis*; B—Dusky Shiner, *Pteronotropis cummingsae*.

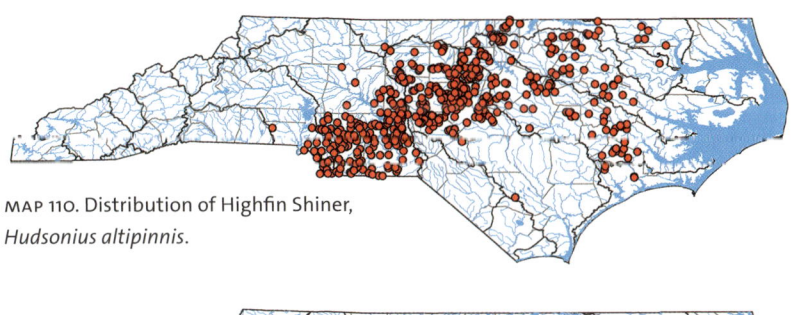

MAP 110. Distribution of Highfin Shiner, *Hudsonius altipinnis*.

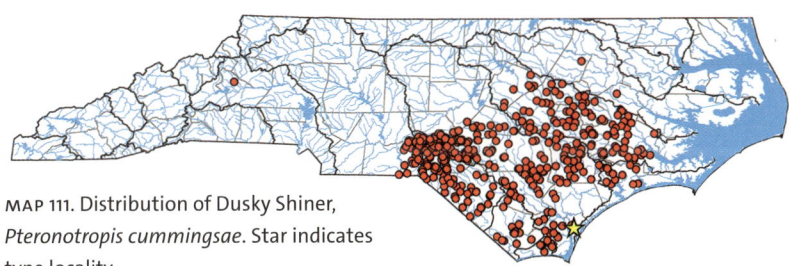

MAP 111. Distribution of Dusky Shiner, *Pteronotropis cummingsae*. Star indicates type locality.

50a. Predorsal scales large, 13–16; mid-row straight and distinct. Predorsal circumferential scales above lateral line 10 or 11. Two dark lines on head between the eyes. Dorsolateral scales form two or three horizontal stripes bordered with brown or black (figure 157) (map 112)......................
........................Telescope Shiner, *Notropis telescopus* (Cope, 1868)

50b. Predorsal scales small, 17–29; mid-row seldom distinct anteriorly. Predorsal circumferential scales above lateral line 13–17. Two dark lines on head between the eyes absent. Dorsolateral scales do not form distinct horizontal stripes...51

FIGURE 157. Telescope Shiner, *Notropis telescopus*.

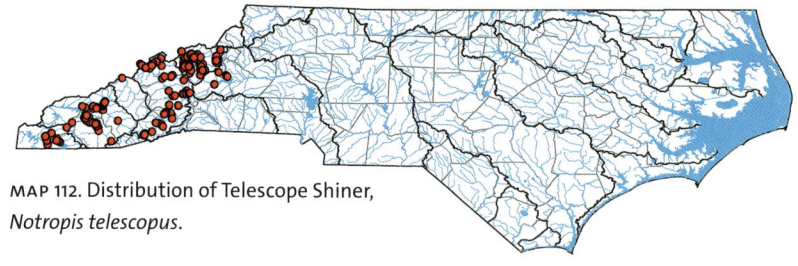

MAP 112. Distribution of Telescope Shiner, *Notropis telescopus*.

51a. Eye large; diameter greater than length of snout. Predorsal profile curves down above nostrils (figure 158) (map 113)...................................
............Sandbar Shiner, *Notropis scepticus* (Jordan and Gilbert, 1883)

51b. Eye diameter equal to or less than length of snout. Predorsal profile does not curve abruptly down above nostrils...................................52

FIGURE 158. Sandbar Shiner, *Notropis scepticus*.

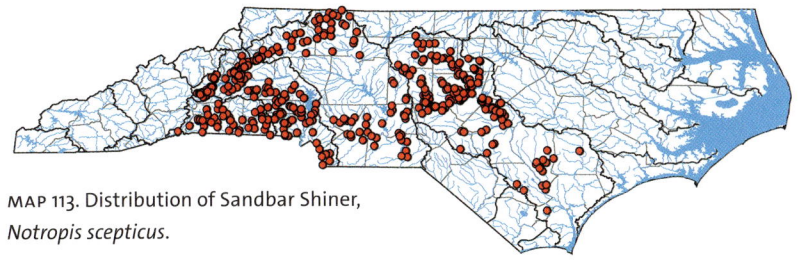

MAP 113. Distribution of Sandbar Shiner,
Notropis scepticus.

52a. Predorsal circumferential scales above lateral line 15. Anal fin rays (10) 11 (12). Predorsal profile straight in adults (figure 159). Range restricted to Atlantic slope basins (map 114).. ....Comely Shiner, *Notropis amoenus* (Abbott, 1874)
52b. Predorsal circumferential scales above lateral line (11) 13. Anal fin rays (9) 10 (12). Predorsal profile slightly rounded in adults. Range restricted to New basin and basins west of the Appalachian Mountains.............53

FIGURE 159. Comely Shiner, *Notropis amoenus.*

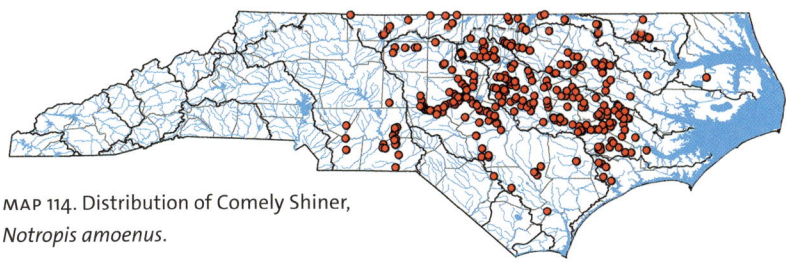

MAP 114. Distribution of Comely Shiner,
Notropis amoenus.

53a. Dorsal fin origin 1.5–2.5 scales behind pelvic fin origin (figure 160). Distance from dorsal fin origin to hypural plate greater than distance from dorsal fin origin to center of pupil. Pigment extends below lateral line. Dark crescents often visible between nostrils. Pelvic fin rays 9 or 10 (map 115) ....Silver Shiner, *Notropis photogenis* (Cope, 1865)

53b. Dorsal fin origin 2.5–4.0 scales behind pelvic fin origin (figure 160). Distance from dorsal fin origin to hypural plate less than distance from dorsal fin origin to center of pupil. No pigment on sides below lateral line. No dark crescents visible between nostrils. Pelvic fin rays 8.................54

FIGURE 160. Dorsal fin origin relative to pelvic fin origin. A—Silver Shiner, *Notropis photogenis*; B—"Kanawha" Rosyface Shiner, *Notropis* sp.

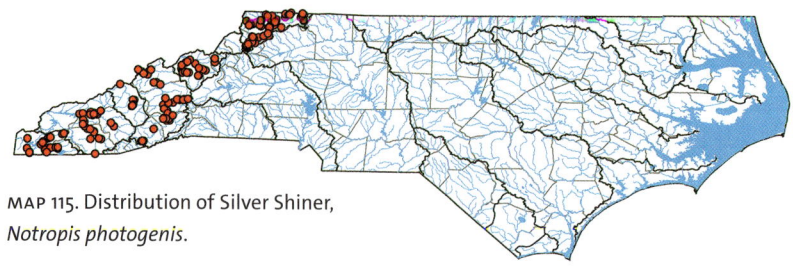

MAP 115. Distribution of Silver Shiner, *Notropis photogenis*.

54a. Range restricted to New basin (figure 161) (map 116).......................
......................................"Kanawha" Rosyface Shiner, *Notropis* sp.

54b. Range restricted to Hiwassee, Little Tennessee, Pigeon, French Broad, and Nolichucky basins (figure 161) (map 117)...............................
.......................Highland Shiner, *Notropis micropteryx* (Cope, 1868)

FIGURE 161. A—"Kanawha" Rosyface Shiner, *Notropis* sp.; B—Highland Shiner, *Notropis micropteryx*.

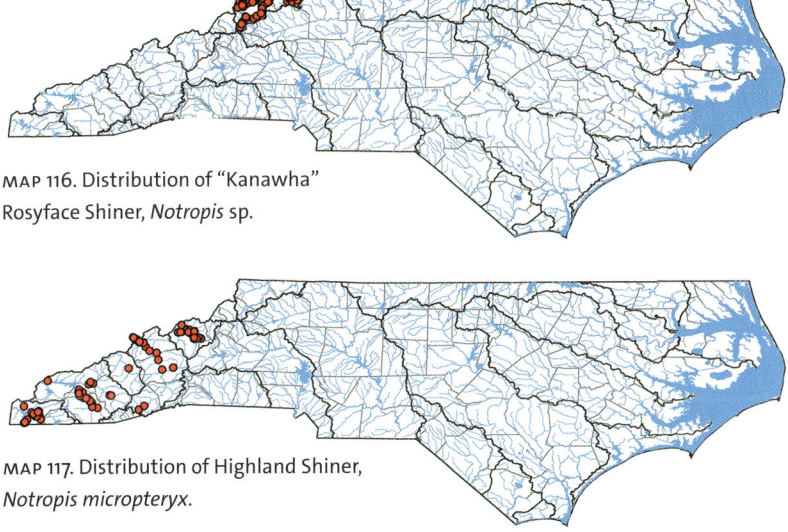

MAP 116. Distribution of "Kanawha" Rosyface Shiner, *Notropis* sp.

MAP 117. Distribution of Highland Shiner, *Notropis micropteryx*.

55a. Anal fin rays modally 7; pronounced black lateral and preorbital stripes; breeding males not brightly colored with reds or oranges. Range restricted to lower Piedmont and/or Coastal Plain..................................56

55b. Anal fin rays modally 8; pronounced black lateral or preorbital stripes present or not; breeding males may or may not be brightly colored with reds or oranges. Distributed within the Mountains, Piedmont, and/or Coastal Plain..59

56a. Pharyngeal teeth 2,4–4,2. Lateral stripe absent just behind eye. Mouth subterminal, distinctly oblique (figure 162) (map 118).....................
........................Coastal Shiner, *Alburnops petersoni* (Fowler, 1942)

56b. Pharyngeal teeth 4–4. Lateral stripe extending to the eye. Mouth subterminal, almost horizontal (except in the Bridle Shiner, *Notropis bifrenatus*, which has an extremely small mouth that is subterminal and distinctly oblique)..57

FIGURE 162. Coastal Shiner, *Alburnops petersoni*.

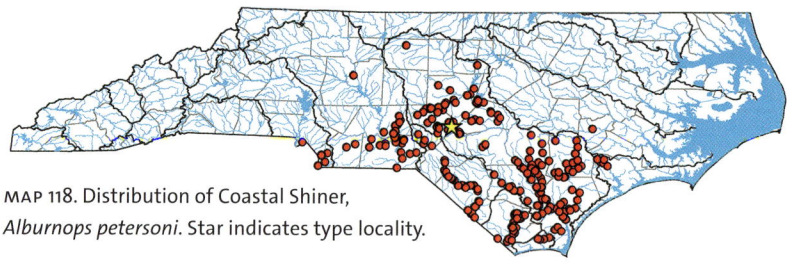

MAP 118. Distribution of Coastal Shiner, *Alburnops petersoni*. Star indicates type locality.

57a. Breast often entirely naked, half-scaled at most. Caudal fin spot usually not connected to lateral stripe. Smudge of dark pigment anterior to dorsal fin. Dark lateral stripe does not encircle snout (figure 163a). Light area absent on snout above dark stripe (figure 164) (map 119)..................
..........................Swallowtail Shiner, *Miniellus procne* (Cope, 1865)[1]

57b. Breast usually half- to fully scaled. Caudal fin spot connected to lateral stripe. No dark smudge anterior to dorsal fin. Dark lateral stripe encircling the snout. Light area present on snout above dark stripe (figure 163b and c)
....58

FIGURE 163. A—Swallowtail Shiner, *Miniellus procne*, with dark lateral stripe not encircling the snout; B—Whitemouth Shiner, *Miniellus alborus*, with dark lateral stripe encircling snout and lips unpigmented; C—Bridle Shiner, *Notropis bifrenatus*, with dark lateral strip encircling snout and upper lip pigmented.

FIGURE 164.
Swallowtail Shiner,
Miniellus procne.

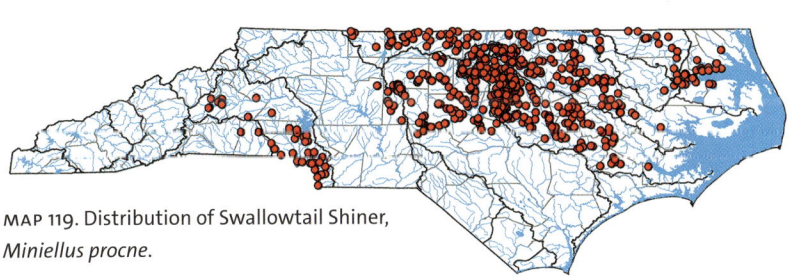

MAP 119. Distribution of Swallowtail Shiner,
Miniellus procne.

1. Populations of Swallowtail Shiner, *Miniellus procne*, in the lower Catawba basin appear different from their eastern counterparts. They resemble the Sand Shiner, *Miniellus stramineus* (Cope, 1865), in terms of overall body color and physical appearance (B. H. Tracy, personal observation). Raney (1947) recognized two subspecies of the Swallowtail Shiner, *M. procne procne* (Cope), found north of the James River in Virginia, and *M. procne longiceps* (Cope), found from the Roanoke River in Virginia south to the Santee drainage in South Carolina, but the validity of these subspecific designations was dismissed by Snelson (1971) and Jenkins and Sorensen (1980). Rohde et al. (2009) acknowledged the results of some preliminary genetic work from different drainages in South Carolina, but the results were never published, and the subject continues to require further study.

58a. Band of dark pigment encircling snout (figure 163b), ca. uniform in width or slightly narrowed at snout tip. Front of upper lip usually unpigmented. Mouth subterminal, almost horizontal. Often with red pigment in upper portion of iris (figure 165). Range restricted to Yadkin, Cape Fear, and Roanoke basins; introduced into the lower Catawba basin (map 120).........
.........Whitemouth Shiner, *Miniellus alborus* (Hubbs and Raney, 1947)

58b. Band of dark pigment encircling the snout, markedly constricted at snout tip (figure 163c). Front of upper lip liberally pigmented. Mouth subterminal, distinctly oblique (figures 163c and 165). Range restricted to lower Neuse and Chowan basins (map 121)......................................
..........................Bridle Shiner, *Notropis bifrenatus* (Cope, 1867)

FIGURE 165. A—Whitemouth Shiner, *Miniellus alborus*; B—Bridle Shiner, *Notropis bifrenatus*.

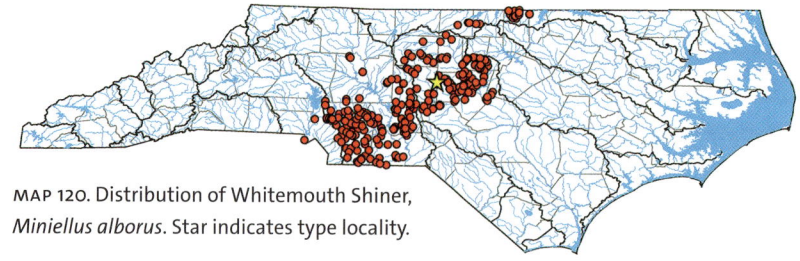

MAP 120. Distribution of Whitemouth Shiner, *Miniellus alborus*. Star indicates type locality.

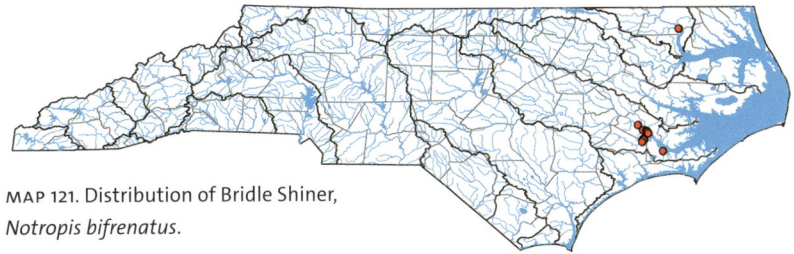

MAP 121. Distribution of Bridle Shiner,
Notropis bifrenatus.

59a. Lateral line incomplete. Caudal fin spot black, larger than pupil, surrounded by distinct light area (figure 166). Large black blotch along front
of dorsal fin (map 122)..
...........................Taillight Shiner, *Notropis maculatus* (Hay, 1881)
59b. Lateral line usually complete. Caudal fin spot, if present, not larger than
pupil and not surrounded by a distinct light area. Large black blotch along
front of dorsal fin absent...60

FIGURE 166. Taillight Shiner, *Notropis maculatus*.

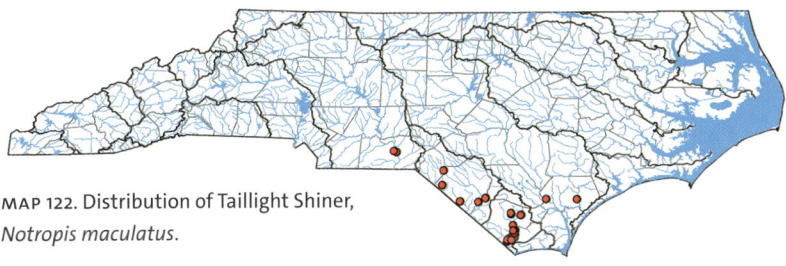

MAP 122. Distribution of Taillight Shiner,
Notropis maculatus.

60a. Dorsal fin rounded. Caudal spot triangular, much darker than lateral stripe. Lateral stripe faint to absent; light streak over lateral stripe absent. Head flat between the eyes. Body cylindrical. Scales on the nape small and crowded (figure 167) (map 123)..
.................Mirror Shiner, *Paranotropis spectrunculus* (Cope, 1868)

60b. Dorsal fin pointed or truncate. Caudal fin spot variable. Lateral stripe medium to well developed with a light streak usually present over dark lateral stripe. Head shape variable. Body compressed. Scales on nape not crowded, clearly visible...61

FIGURE 167. Mirror Shiner, *Paranotropis spectrunculus*.

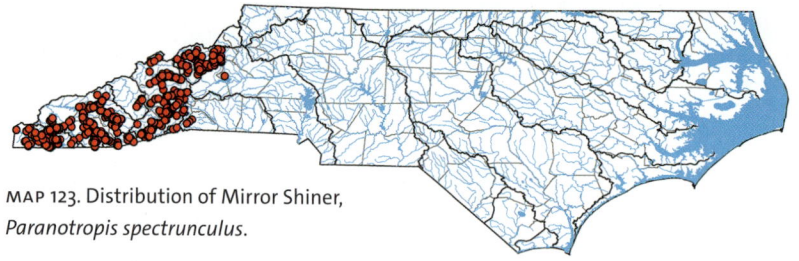

MAP 123. Distribution of Mirror Shiner, *Paranotropis spectrunculus*.

61a. Body along anal fin base black with pigment extending under caudal peduncle. Mouth small, not reaching to front of eye. Roof of mouth black. Anal and pelvic fins usually bordered with black (figure 168) (map 124)
......................Ironcolor Shiner, *Alburnops chalybaeus* (Cope, 1867)

61b. Body along anal fin base with little or no pigment. Mouth size variable. Roof of mouth unpigmented or with very diffuse pigment. Anal and pelvic fin rays not bordered with black....................................62

FIGURE 168. Ironcolor Shiner, *Alburnops chalybaeus*.

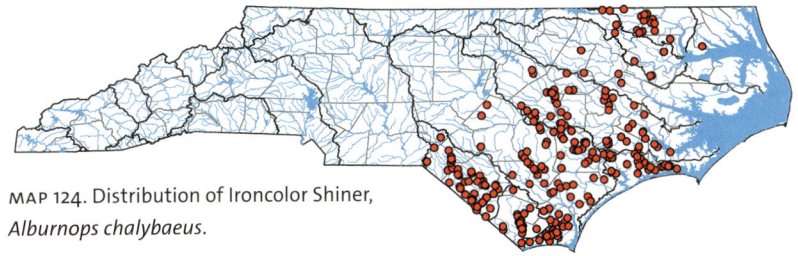

MAP 124. Distribution of Ironcolor Shiner,
Alburnops chalybaeus.

62a. Dorsal fin pointed, first ray usually reaching two or three scales behind last ray. Dorsal fin origin over or anterior to pelvic fin origin. Mouth nearly horizontal (figure 169) (map 125) . . . . .Spottail Shiner, *Hudsonius hudsonius* (Clinton, 1824)

62b. Dorsal fin not strongly pointed, first ray reaching zero or one scale behind last ray. Dorsal fin origin posterior to pelvic fin origin (except in Cape Fear Shiner, *Miniellus mekistocholas*). Mouth usually oblique 63

FIGURE 169. Spottail Shiner, *Hudsonius hudsonius*.

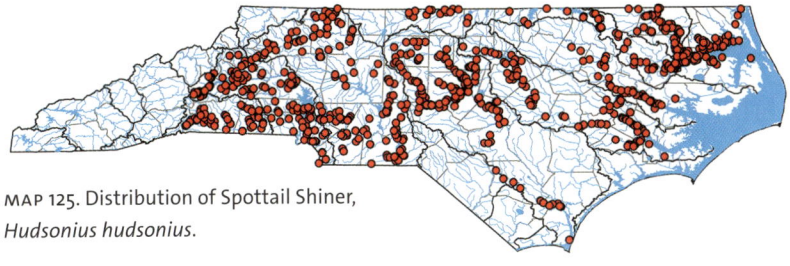

MAP 125. Distribution of Spottail Shiner,
Hudsonius hudsonius.

63a. Punctate lateral line. Caudal fin spot rectangular. Light stripe present over dark lateral stripe (figure 170) (map 126).............................
.....................Tennessee Shiner, *Paranotropis leuciodus* (Cope, 1868)

63b. Lateral line not punctate (except in New River Shiner, *Miniellus scabriceps*, which sometimes has a punctate lateral line on the anterior portion). Caudal fin spot, if present, round. Light stripe usually absent over dark lateral stripe...64

FIGURE 170. Tennessee Shiner, *Paranotropis leuciodus*. A—female; B—male in breeding colors.

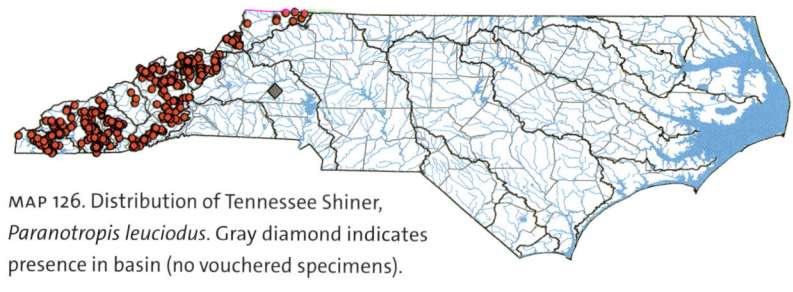

MAP 126. Distribution of Tennessee Shiner, *Paranotropis leuciodus*. Gray diamond indicates presence in basin (no vouchered specimens).

64a. Eye large, bulging outward, directed upward more than usual. Pre-dorsolateral scales arranged in clearly defined diagonal rows (figure 171). Range restricted to New basin (map 127).....................................
........................New River Shiner, *Miniellus scabriceps* (Cope, 1868)

64b. Eye not large or bulging outward, usually directed sideways. Pre-dorsolateral scales not usually arranged in clearly defined diagonal rows.............65

FIGURE 171. New River Shiner, *Miniellus scabriceps*.

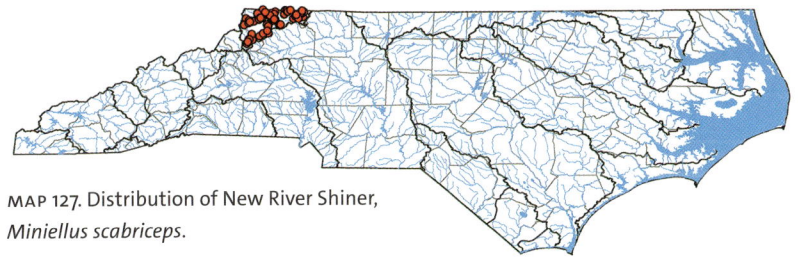

MAP 127. Distribution of New River Shiner, *Miniellus scabriceps*.

65a. Anterior lateral line scales elevated, exposed width 2.1–2.9 times height (best observed in third to seventh scales from the head) (figure 151). Extensive development of neuromasts on anterior half of the head: dorsum, snout, and subnasal area; around the orbit, particularly on the cheek; and on the anterior portion of lateral line scales. Lateral stripe diffuse, not entering snout. Caudal fin spot faint to absent (figure 172) (map 128).....
........................Mimic Shiner, *Paranotropis volucellus* (Cope, 1865)

65b. Lateral line scales not elevated anteriorly, width less than two times height (figure 151). Development of neuromasts variable. Lateral stripe black, sometimes diffuse anteriorly, often entering snout. Caudal fin spot black, often joined to lateral stripe..66

FIGURE 172. Mimic Shiner, *Paranotropis volucellus*.

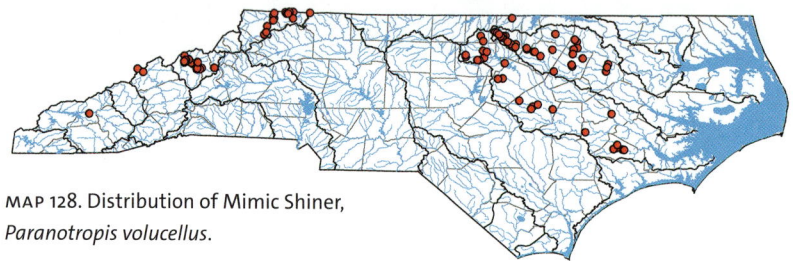

MAP 128. Distribution of Mimic Shiner, *Paranotropis volucellus*.

66a. Dorsal fin origin anterior or above pelvic fin origin. Mostly scaled anterior breast area. Lower lip lined with dark pigment. Breeding males bright brassy yellow-gold (figure 173). Range restricted to Cape Fear basin (map 129)......Cape Fear Shiner, *Miniellus mekistocholas* (Snelson, 1971)

66b. Dorsal fin origin posterior to pelvic fin origin. Anterior breast mostly naked (except in Yellowfin Shiner, *Hydrophlox lutipinnis*, in which scalation of anterior breast is variable). Lower lip not lined with dark pigment. Breeding males brightly colored with yellow, red, or white-colored fins ....*Hydrophlox*, Couplet 67

FIGURE 173. Cape Fear Shiner, *Miniellus mekistocholas*. A—bar indicating dorsal fin anterior to or above pelvic fin origin; B—arrow indicating pigmented lower lip.

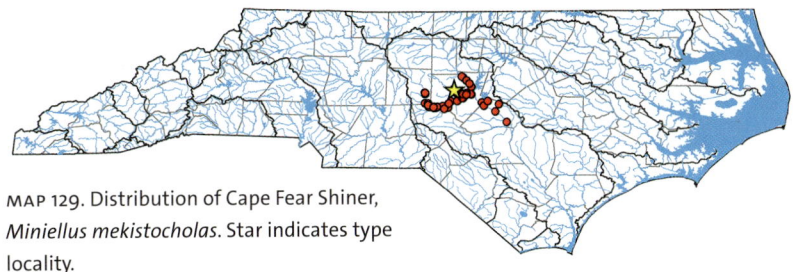

MAP 129. Distribution of Cape Fear Shiner, *Miniellus mekistocholas*. Star indicates type locality.

67a. Lateral stripe faint anteriorly; does not enter head (if enters head, passes above eye). Lateral line located one or two scales under top of diffuse lateral stripe in front of dorsal fin. Humeral bar usually dark.............68
67b. Lateral stripe dark anteriorly; passes uninterrupted from hypural plate onto opercle; crosses opercle level with eye; often passes through eye. Lateral line located in, or at lower edge of, lateral stripe in front of dorsal fin. Humeral bar faint...69

68a. Small pale spot at each end of dorsal fin base; dorsal fin with melanophores covering one-fourth to one-third of the interradial membranes from the base (figure 174). Ventral portion of the midlateral stripe fades anteriorly, especially along lateral line and beginning at posterior end of dorsal fin. Scattered blotches present on sides of body (figure 175) (map 130) ....Redlip Shiner, *Hydrophlox chiliticus* (Cope, 1870)
68b. Small pale spot absent at each end of dorsal fin base; dorsal fin with melanophores covering only the base of the interradial membranes (figure 174). Midlateral stripe usually extending from head to tail. Sides with no scattered dark blotches (figure 175) (map 131)...................................
....................Saffron Shiner, *Hydrophlox rubricroceus* (Cope, 1868)

A 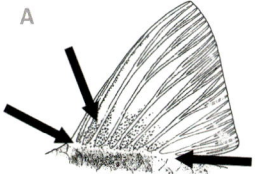   B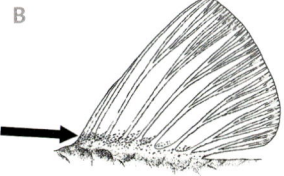

FIGURE 174. Arrows indicating pigmentation of dorsal fin. A—Redlip Shiner, *Hydrophlox chiliticus*, with pale spot at both ends of fin and melanophores on interradial membranes; B—Saffron Shiner, *Hydrophlox rubricroceus*, with melanophores at fin's base. Illustration adapted from Jenkins and Burkhead (1994).

FIGURE 175. A—Redlip Shiner, *Hydrophlox chiliticus*; B—Saffron Shiner, *Hydrophlox rubricroceus*.

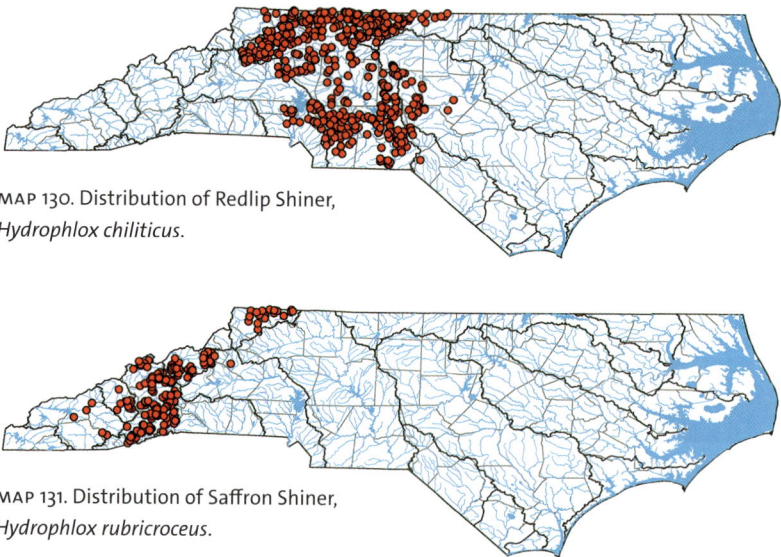

MAP 130. Distribution of Redlip Shiner, *Hydrophlox chiliticus*.

MAP 131. Distribution of Saffron Shiner, *Hydrophlox rubricroceus*.

69a. Range restricted to Savanah basin with introduced populations in Little Tennessee basin (map 132). Lower margin of lateral stripe on the same level on opercle as on cheek. Lateral line occurs in bottom half of lateral stripe below dorsal fin origin (figure 176). Pigment usually absent from second scale row below lateral line anterior to dorsal fin. Pharyngeal teeth 2,4−4,2..
......Yellowfin Shiner, *Hydrophlox lutipinnis* (Jordan and Brayton, 1878)

69b. Range restricted to Broad or Catawba basins. Lower margin of lateral stripe lower on opercle than on cheek. Lateral line occurs in bottom half or in middle of lateral stripe (including all pigmented areas) below dorsal fin origin. Pigment present or absent on second scale row below lateral line anterior of dorsal fin. Pharyngeal teeth variable....................70

FIGURE 176. Yellowfin Shiner, *Hydrophlox lutipinnis*.

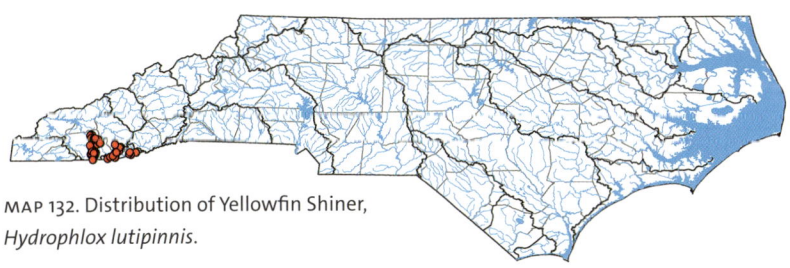

MAP 132. Distribution of Yellowfin Shiner, *Hydrophlox lutipinnis*.

70a. Range restricted to Broad basin (map 133). Pelvic, pectoral, and anal fins yellow to yellow and white, rarely red (figure 177)..........................
...."Piedmont" Shiner, *Hydrophlox* sp.

70b. Range restricted to Catawba basin (map 134). Pelvic, pectoral, and anal fins white (figure 177)...
..............Greenhead Shiner, *Hydrophlox chlorocephalus* (Cope, 1870)

FIGURE 177. A—"Piedmont" Shiner, *Hydrophlox* sp.; B—Greenhead Shiner, *Hydrophlox chlorocephalus*.

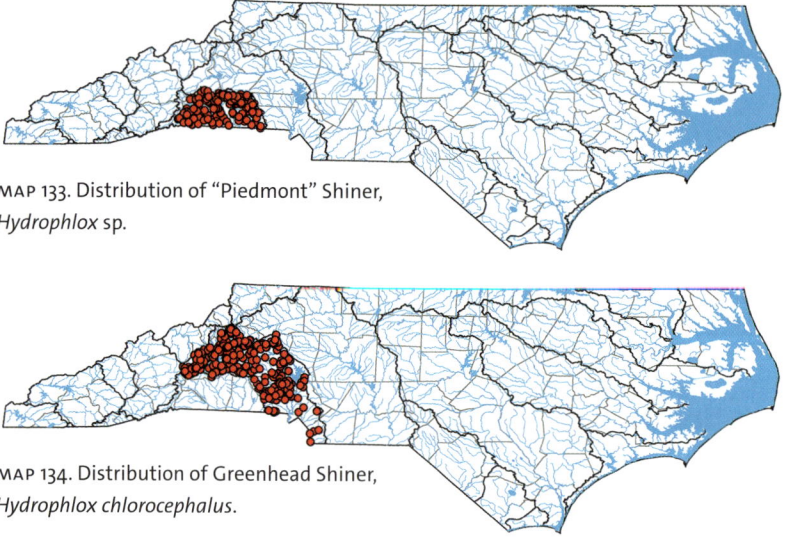

MAP 133. Distribution of "Piedmont" Shiner, *Hydrophlox* sp.

MAP 134. Distribution of Greenhead Shiner, *Hydrophlox chlorocephalus*.

9

North American Catfishes
Family Ictaluridae

There are eighteen species of catfishes in North Carolina, including three undescribed species (tables 2 and 13; Tracy, Rohde, and Hogue, 2020; Tracy et al., 2021). Colloquial names for catfishes include blue channel cats, bullheads, butter balls, mud cats, madtoms, squealers, and many more.

In North Carolina, catfishes range in size from the diminutive "Broadtail" Madtoms of just a few centimeters in TL to the behemoth Blue Catfish with a maximum TL greater than 1.5 meters. Similarly, they may weigh just a few grams for the smaller madtoms upwards to ca. sixty-eight kilograms for the Blue Catfish. Many species are recreationally and commercially important as delectable table fare, such as the Blue Catfish, Channel Catfish, and Flathead Catfish. Game species include the Blue Catfish, Channel Catfish, Flathead Catfish, and the bullheads, *Ameiurus* spp.; whereas the smaller madtom species, *Noturus* spp., are considered nongame species (NCWRC, 2022).

Catfishes are found throughout North Carolina in streams, swamps, big rivers, and reservoirs, from Cherokee County in the mountains to Dare County along the Albemarle Sound (Tracy, Rohde, and Hogue, 2020). The following three species are found in only one basin: the Mountain Madtom in the lower French Broad; "Cape Fear Broadtail" Madtom in the Cape Fear; and "Lake Waccamaw Broadtail" Madtom in the Waccamaw basin (table 2). The Brown Bullhead is our most widely distributed species, found in nineteen of our twenty-one basins, but there are no records of its occurrence in the Savannah or Shallotte basins (table 2; Tracy, Rohde, and Hogue, 2020). Several species have been introduced, legally or illegally, outside of their historical ranges. For example, the Margined Madtom has been collected and transported outside its

TABLE 13. Species of catfishes in North Carolina

Scientific Name	Common Name	Imperilment Status	Management Status
Ameiurus brunneus	Snail Bullhead		
Ameiurus catus	White Catfish		Managed by NCWRC
Ameiurus melas	Black Bullhead		
Ameiurus natalis	Yellow Bullhead		
Ameiurus nebulosus	Brown Bullhead		
Ameiurus platycephalus	Flat Bullhead		
Ictalurus furcatus	Blue Catfish		Managed by NCWRC
Ictalurus punctatus	Channel Catfish		Managed by NCWRC
Noturus eleutherus[a]	Mountain Madtom	state special concern	
Noturus flavus	Stonecat	state endangered	
Noturus furiosus	Carolina Madtom	federally endangered	
Noturus gilberti[a]	Orangefin Madtom	state endangered	
Noturus gyrinus	Tadpole Madtom		
Noturus insignis	Margined Madtom		
Noturus sp. "Cape Fear Broadtail" Madtom[a]		state special concern	
Noturus sp. "Lake Waccamaw Broadtail" Madtom[a]		state special concern	
Noturus sp. "Pee Dee Broadtail" Madtom		state special concern	
Pylodictis olivaris	Flathead Catfish		

Note: Names in quotation marks denote scientifically undescribed species.
[a] Species found in only one river basin.

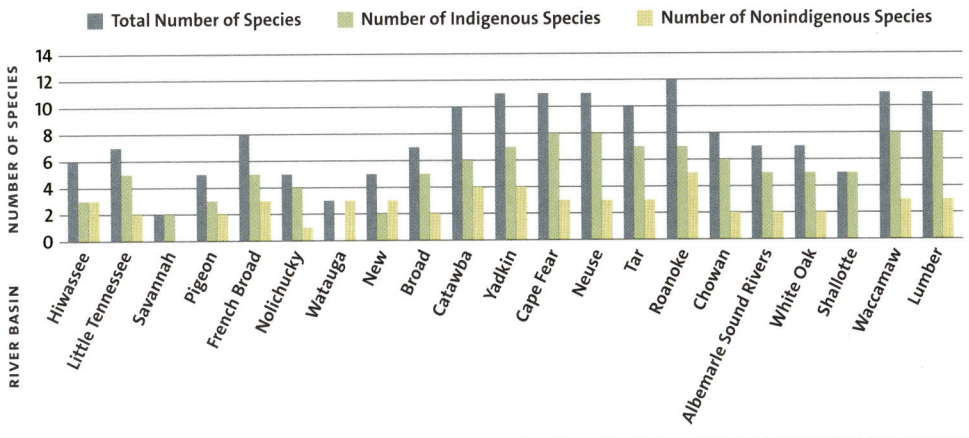

GRAPH 3. Diversity of catfishes across North Carolina's river basins.

native range east of the Appalachian Mountains and used as bait for catching Smallmouth Bass in the Watauga and New basins.

Our least speciose basin is the Savannah, which consists of small headwater tributaries of the Savannah River in North Carolina. Only the Margined Madtom and Snail Bullhead are found there (graph 3; table 2; Tracy, Rohde, and Hogue, 2020). More species of catfishes, twelve, are found in the Roanoke basin than in any of the other twenty-one basins (graph 3). Those twelve species include five species, the most of any basin, that have been introduced from other basins within or outside of North Carolina—the Black Bullhead, Blue Catfish, Channel Catfish, Flathead Catfish, and Snail Bullhead. The seven indigenous species in the Roanoke basin include the Margined, Orangefin, and Tadpole Madtoms, White Catfish, and Brown, Flat, and Yellow Bullheads. No catfish species is native to the Watauga basin, and no nonindigenous species has been found in the Savannah or Shallotte basins. Seven species are considered imperiled in North Carolina (table 13; NCAC, 2023; NCNHP, 2022; NCWRC, 2021).

Key characteristics for the proper identification of catfishes include the shape of the caudal and anal fins; the color of the barbels surrounding the mouth; and body and fin coloration (NCWRC, n.d. [a]). Most species can easily be distinguished from one another, with the possible exceptions of the Flat Bullhead versus Snail Bullhead, Black Bullhead versus Brown Bullhead, and, in the upper Roanoke basin in Stokes County, the Margined Madtom versus Orangefin Madtom.

Identification Key to the Species of North American Catfishes (Family Ictaluridae)

1a. Adipose fin attached to caudal fin (figure 178). Maximum TL less than 200 mm (except for the Stonecat, *Noturus flavus*)....*Noturus*, Couplet 2

1b. Adipose fin free from caudal fin (figure 178). Maximum TL far exceeding 200 mm..9

FIGURE 178. Arrows indicating adipose fin. A—attached in Orangefin Madtom, *Noturus gilberti*; B—free in Snail Bullhead, *Ameiurus brunneus*.

2a. Band of teeth in upper jaw with backward lateral extensions (figure 179). Light blotch on nape (figure 180). Range restricted to Little Tennessee, lower French Broad, and Nolichucky basins (map 135)......................
....................................Stonecat, *Noturus flavus* Rafinesque, 1818

2b. Band of teeth in upper jaw without backward lateral extensions. No light blotch on nape (except in the Carolina Madtom, *Noturus furiosus*)......3

FIGURE 179. Premaxillary band of teeth with backward lateral extensions. Stonecat, *Noturus flavus*.

FIGURE 180. Stonecat, *Noturus flavus*. Minnesota, with arrow indicating light blotch on nape.

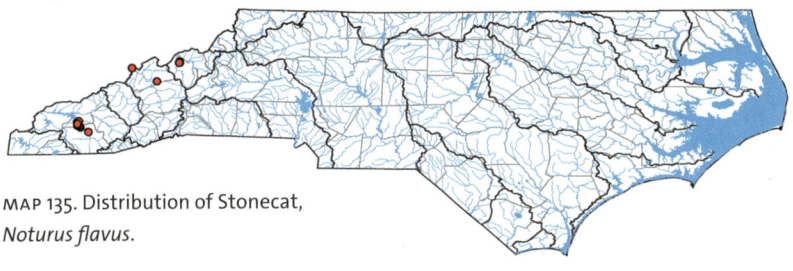

MAP 135. Distribution of Stonecat, *Noturus flavus*.

3a. Body with dorsal saddles and lateral blotches (figure 181)................4
3b. Body without dorsal saddles and lateral blotches (figure 181)............5

FIGURE 181. Body. A—with dorsal saddles in Carolina Madtom, *Noturus furiosus*; B—without dorsal saddles in Margined Madtom, *Noturus insignis*.

4a. Range restricted to Neuse and Tar basins (figure 182) (map 136)...........
.............Carolina Madtom, *Noturus furiosus* Jordan and Meek, 1889
4b. Range restricted to lower French Broad basin (figure 182) (map 137)......
...................Mountain Madtom, *Noturus eleutherus* Jordan, 1877

FIGURE 182. A—Carolina Madtom, *Noturus furiosus*; B—Mountain Madtom, *Noturus eleutherus*.

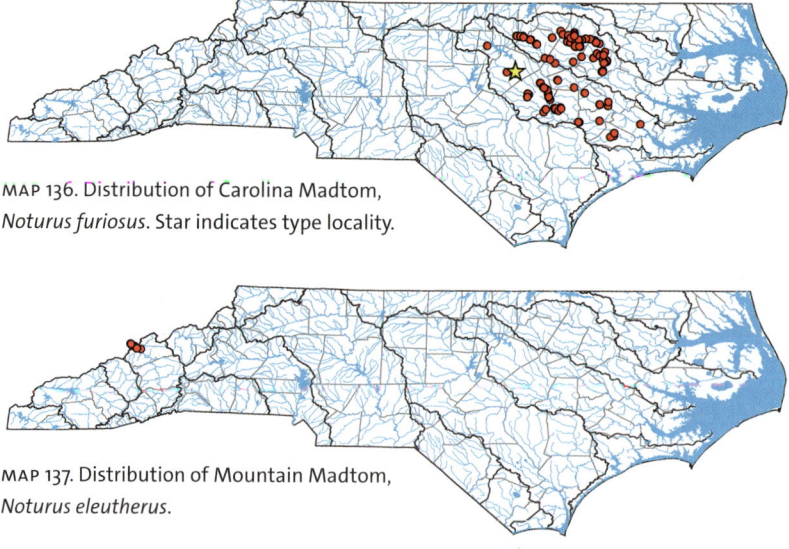

MAP 136. Distribution of Carolina Madtom, *Noturus furiosus*. Star indicates type locality.

MAP 137. Distribution of Mountain Madtom, *Noturus eleutherus*.

5a. Caudal fin large or caudal fin small with a dark blotch at its base. Range widespread across the Sand Hills and Coastal Plain (Tadpole Madtom, *Noturus gyrinus*) or restricted to lower Cape Fear, Waccamaw, and Lumber basins ("Broadtail" Madtoms, *Noturus* spp.)..........................6

5b. Caudal fin not large. Range widespread across North Carolina (Margined Madtom, *Noturus insignis*) or restricted to Roanoke basin (Orangefin Madtom, *Noturus gilberti*)...8

6a. Caudal fin large. Prominent, narrow, and dark midlateral streak present. Mouth terminal or very slightly subterminal; jaws of equal length (figure 183). Nasal barbel when pulled backward extends very far behind the eye. Area at the corner of mouth posterior to maxillary barbel pigmented (map 138).............Tadpole Madtom, *Noturus gyrinus* (Mitchill, 1817)

6b. Caudal fin small with a dark blotch at its base. Thin, dark midlateral streak usually absent or indistinct. Mouth subterminal (figure 183); upper jaw projects forward of lower jaw. Nasal barbel when pulled backward extends only to posterior edge of eye or slightly beyond. Area at the corner of mouth posterior to maxillary barbel unpigmented (appears white)...7

FIGURE 183. Arrows indicating nasal barbels and mouth location. A—terminal mouth in Tadpole Madtom, *Noturus gyrinus*; B—subterminal mouth in "Lake Waccamaw Broadtail" Madtom, *Noturus* sp.

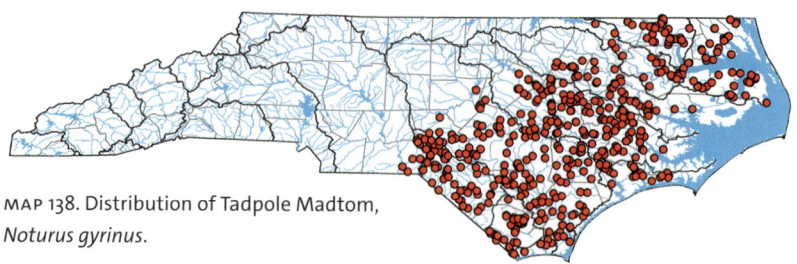

MAP 138. Distribution of Tadpole Madtom, *Noturus gyrinus*.

7a. Modally fourteen anal rays and fifty-five caudal fin rays. Ratio of caudal fin length to caudal fin height 1:11. Broad head (figure 184). Range restricted to Lake Waccamaw (Waccamaw basin) (map 139)................
......................"Lake Waccamaw Broadtail" Madtom, *Noturus* sp.

7b. Modally sixteen anal rays and fifty-nine to sixty caudal fin rays. Ratio of caudal fin length to caudal fin height 1:34 (figure 184). Range restricted to Lumber and Waccamaw basins (map 140)..............................
...................................."Pee Dee Broadtail" Madtom, *Noturus* sp.

7c. Modally sixteen anal rays and sixty caudal fin rays. Ratio of caudal fin length to caudal fin height 1:42 (figure 184). Range restricted to lower Cape Fear basin (map 141)...
..............................."Cape Fear Broadtail" Madtom, *Noturus* sp.

FIGURE 184. *Noturus* sp. A—"Lake Waccamaw Broadtail" Madtom; B—"Pee Dee Broadtail" Madtom (South Carolina); C—"Cape Fear Broadtail" Madtom.

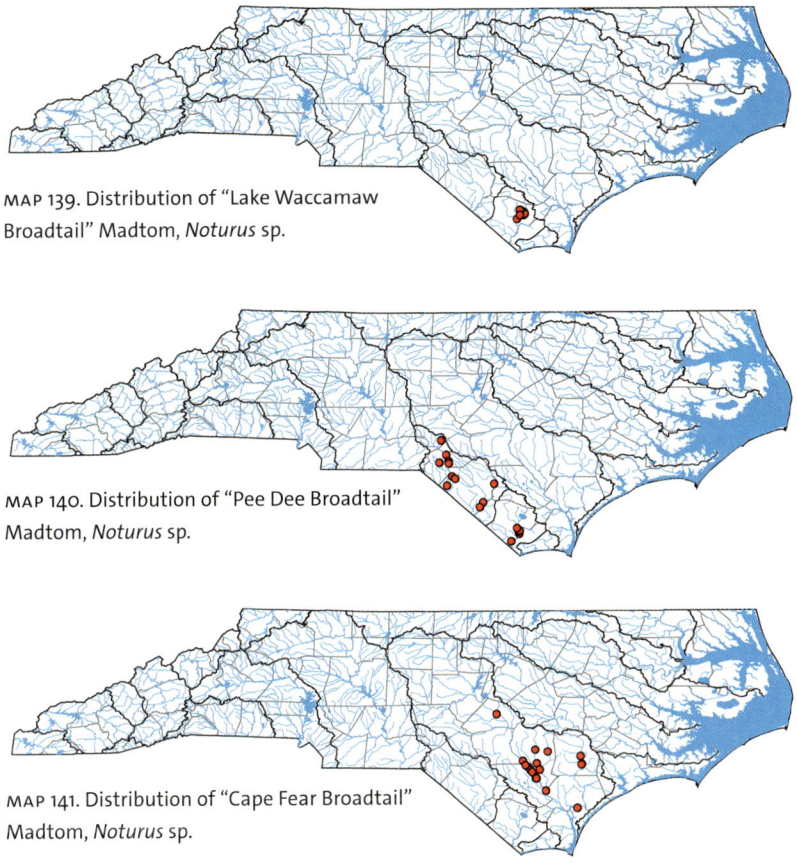

MAP 139. Distribution of "Lake Waccamaw Broadtail" Madtom, *Noturus* sp.

MAP 140. Distribution of "Pee Dee Broadtail" Madtom, *Noturus* sp.

MAP 141. Distribution of "Cape Fear Broadtail" Madtom, *Noturus* sp.

8a. Caudal fin margin pale with the pale portion slightly to distinctly wider on upper lobe (often forming a somewhat triangular pale area) than on lower lobe. Caudal fin distinctly darker on lower lobe than upper lobe submarginally. Chin behind barbels strongly papillose (figure 185). Range restricted to upper Dan River watershed (Roanoke basin) (map 142).......
.......Orangefin Madtom, *Noturus gilberti* Jordan and Evermann, 1889

8b. Caudal fin margin dusky or black, equally developed on both lobes, or if margin is pale, the pale portion is narrow and equal in width on upper and lower lobes. Chin behind barbels weakly papillose (figure 185) (map 143)............Margined Madtom, *Noturus insignis* (Richardson, 1836)

FIGURE 185. A—Orangefin Madtom, *Noturus gilberti*; B—Margined Madtom, *Noturus insignis*.

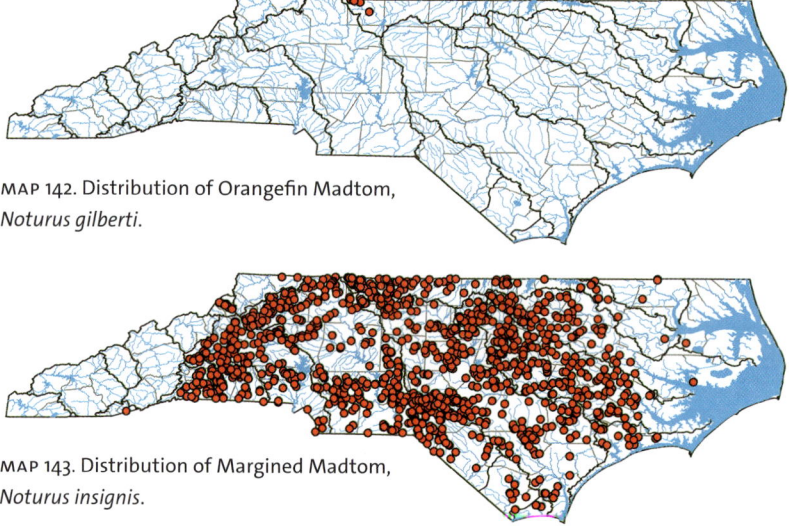

MAP 142. Distribution of Orangefin Madtom, *Noturus gilberti*.

MAP 143. Distribution of Margined Madtom, *Noturus insignis*.

9a. Premaxillary band of teeth with backward lateral extensions (figure 179). Anal fin rays 12–16. Entire body mottled. Tip of dorsal lobe of caudal fin white (except in very large specimens) (figure 186) (map 144)............ ....Flathead Catfish, *Pylodictis olivaris* (Rafinesque, 1818)

9b. Premaxillary band of teeth nearly straight, without backward lateral extensions. Anal fin rays 18–36. Entire body not mottled. Tip of dorsal lobe of caudal fin not white...10

FIGURE 186. Flathead Catfish, *Pylodictis olivaris*.

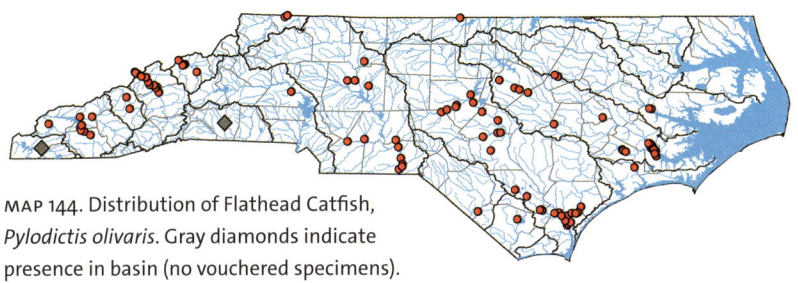

MAP 144. Distribution of Flathead Catfish, *Pylodictis olivaris*. Gray diamonds indicate presence in basin (no vouchered specimens).

10a. Caudal fin deeply forked.............................*Ictalurus*, Couplet 11
10b. Caudal fin moderately forked, emarginate, or rounded....................
....*Ameiurus*, Couplet 12

11a. Anal fin straight (figure 187); anal fin rays (27) 30–36 (38). Body with no spots (map 145).....Blue Catfish, *Ictalurus furcatus* (Valenciennes, 1840)
11b. Anal fin rounded (figure 187); anal fin rays (23) 25–30 (32). Young to small adults with few to many dark spots (map 146)............................
...................Channel Catfish, *Ictalurus punctatus* (Rafinesque, 1818)

A

FIGURE 187.
A—Blue Catfish, *Ictalurus furcatus*;
B—Channel Catfish, *Ictalurus punctatus*.

B

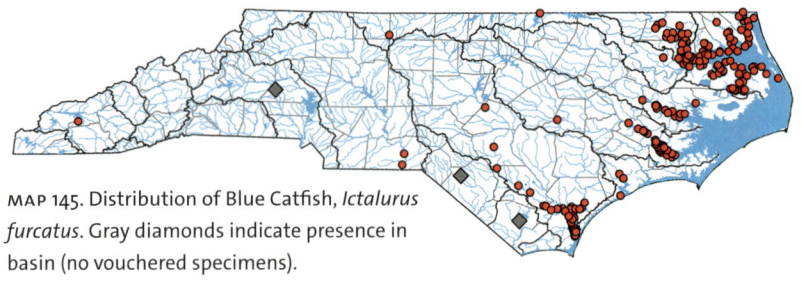

MAP 145. Distribution of Blue Catfish, *Ictalurus furcatus*. Gray diamonds indicate presence in basin (no vouchered specimens).

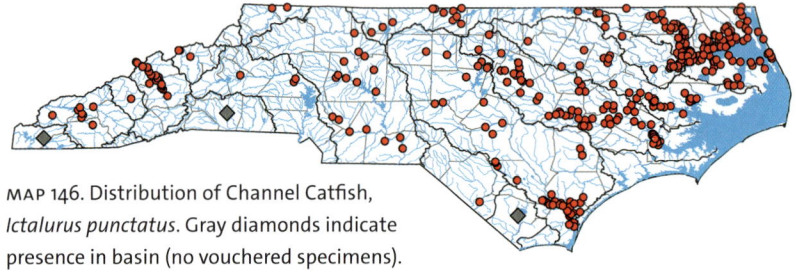

MAP 146. Distribution of Channel Catfish, *Ictalurus punctatus*. Gray diamonds indicate presence in basin (no vouchered specimens).

12a. Dorsal fin with a dark basal blotch (figure 188). Eye size moderate......13
12b. Dorsal fin without a dark basal blotch (figure 188). Eye size small (except in White Catfish, *Ameiurus catus*, in which eye is fairly large in young, moderate in adults)...14

FIGURE 188. Dark basal blotch on dorsal fin. A—with blotch in Snail Bullhead, *Ameiurus brunneus*; B—without blotch in White Catfish, *Ameiurus catus*.

13a. Chin barbels usually profusely pigmented (in small specimens pigment occasionally only developed basally). Maxillary barbels uniformly dark. When viewed from the side, upper jaw with an extreme "overbite" (figure 189). Premaxillary tooth patch in large juveniles and adults uniformly wide, lateral ends indented, and in adults, anterior teeth larger than posterior teeth (figure 190). Anal rays usually 17–20 (22) (figure 191) (map 147) ....Snail Bullhead, *Ameiurus brunneus* Jordan, 1877

13b. Chin barbels usually without pigment (pigment may be present in large specimens on lateral barbels, rarely on medial) (figure 189). Leading edge of maxillary barbel pale, appearing bicolored. When viewed from the side, upper jaw without an "overbite" (figure 189). Premaxillary tooth patch in large juveniles and adults narrower medially, lateral ends not indented, and teeth of uniform size (figure 190). Anal rays usually (19) 20–24 (26) (figure 191) (map 148)..
......................Flat Bullhead, *Ameiurus platycephalus* (Girard, 1859)

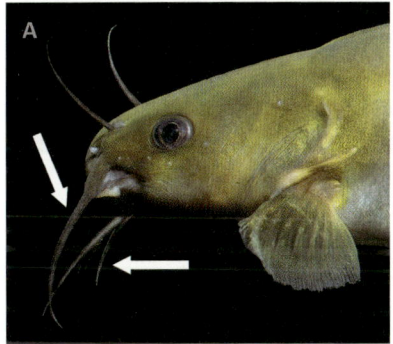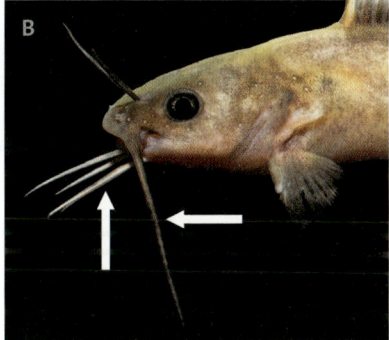

FIGURE 189. Maxillary and chin barbels. A—Snail Bullhead, *Ameiurus brunneus*; B—Flat Bullhead, *Ameiurus platycephalus*.

FIGURE 190. Premaxillary tooth patches. A—Snail Bullhead, *Ameiurus brunneus*; B—Flat Bullhead, *Ameiurus platycephalus*. Illustration adapted from Jenkins and Burkhead (1994).

FIGURE 191. A—Snail Bullhead, *Ameiurus brunneus*; B—Flat Bullhead, *Ameiurus platycephalus*.

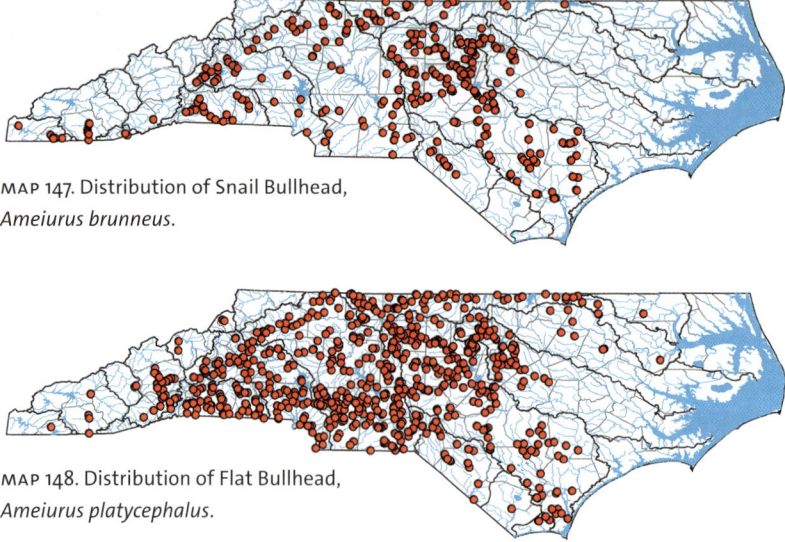

MAP 147. Distribution of Snail Bullhead, *Ameiurus brunneus*.

MAP 148. Distribution of Flat Bullhead, *Ameiurus platycephalus*.

14a. Caudal fin moderately forked or rounded. Chin barbels pale...........15

14b. Caudal fin slightly emarginate. Chin barbels dark......................16

15a. Caudal fin moderately forked (figure 192). Anal margin rounded; anal fin rays (21) 22–24 (25) (map 149)..
.........................White Catfish, *Ameiurus catus* (Linnaeus, 1758)

15b. Caudal fin rounded (figure 192). Anal margin nearly truncate; anal fin rays (23) 25–28 (29) (map 150)...
........................Yellow Bullhead, *Ameiurus natalis* (Lesueur, 1819)

FIGURE 192. A—White Catfish, *Ameiurus catus*; B—Yellow Bullhead, *Ameiurus natalis*.

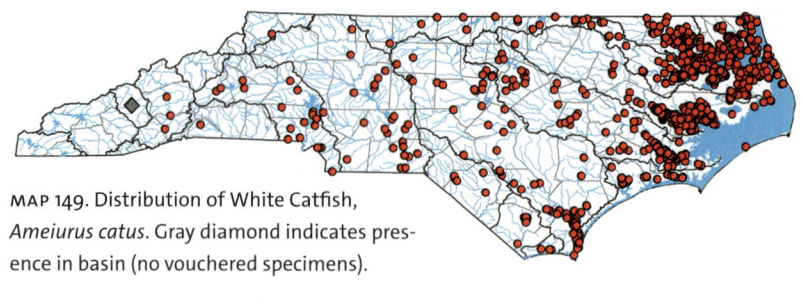

MAP 149. Distribution of White Catfish, *Ameiurus catus*. Gray diamond indicates presence in basin (no vouchered specimens).

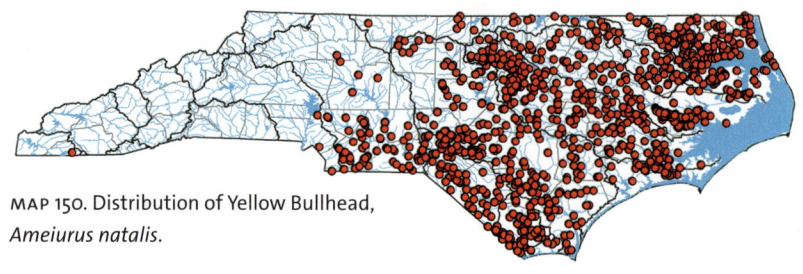

MAP 150. Distribution of Yellow Bullhead, *Ameiurus natalis*.

16a. Caudal fin base uniformly dusky or dark in large juveniles and adults (figure 193). Body often mottled. Total gill rakers on first gill arch typically 13–15 (map 151).....Brown Bullhead, *Ameiurus nebulosus* (Lesueur, 1819)

16b. Caudal fin base with a rectangular pale area often present in large juveniles and adults (figure 193). Body not mottled. Total gill rakers on first gill arch typically 17–29 (map 152)..
......................Black Bullhead, *Ameiurus melas* (Rafinesque, 1820)

FIGURE 193. A—Brown Bullhead, *Ameiurus nebulosus*; B—Black Bullhead, *Ameiurus melas.* South Carolina.

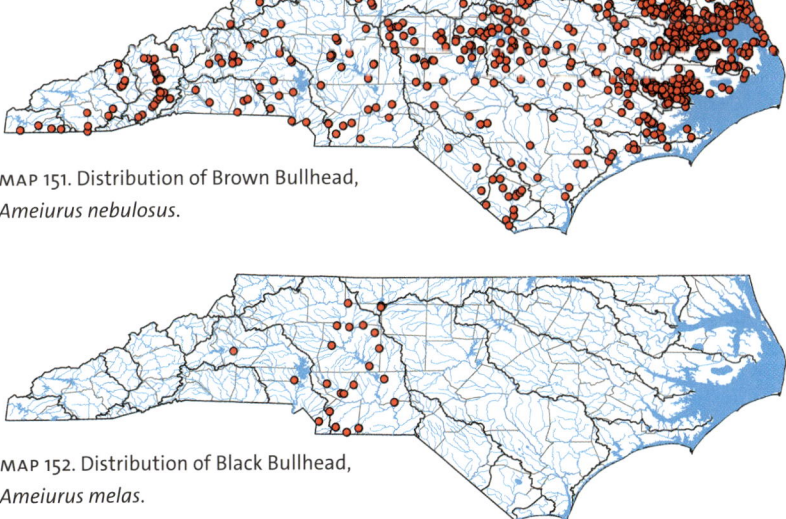

MAP 151. Distribution of Brown Bullhead, *Ameiurus nebulosus*.

MAP 152. Distribution of Black Bullhead, *Ameiurus melas*.

10

Pikes
Family Esocidae

North Carolina is home to three of the four species of pikes known to occur in North America: the Redfin Pickerel, Chain Pickerel, and Muskellunge (tables 2 and 14). There are two subspecies of the Redfin Pickerel, *Esox americanus*, recognized: the Redfin Pickerel, *E. americanus americanus*; and Grass Pickerel, *E. americanus vermiculatus*. The Grass Pickerel does not occur in North Carolina. Decades ago, the Northern Pike, *E. lucius*, was introduced into several river basins (Yadkin and Roanoke), but those introductions were unsuccessful. Colloquial names for the pikes include jack, jackfish, little pickerel, musky (muskie), pickerel, and pike.

Both the Redfin Pickerel and Chain Pickerel are native (indigenous) east of the mountains (table 2; Tracy, Rohde, and Hogue, 2020). The Redfin Pickerel is primarily a Coastal Plain species, but there are records from the western Piedmont in the Catawba basin and from the upper Roanoke basin. The Chain Pickerel, whose distribution overlaps that of the Redfin Pickerel, has been introduced (nonindigenous) in the French Broad basin (Buncombe, Henderson, and Transylvania Counties), where it is likely to be confused with the indigenous Muskellunge. By comparison, the Muskellunge is indigenous to the Tennessee drainage (Hiwassee, Little Tennessee, French Broad, and Nolichucky basins), but it has been stocked as a sport fish in the New River (New basin) and in the Broad River and Lake Adger (Broad basin). All three species are considered inland game fishes, and none are considered imperiled in North Carolina (NCNHP, 2022; NCWRC, 2021; NCWRC, 2022). None of the species are endemic to a specific river basin in North Carolina (tables 2 and 14).

North Carolina's pickerels are common inhabitants of the Coastal Plain's swamps, sloughs, creeks, and mill ponds and of our natural lakes. The Redfin

TABLE 14. Species of pikes in North Carolina

Scientific Name	Common Name
Esox americanus	Redfin Pickerel
Esox masquinongy	Muskellunge
Esox niger	Chain Pickerel

Note: All species are managed by the NCWRC.

and Chain Pickerels are frequently found in tannin-stained blackwater and acidic Sand Hills and Coastal Plain streams wherever there is cover such as vegetation, brush piles, and logs (Rohde et al., 2009). In the Piedmont, pickerels are less common and are found in creeks, rivers, and reservoirs. The Muskellunge tends to inhabit our bigger, deeper, and slower moving rivers in the Mountains such as the French Broad and New, along with reservoir populations in Fontana Reservoir (Little Tennessee basin) and Lake Adger (Broad basin).

Key characteristics for the proper identification of pikes include the presence/absence of a subocular bar; color and spotting of the fins; scalation of the opercle; and the relative length of the snout. Additionally, size can aid in their identification: the maximum size of the Redfin Pickerel is ca. 380 mm TL, ca. 990 mm TL for the Chain Pickerel, and up to 1,830 mm TL for the Muskellunge (Rohde et al., 2009). Hybridization is known to occur between the Redfin Pickerel and Chain Pickerel, which can make identification problematic.

Identification Key to the Species of Pikes (Family Esocidae)

1a. Black bar beneath eye. Dorsal and anal fins unmarked. Opercle completely scaled. Four mandibular pores on the underside of each jaw.....2
1b. No black bar beneath eye (figure 194). Dorsal and anal fins with spots or streaks. Opercle scaled only on upper half. Usually six or more mandibular pores on the underside of each jaw (map 153)...........................
...........................Muskellunge, *Esox masquinongy* Mitchill, 1824

FIGURE 194. Muskellunge, *Esox masquinongy*. Photograph courtesy of Luke Etchison.

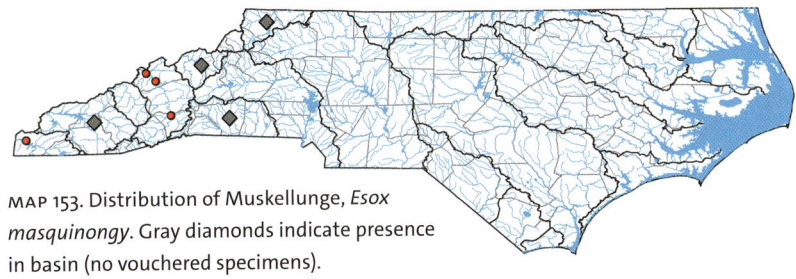

MAP 153. Distribution of Muskellunge, *Esox masquinongy*. Gray diamonds indicate presence in basin (no vouchered specimens).

2a. Sides of adults longer than 100 mm TL with dark vertical bars; bars are faint in individuals less than 50 mm TL (figures 195 and 196). Distance from center of eye to tip of snout less than or equal to distance from center of eye to upper end of gill opening, giving the appearance of a "duck" bill (figure 197). Subocular bar usually slanted posteriorly (figure 197). Some of the fins with red or reddish color (figure 195) (map 154)................
..........................Redfin Pickerel, *Esox americanus* Gmelin, 1789

2b. Sides of adults longer than 200 mm TL with numerous elongate light spots (chain-like pattern); young with bars (figures 196 and 198). Snout longer; distance from center of eye to tip of snout greater than or equal to distance from center of eye to upper end of gill opening (figure 199). Subocular bar usually vertical (figure 198). Fins not red (figure 198) (map 155).........................Chain Pickerel, *Esox niger* Lesueur, 1818

FIGURE 195. Redfin Pickerel, *Esox americanus*.

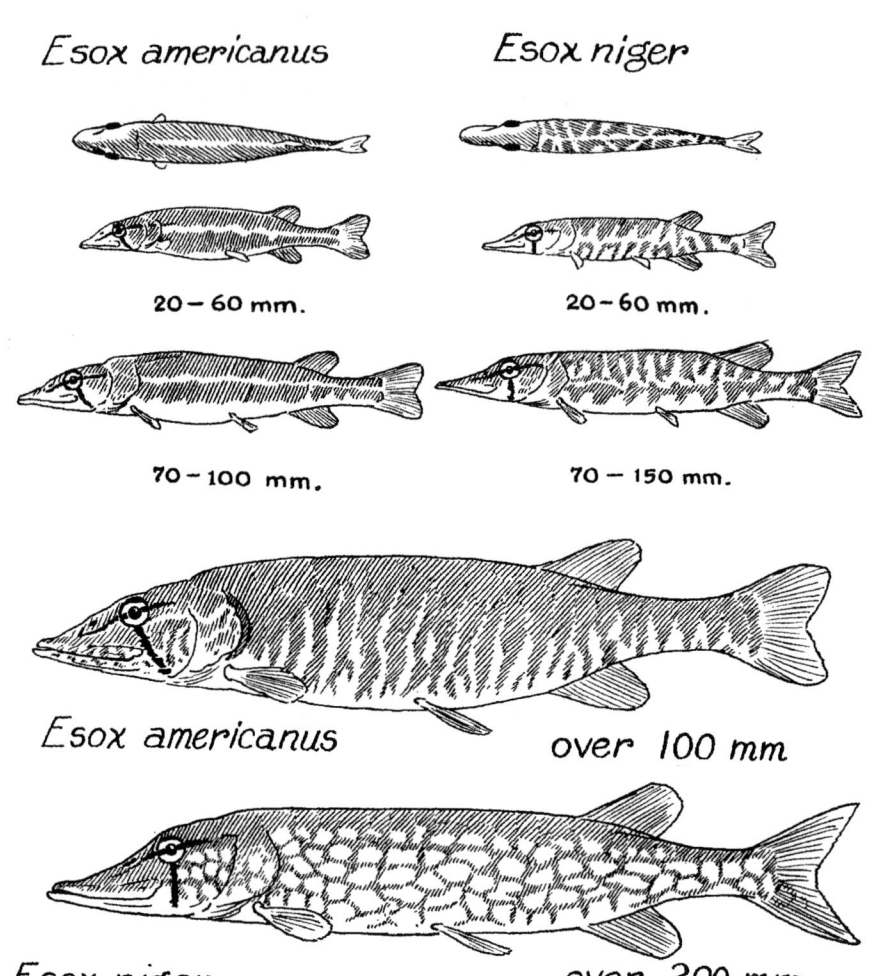

FIGURE 196. Comparison of body shape and color pattern of Redfin Pickerel, *Esox americanus*, and Chain Pickerel, *Esox niger*, from 20 to 200 mm TL. Illustration adapted from Crossman (1962).

A ≤ B

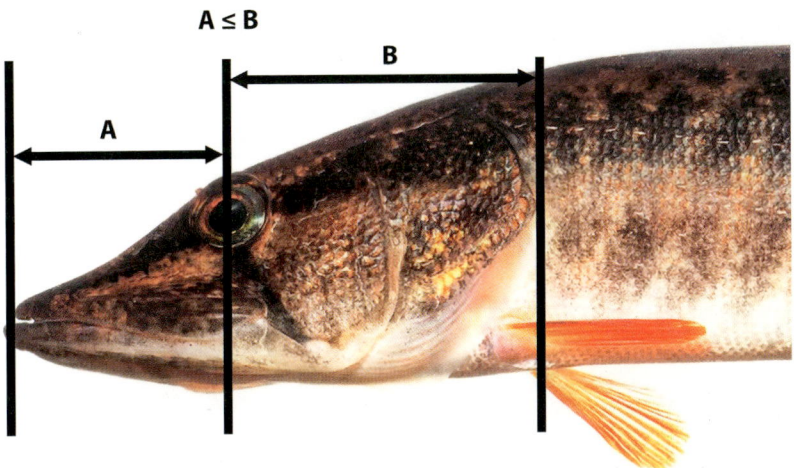

FIGURE 197. Redfin Pickerel, *Esox americanus*, with arrows and bars indicating distance from center of eye to tip of snout less than or equal to distance from center of eye to upper end of gill opening.

FIGURE 198. Chain Pickerel, *Esox niger*.

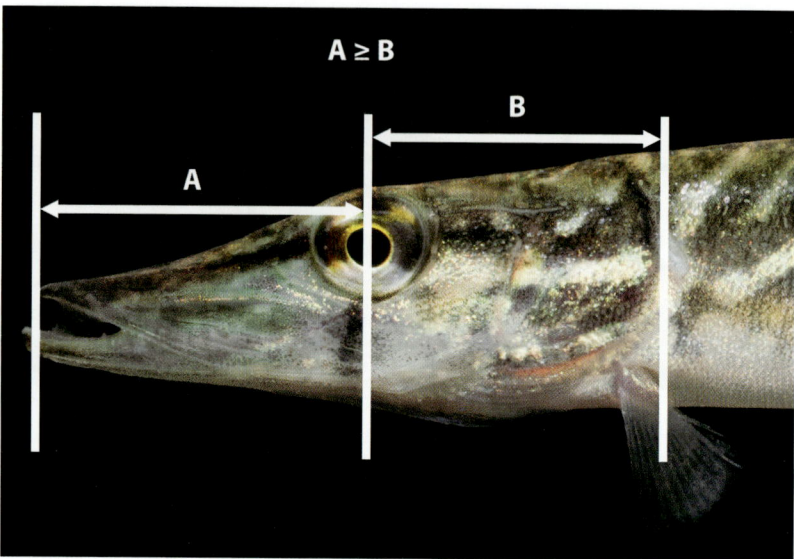

FIGURE 199. Chain Pickerel, *Esox niger*, with arrows and bars indicating distance from center of eye to tip of snout greater than or equal to distance from center of eye to upper end of gill opening.

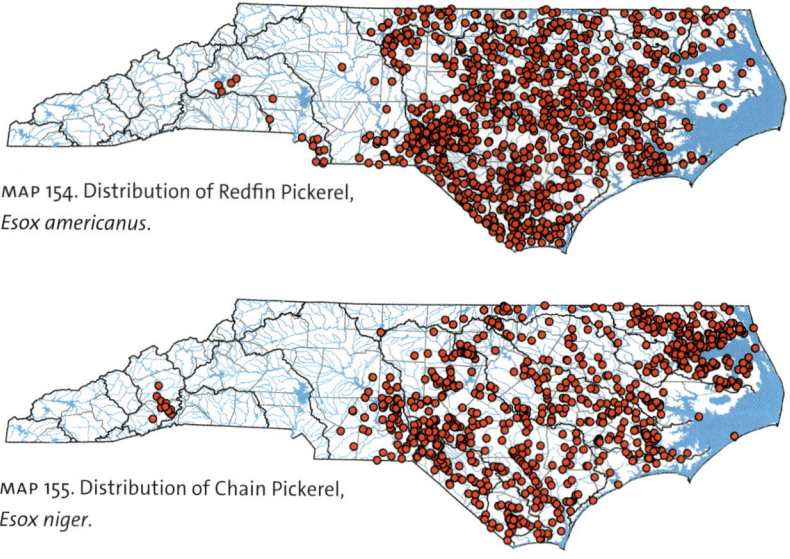

MAP 154. Distribution of Redfin Pickerel, *Esox americanus*.

MAP 155. Distribution of Chain Pickerel, *Esox niger*.

11

Trouts and Salmons
Family Salmonidae

There are three species of trout and one species of salmon in North Carolina: the Brook Trout, Brown Trout, Rainbow Trout, and Sockeye Salmon (tables 2 and 15). Even though the family Salmonidae is made up of trouts and salmons, in North Carolina they are known collectively as trout. Other colloquial names include brookies, speckled trout, or specks. In North Carolina and elsewhere, Sockeye Salmon populations that are not anadromous (that is, land-locked populations that do not migrate from the ocean to fresh water to spawn) are called Kokanee.

North Carolina's only indigenous (native) trout species is the Brook Trout, specifically the Southern Appalachia strain, which was historically found throughout the Appalachian Mountains on both sides of the Eastern Continental Divide (NCWRC n.d. [b]). The onset of large-scale industrial logging resulted in habitat and water quality degradation of many mountain streams, and so the northern strain of the Brook Trout along with the Rainbow Trout and the Brown Trout were introduced in the 1870s and 1880s to offset the dwindling populations and numbers of the Southern Appalachia strain Brook Trout.

In the later 1950s to early 1960s, the Sockeye Salmon was stocked as a forage fish for the Rainbow Trout and as a sport fish in Nantahala Lake (Little Tennessee basin) (Tracy, Rohde, and Hogue, 2020). The population in Nantahala Lake is not self-sustaining and persists only through periodic stockings by the North Carolina Wildlife Resources Commission (NCWRC).

The Rainbow Trout was originally stocked in the Pigeon (Pigeon basin), French Broad (French Broad basin), Broad and Green (Broad basin), Catawba, Johns, and Linville (Catawba basin), Yadkin (Yadkin basin), and Dan (Roanoke

TABLE 15. Species of trouts and salmons in North Carolina

Scientific Name	Common Name
Oncorhynchus mykiss	Rainbow Trout
Oncorhynchus nerka	Sockeye Salmon
Salmo trutta	Brown Trout
Salvelinus fontinalis	Brook Trout

Note: All species are managed by the NCWRC.

basin) Rivers (Tracy, Rohde, and Hogue, 2020). Today, wild (naturalized) and stocked populations of the Rainbow Trout are found throughout all Mountain basins and in the montane reaches of Atlantic slope basins, including the Savannah, Broad, Catawba, Yadkin, and Roanoke (table 2).

The Brown Trout has been widely stocked as a sport fish since the late 1880s and is now found as wild (naturalized) and stocked populations throughout all Mountain basins and in the montane reaches of Atlantic slope basins, including the Savannah, Broad, Catawba, Yadkin, and Roanoke (table 2; Tracy, Rohde, and Hogue, 2020).

The Brook Trout is found in all Mountain basins and in the headwater montane streams of the Savannah, Catawba, and Yadkin. This species has been stocked in the Broad and Roanoke basins, where it is not indigenous (table 2; Tracy, Rohde, and Hogue, 2020). Today most wild populations are restricted to remote, clear, cold, well-oxygenated, high-elevation, high-gradient, and turbulent streams beneath a canopy of rhododendron, eastern hemlock, yellow poplar, and other native trees of the Appalachian Mountains.

Most individuals of wild Brook Trout are only ca. 130–200 mm TL, although hatchery-reared and stocked Brook Trout may approach ca. 700 mm TL. The Rainbow and Brown Trout can reach ca. 825 mm TL, and the North Carolina record for the largest Kokanee was ca. 610 mm TL.

None of the species are considered imperiled in North Carolina because all are classified and managed as game species by the NCWRC (NCAC, 2023; NCNHP, 2022; NCWRC, 2021, 2022).

Key characteristics for the proper identification of trouts and salmons include the presence (and location) or absence of spots on the body and fins and pigmentation patterns on the pectoral, anal, and adipose fins.

Identification Key to the Species of Trouts and Salmons (Family Salmonidae)

1a. Body and fins with no distinct spots in adults (figure 200). Anal rays 13–19. Gill rakers 30–50. Range restricted to Nantahala Reservoir (Little Tennessee basin) (map 156)...
.......Sockeye Salmon (Kokanee), *Oncorhynchus nerka* (Walbaum, 1792)

1b. Body and fins with distinct spots. Anal rays 9–12. Gill rakers fewer than 20...2

FIGURE 200. Sockeye Salmon, *Oncorhynchus nerka*. Photograph courtesy of David A. Neely.

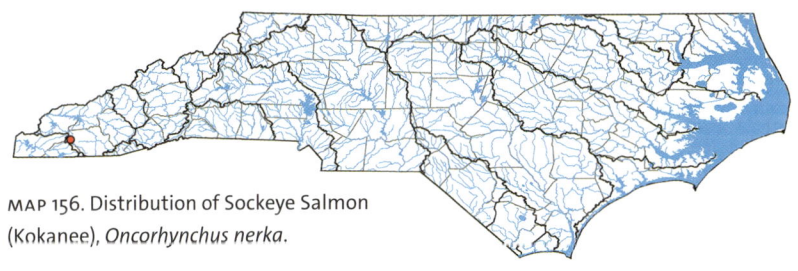

MAP 156. Distribution of Sockeye Salmon (Kokanee), *Oncorhynchus nerka*.

2a. Side with pale spots on a dark background. Dorsum with a wormlike pattern. Front edge of pectoral fin margined with white (figure 201) (map 157)
........................Brook Trout, *Salvelinus fontinalis* (Mitchill, 1814)

2b. Side with pale spots on a light background. Dorsum without wormlike markings. Front edge of pectoral fin not margined with white (figure 200)
....3

FIGURE 201. Brook Trout, *Salvelinus fontinalis*, Southern Appalachian strain.

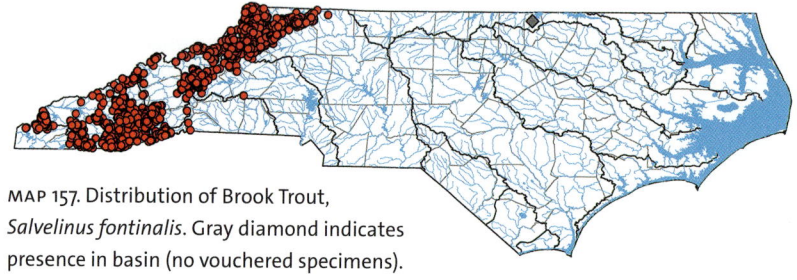

MAP 157. Distribution of Brook Trout,
Salvelinus fontinalis. Gray diamond indicates
presence in basin (no vouchered specimens).

3a. Caudal fin covered with black spots. Side in life with a pinkish stripe. Adipose fin black-edged. Anterior border of pelvic and anal fins without white front edge and with no black submarginal band (figure 202) (map 158)
.................Rainbow Trout, *Oncorhynchus mykiss* (Walbaum, 1792)

3b. Caudal fin lacking black spots. Side in life lacking a pinkish stripe. Side with scattered red spots surrounded by halos. Adipose fin pale-edged. Anterior border of pelvic and anal fins with white front edge and black submarginal band (figure 203) (map 159).................................
................................Brown Trout, *Salmo trutta* Linnaeus, 1758

FIGURE 202. Rainbow Trout, *Oncorhynchus mykiss*.

FIGURE 203. Brown Trout, *Salmo trutta*.

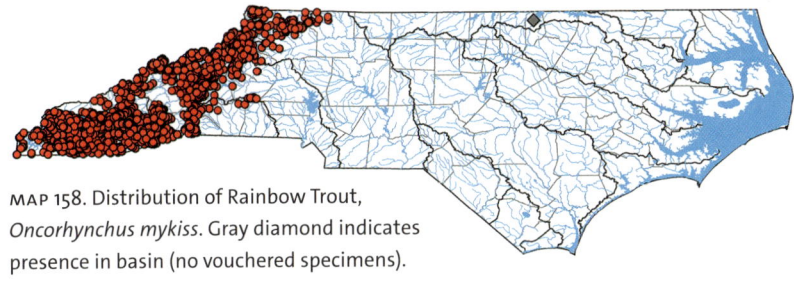

MAP 158. Distribution of Rainbow Trout,
Oncorhynchus mykiss. Gray diamond indicates
presence in basin (no vouchered specimens).

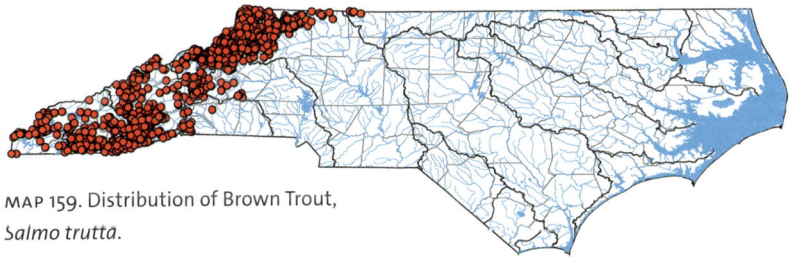

MAP 159. Distribution of Brown Trout,
Salmo trutta.

Sleepers
Family Eleotridae

Inhabiting the shallow coastal waters of North Carolina, sleepers constitute a small family with just two species, the Fat Sleeper and Largescaled Spinycheek Sleeper (table 2; Tracy, Rohde, and Hogue, 2020). Neither species ventures very far inland, but both can be found in some of the smaller tributaries along the coast from the Albemarle Sound to the Lockwoods Folly River (Brunswick County, Shallotte basin). In addition, neither species is listed as an imperiled species (NCAC, 2023; NCNHP, 2022; NCWRC, 2021) or as endemic to a specific river basin in North Carolina (table 2). A third species, the Emerald Sleeper, *Erotelis smaragdus*, is known only from two specimens collected in 1987 and 1993 (Ross and Rohde, 2004), but no other specimens have been collected since then.

The Fat Sleeper appears restricted to shallow fresh and estuarine waters including coastal streams, ponds, and ditches in the Cape Fear, Neuse, Tar, and White Oak basins, and as far north as Ocracoke Island on the Cape Hatteras National Seashore, Dare County (Ross and Rohde, 2004; Tracy, Rohde, and Hogue, 2020). It may reach 380 mm TL but is seldom longer than 250 mm TL (Rohde et al., 2009). The Largescaled Spinycheek Sleeper is found in the Shallotte, Cape Fear, White Oak, and Albemarle Sound basins and as far north as off Roanoke Island in the Croatan Sound, Dare County, and reaches 250 mm TL (Rohde et al., 2009).

Key characteristics for the proper identification of sleepers include the number of dorsal fin spines and pectoral fin rays; the lateral scale count; the presence or absence of a preopercular spine; and body coloration.

Identification Key to the Freshwater and Estuarine Species of Sleepers (Family Eleotridae)

1a. First dorsal fin with seven spines. Preopercular spine absent. Pectoral fin rays 14. Scales large, 33–36 in lateral series. Blotch just above opercle ringed in light blue with bluish bars on anal fin (figure 204) (map 160)...........
...................Fat Sleeper, *Dormitator maculatus* (Bloch, 1792)[1]

1b. First dorsal fin with six spines. Preopercle with sharp concealed spine. Pectoral fin rays 16–18 (usually 17). Scales small, 40–65 in lateral series. No bluish spot above opercle, and anal fin without bluish bars (figure 205) (map 161)..
.........Largescaled Spinycheek Sleeper, *Eleotris amblyopsis* (Cope, 1871)

FIGURE 204. Fat Sleeper, *Dormitator maculatus*. A—male; B—female.

1. The Fat Sleeper may be confused with the co-occurring Banded Pygmy Sunfish, *Elassoma zonatum* (family Elassomatidae). However, the Fat Sleeper has two separated dorsal fins and a scaled head, whereas the Banded Pygmy Sunfish has a single dorsal fin and an unscaled head (figure 206) (Rohde et al., 2009).

FIGURE 205. Largescaled Spinycheek Sleeper, *Eleotris amblyopsis.*

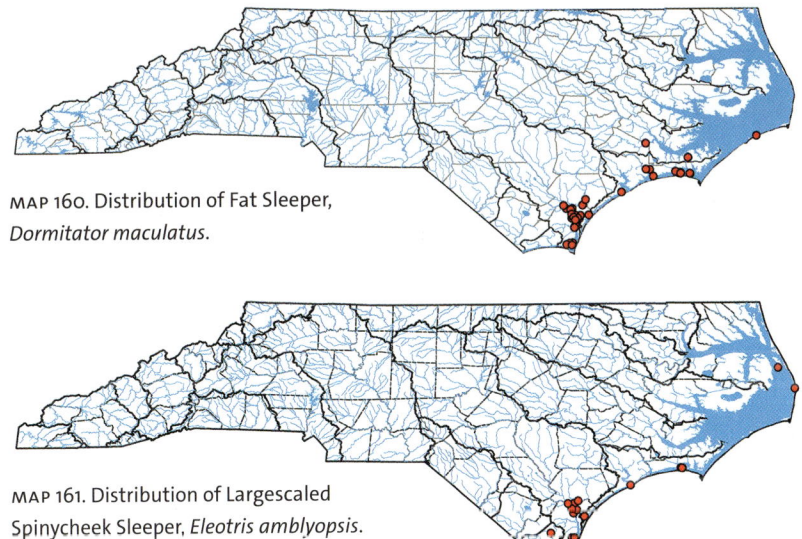

MAP 160. Distribution of Fat Sleeper,
Dormitator maculatus.

MAP 161. Distribution of Largescaled
Spinycheek Sleeper, *Eleotris amblyopsis.*

FIGURE 206. A—Fat Sleeper juvenile, *Dormitator maculatus*; B—Banded Pygmy Sunfish, *Elassoma zonatum*. South Carolina. Photograph courtesy of Zach Alley (Banded Pygmy Sunfish).

Gobies
Family Gobiidae

Gobies are some of the most brightly colored and diverse fishes found around coral reefs. There are at least 220 genera and 2,712 species worldwide primarily inhabiting shallow tropical and subtropical waters (Froese and Pauly, 2022; Murdy and Hoese, 2002). Along and off the North Carolina shoreline, there are twenty-five indigenous species and one nonindigenous species (Ross and Rohde, 2004; Tracy, Rohde, and Hogue, 2020). However, only seven of these species are found in coastal freshwater or estuarine habitats (tables 2 and 16). They range in size from 40 to 300 mm TL, but most are between 40 and 80 mm TL (Murdy and Hoese, 2002). None of the species are endemic to a specific river basin or are listed as an imperiled species in North Carolina (table 2; NCAC, 2023; NCNHP, 2022; NCWRC, 2021).

Most species can be found in coastal river mouths; in estuaries and bays with muddy, sandy, or grassy bottoms; amongst submerged vegetation and oyster beds; or atop rock and rubble bottoms. The Lyre Goby and Freshwater Goby have been found in the Cape Fear basin as far upstream as near Riegelwood (Columbus and Pender Counties) (Tracy, Rohde, and Hogue, 2020). The Highfin Goby has occasionally strayed into fresh waters (Tracy, Rohde, and Hogue, 2020) but spends most of its life in marine waters.

The River Goby is our only nonindigenous species. It was unknown from North Carolina until a single specimen was discovered in a fish kill in 1996 from Burnt Mill Creek in Wilmington (New Hanover County, Cape Fear basin). In 2015 an apparently self-sustaining population was discovered in a stormwater retention pond in Morehead City (Carteret County, White Oak basin). This population has been present for at least four years, despite mortalities from cold temperatures. In 2017 another population was discovered in an

TABLE 16. Species of gobies found in or along the coast of North Carolina

Scientific Name	Common Name
Awaous banana	River Goby
Ctenogobius boleosoma	Darter Goby
Ctenogobius shufeldti	Freshwater Goby
Evorthodus lyricus	Lyre Goby
Gobiosoma bosc	Naked Goby
Gobiosoma ginsburgi	Seaboard Goby
Microgobius thalassinus	Green Goby

unnamed creek near the Visitors' Center in Morehead City (Scott A. Smith, personal communication). All three locations are near North Carolina's two shipping ports, and this species may have been introduced from the release of ballast water (Tracy, Rohde, and Hogue, 2020).

Key characteristics for the proper identification of gobies include color and color patterns; presence or absence of scales; the opercle being scaled or not; lateral scale row counts; dorsal fin and anal fin ray counts; and dorsal fin spine counts.

Identification Key to the Freshwater and Estuarine Species of Gobies (Family Gobiidae)

1a. Shoulder (beneath gill cover) with two or three distinct fleshy lobes (figure 207). Lateral scales greater than 57 (figure 208) (map 162)............ ....River Goby, *Awaous banana* (Valenciennes, 1837)

1b. No lobes on shoulder girdle. Lateral scales, if present, usually less than 57 2

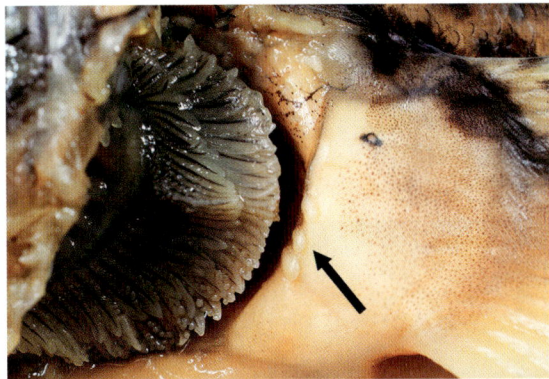

FIGURE 207. Arrow indicating location of fleshy lobes beneath the operculum in River Goby, *Awaous banana*.

FIGURE 208. River Goby, *Awaous banana*.

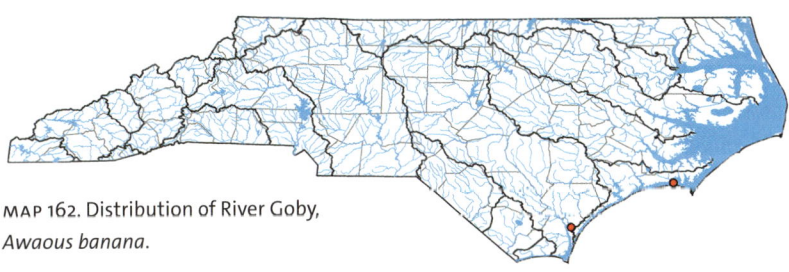

MAP 162. Distribution of River Goby, *Awaous banana*.

2a. Body without scales, except possibly two small scales on caudal fin near base...*Gobiosoma*, Couplet 3

2b. Body mostly covered with scales..4

3a. Two small ctenoid scales on the caudal fin near the base. Lateral line usually with vertically elongated dark spots (figure 209)......................
.....Seaboard Goby, *Gobiosoma ginsburgi* Hildebrand and Schroeder, 1928
3b. No scales anywhere. Lateral line without vertically elongated dark spots (figure 209)...............Naked Goby, *Gobiosoma bosc* (Lacepède, 1800)

FIGURE 209. A—Seaboard Goby, *Gobiosoma ginsburgi*; B—Naked Goby, *Gobiosoma bosc.*

4a. Dorsal fin with seven spines (figure 210)...................................
.........Green Goby, *Microgobius thalassinus* (Jordan and Gilbert, 1883)
4b. Dorsal fin with usually fewer than seven spines...........................5

FIGURE 210. Green Goby, *Microgobius thalassinus.*

5a. Opercle entirely or partially covered with scales (figure 211) (map 163)....
.................................Lyre Goby, *Evorthodus lyricus* (Girard, 1858)
5b. Opercle naked......................................*Ctenogobius*, Couplet 6

FIGURE 211. Lyre Goby, *Evorthodus lyricus*.

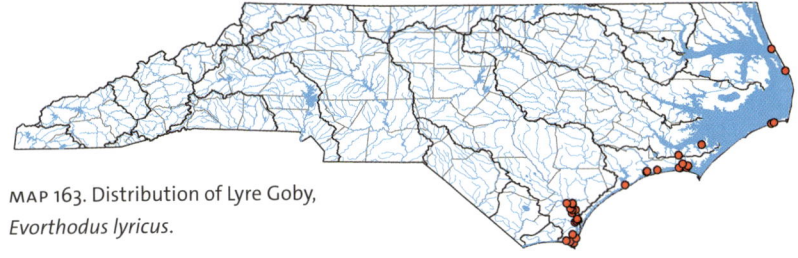

MAP 163. Distribution of Lyre Goby, *Evorthodus lyricus*.

6a. Dorsal and anal fin rays usually 11 and 12, respectively. Distinct spot at upper pectoral fin base present (sometimes faded in preserved specimens). V-shaped saddles on body often present below second dorsal fin (figure 212)..........
.........Darter Goby, *Ctenogobius boleosoma* (Jordan and Gilbert, 1882)
6b. Dorsal and anal fin rays usually 12 and 13, respectively. Spot at upper pectoral fin base usually absent. Four square-like blotches along midline (figure 213) (map 164)................Freshwater Goby, *Ctenogobius shufeldti* (Jordan and Eigenmann, 1887)

FIGURE 212. Darter Goby, *Ctenogobius boleosoma*.

FIGURE 213. Freshwater Goby, *Ctenogobius shufeldti.*

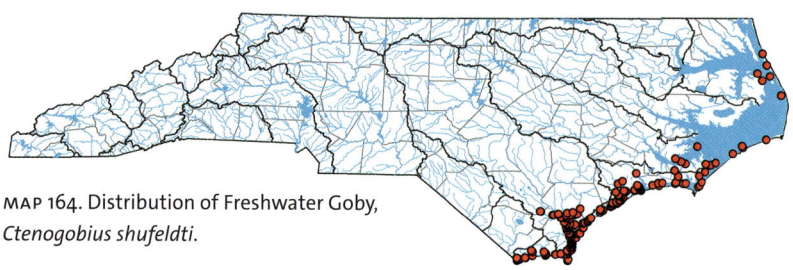

MAP 164. Distribution of Freshwater Goby, *Ctenogobius shufeldti.*

Sand Flounders
Family Paralichthyidae

In the waters along and off the coast of North Carolina, twenty species within the family Paralichthyidae can be found, but only three of these species are found in coastal freshwater or estuarine streams (tables 2 and 17; Tracy, Rohde, and Hogue, 2020). Colloquial names for sand flounders include flounder, fluke, mud flounder, plaice, and summer flounder. Often, any flatfish is simply called a flounder, regardless of its species or to which family it belongs. The Southern Flounder is the largest of the three species and may reach ca. 914 mm TL. Both the Gulf Flounder and Summer Flounder may reach more than 610 mm TL. Sand flounders are demersal fish, meaning they live on or buried beneath the bottom substrate. These bottoms can be hard or soft sand; coarse, shelly debris; or mud in lower coastal river channels and estuaries, inlets, and seagrass beds. They often bury themselves into the sediment as they wait to ambush their prey.

The Southern Flounder may be encountered as a seasonal inhabitant in freshwater habitats along the coast from the Albemarle Sound to Shallotte basins (table 2). This species has been found in the Cape Fear River upstream near Lock and Dam No. 1 (Bladen County, Cape Fear basin), in the Neuse River as far upstream as near the town of LaGrange (Lenoir County, Neuse basin), in the Roanoke River as far upstream as at the town of Weldon (Halifax County, Roanoke basin), and in the Chowan River as far upstream as at Arrowhead Beach (Chowan County, Chowan basin) (Tracy, Rohde, and Hogue, 2020).

The recreational and commercial harvesting of all three species is regulated by the North Carolina Division of Marine Fisheries and the North Carolina Wildlife Resources Commission (NCDMF, 2020; NCWRC, 2022). None of the species are endemic to a specific river basin in North Carolina (table 2).

TABLE 17. Species of sand flounders found in or along the coast of North Carolina

Scientific Name	Common Name
Paralichthys albigutta	Gulf Flounder
Paralichthys dentatus	Summer Flounder
Paralichthys lethostigma	Southern Flounder

Note: All species are managed by the NCDMF and NCWRC.

Key characteristics for the proper identification of sand flounders include the presence or absence of pigmented spots (ocelli) and their placement, dorsal and anal fin ray counts, and gill raker counts. The identification of sand flounders can be challenging. Complicating that fact is that specimens captured during trawling often sustain fin damage and scale loss, making the use of scale pigmentation patterns for identification impractical because all that remain are the scale pockets.

Identification Key to the Freshwater and Estuarine Species of Sand Flounders (Family Paralichthyidae)

1a. Prominent ocelli on eyed side (figure 214)................................2
1b. No prominent ocelli on eyed side (figure 214) (map 165)..................
...Southern Flounder, *Paralichthys lethostigma* Jordan and Gilbert, 1884

2a. Many ocelli on eyed side, but with five prominent ocellated dark spots on posterior half of body (figure 214). Gill rakers on lower limb of first arch 14 or more (rarely 13). Dorsal fin rays 80–96. Anal fin rays 61–73. Lateral line scales 91–106..
..............Summer Flounder, *Paralichthys dentatus* (Linnaeus, 1766)
2b. Three prominent ocellated dark spots on body arranged in a triangle with two (one above the other) in midbody and one on the lateral line in the posterior part of body (figure 214). Gill rakers on lower limb of first arch 9–12. Dorsal fin rays 71–85. Anal fin rays 53–63. Lateral line scales 78–81
..........Gulf Flounder, *Paralichthys albigutta* Jordan and Gilbert, 1882

A

B

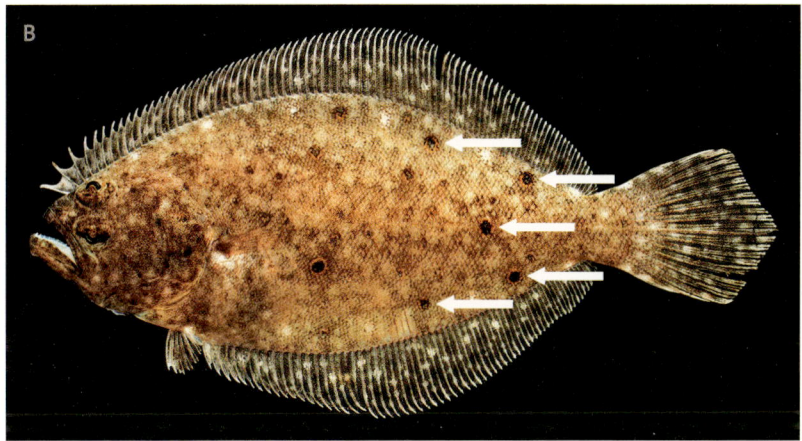

C

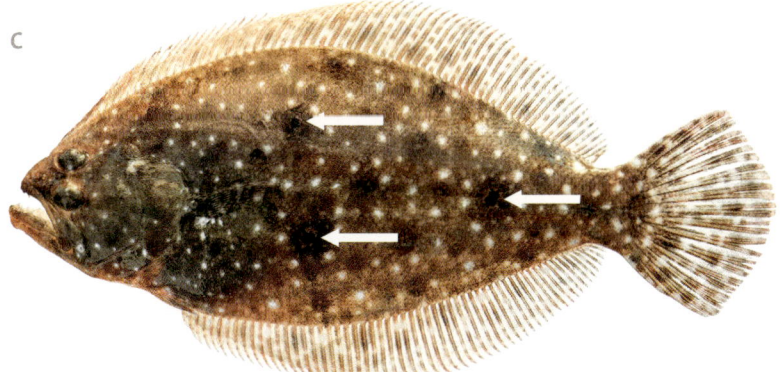

FIGURE 214. A—Southern Flounder, *Paralichthys lethostigma*; B—Summer Flounder, *Paralichthys dentatus*; C—Gulf Flounder, *Paralichthys albigutta*. Arrows indicating ocellated spots.

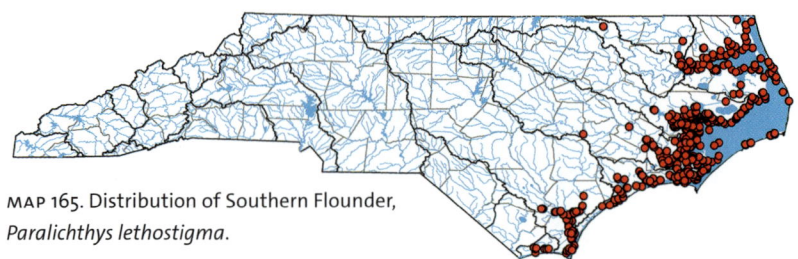

MAP 165. Distribution of Southern Flounder,
Paralichthys lethostigma.

15

Cichlids
Family Cichlidae

The family Cichlidae, known collectively as cichlids, is a very diverse family with more than 2,500 species indigenous to tropical and subtropical fresh and brackish waters of Mexico, Central and South America, the West Indies, Africa, the Middle East, and the Indian subcontinent (Froese and Pauly, 2022). In the United States cichlids are popular in the aquarium trade and in aquaculture, ultimately for human consumption. Unwanted or overgrown aquarium fishes are often dumped illegally into local ponds, lakes, and other waterways. More than 60 nonindigenous species have been found in waters of the United States (US Geological Survey, 2021). Only one species, the Rio Grande Cichlid, *Herichthys cyanoguttatus*, is native to the United States and is found in Texas (Fuller, Nico, and Williams, 1999). There are two species in North Carolina with established, reproducing, and (until recently) persistent populations: the Redbelly Tilapia and Blue Tilapia (table 2; Tracy, Rohde, and Hogue, 2020), but many other species and possible hybrids may be encountered. The Redbelly Tilapia can reach ca. 320 mm TL and Blue Tilapia ca. 370 mm TL (Page and Burr, 2011). Neither species is endemic to a specific river basin in North Carolina (table 2).

The Redbelly Tilapia was originally stocked in Duke Energy's Sutton Lake (New Hanover County, Cape Fear basin), in PCS Phosphate Company's ponds near Aurora (Beaufort County, Tar basin), and in Weatherspoon cooling pond near Lumberton (Robeson County, Lumber basin) in attempts to manage aquatic macrophytes. It was inadvertently introduced into Hyco Reservoir (Caswell and Person Counties, Roanoke basin) from an on-site aquacultural study (Fuller, Nico, and Williams, 1999; Tracy, Rohde, and Hogue, 2020). Except for Hyco Reservoir, sustaining populations are no longer evident in the other three localities because of the species' intolerance to ambient low water temperatures during the winter. Its future persistence in Hyco Reservoir is in doubt because the power generating plants are no longer discharging as much heated effluent as before.

The Blue Tilapia was stocked in 1965 as a forage fish and to control aquatic macrophytes in Lake Julian (Buncombe County, French Broad basin), introduced into Lake Norman for unknown reasons (Catawba, Iredell, Lincoln, and Mecklenburg Counties, Catawba basin), and, like the Redbelly Tilapia, was accidentally introduced into Hyco Reservoir from an on-site aquacultural study during 1984 (Fuller, Nico, and Williams, 1999; Tracy, Rohde, and Hogue, 2020). Thermal discharge from the coal-fired and gas-fired power-generating plant ceased in February 2020 in Lake Julian, and surveys in September 2021 did not detect the Blue Tilapia (Kyle Hussey, Duke Energy, personal communication). It is unknown whether the record mapped for Lake Norman represents a persistent and reproducing population or whether it was the remnant of a one-time bait bucket release.

One additional species, the Nile Tilapia, *Oreochromis niloticus*, was anecdotally reported by Henson, Aday, and Rice (2018) as occurring in North Carolina. At least two specimens were captured in September 2014 with a cast net near Duke Energy's Allen Steam Station near Belmont on Lake Wylie (Gaston County, Catawba basin) and were preserved frozen to be given to state resource agency personnel or a museum (Corey Oakley and Chris Woods, NCWRC, personal communication). The Nile Tilapia has banding on the caudal fin, which differentiates it from the Blue Tilapia, along with the number of dorsal spines and rays, breeding coloration of males, and color of the dorsal fin margin (Nico, Schofield, and Neilson, 2021). Although there are photographs of the two specimens showing banding on the caudal fin of one of the specimens, indicative of the Nile Tilapia, the whereabouts of the actual specimens are unknown, and a definitive identification is impossible. Since 2014, there have been no records of the Nile Tilapia from waters in the vicinity of the Allen Steam Station or farther down the reservoir near the Catawba Nuclear Plant in South Carolina (Ryan Heise and Nick Wahl, Duke Energy, personal communication). Two other species of cichlids have been reported from North Carolina—the Oscar, *Astronotus ocellatus*, and Mozambique Tilapia, *O. mossambicus*—but surviving and reproducing populations have not been found (Tracy, Rohde, and Hogue, 2020).

Key characteristics for the proper identification of cichlids include color and color patterns and the shape of the lateral profile of the head.

Identification Key to the Species of Cichlids (Family Cichlidae)

1a. Body and fins blue. Pink-red borders on dorsal and caudal fins. No red on belly. No black blotch near posterior base of dorsal fin in adults. Lateral profile in front of eyes straight or concave (figure 215) (map 166).........
............................Blue Tilapia, *Oreochromis aureus* (Steindachner, 1864)

1b. Underside of head, belly, and caudal peduncle pink to blood red. Dark blotch near posterior base of dorsal fin in adults. Lateral profile in front of eyes straight or convex (figure 216) (map 167)...........................
...........................Redbelly Tilapia, *Coptodon zillii* (Gervais, 1848)

FIGURE 215. Blue Tilapia, *Oreochromis aureus*. Florida. Photograph courtesy of Tim Aldridge.

FIGURE 216. Redbelly Tilapia, *Coptodon zillii*. Arizona. Photograph courtesy of Ben Cantrell.

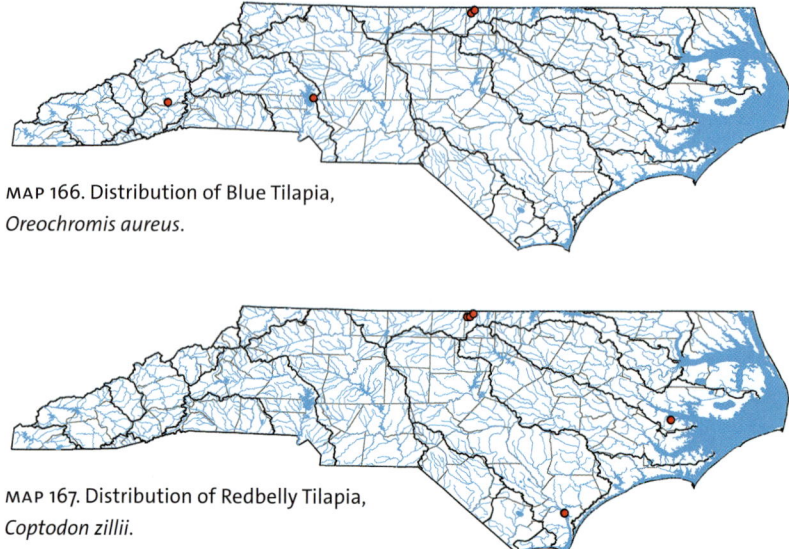

MAP 166. Distribution of Blue Tilapia,
Oreochromis aureus.

MAP 167. Distribution of Redbelly Tilapia,
Coptodon zillii.

New World Silversides
Family Atherinopsidae

Atherinopsidae is a small family comprising six species in North Carolina (tables 2 and 18). The common name, silversides, refers to a distinct silver stripe on the side of the fish, which often reflects sunlight like a mirror when they turn near the surface of the water (Rohde et al., 2009). The Waccamaw Silverside is often referred to as skipjack or glass minnow because of its translucency. Silversides are generally delicate, slender, laterally compressed, and translucent (Rohde et al., 2009). Maximum length of silversides in North Carolina is ca. 125 mm TL, and most live only one year.

Silversides may be found in fresh and saltwater environments from Hot Springs in Madison County to Hatteras Village in Dare County. They are most abundant in coastal rivers and estuaries and offshore (Tracy, Rohde, and Hogue, 2020). The Atlantic Silverside, Inland Silverside, and Rough Silverside are found along the coast, but the Inland Silverside was recently found in Lake Norman (Catawba, Iredell, Lincoln, and Mecklenburg Counties, Catawba basin) and as far upstream as near Elizabethtown (Bladen County, Cape Fear basin), near Greenville (Pitt County, Tar basin), and near Murfreesboro on the Meherrin River (Hertford County, Chowan basin) (Tracy, Rohde, and Hogue, 2020). The Waccamaw Silverside is endemic to Lake Waccamaw, Columbus County (Waccamaw basin) and is not found anywhere else in the world. Because of its limited distribution and the anthropogenic impacts upon its habitat and water quality, the Waccamaw Silverside is listed as federally Threatened (table 18; Krabbenhoft, Rohde, and Quattro, 2005; NCAC, 2023; NCNHP, 2022; NCWRC, 2021).

The Green Silverside is a recent, natural immigrant into North Carolina from South Carolina. It was first discovered in North Carolina in 1995 in the

TABLE 18. Species of silversides in North Carolina

Scientific Name	Common Name	Imperilment Status
Labidesthes sicculus[a]	Brook Silverside	
Labidesthes vanhyningi	Green Silverside	
Membras martinica	Rough Silverside	
Menidia beryllina	Inland Silverside	
Menidia extensa[a]	Waccamaw Silverside	federally threatened
Menidia menidia	Atlantic Silverside	

[a] Species found in only one river basin.

Waccamaw River, Waccamaw basin (Moser et al., 1998). Since then, it has naturally dispersed into Lake Waccamaw, where it now co-occurs with the Waccamaw Silverside. It has further dispersed into the Waccamaw River and throughout the lower Yadkin, lower Cape Fear, and Lumber basins. The Atlantic slope in North Carolina is now the northern limit of the Green Silverside's range (Werneke and Armbruster, 2015).

The Brook Silverside is also a recent, natural immigrant into North Carolina after swimming upstream from Tennessee. It was first discovered in North Carolina in 2012 by North Carolina Wildlife Resources Commission staff in the mainstem of the lower French Broad River in Madison County, French Broad basin (table 2; Tracy, Rohde, and Hogue, 2020). It was collected again in 2022 from the lower French Broad River, its only known locale in the state.

Key characteristics for the proper identification of silversides include the shape of the snout; scale textures; positioning of the origin of the spinous dorsal fin in relation to the origin of the anal fin; the number of anal fin rays; and their geographical distribution. Several species do co-occur, such as the Atlantic Silverside and Inland Silverside, rendering field identification a challenge.

Identification Key to the Species of New World Silversides (Family Atherinopsidae)

1a. Snout triangular when viewed from above; snout length longer than width of the eye. Lateral scales ca. 75. Predorsal scales ca. 33. Found in fresh water...*Labidesthes*, Couplet 2

1b. Snout rounded when viewed from above; snout length shorter than or equal to width of the eye. Lateral scales 34–50. Predorsal scales 14–23. Found in coastal, estuarine, and marine waters (except the Inland Silverside, *Menidia beryllina*, which has been recently found in Lake Norman (Catawba basin))...3

2a. Midlateral stripe narrows in front of first dorsal fin. Ratio of thoracic length to abdominal length greater than two. (Note: ventrally, thoracic length is the distance from the symphysis posterior to the throat [where soft tissue becomes bony] to the origin of the pelvic fin; ventrally, abdominal length is the distance from the origin of the pelvic fin to the urogenital opening.) (figure 217). Range restricted to the lower French Broad River in Madison County (French Broad basin) (map 168)......................
........................Brook Silverside, *Labidesthes sicculus* (Cope, 1865)

2b. Midlateral stripe usually expands in front of first dorsal fin. Ratio of thoracic length to abdominal length less than two (figure 217) (map 169)....
............Green Silverside, *Labidesthes vanhyningi* Bean and Reid, 1930

FIGURE 217. A—Brook Silverside, *Labidesthes sicculus*. Minnesota; B—Green Silverside, *Labidesthes vanhyningi*.

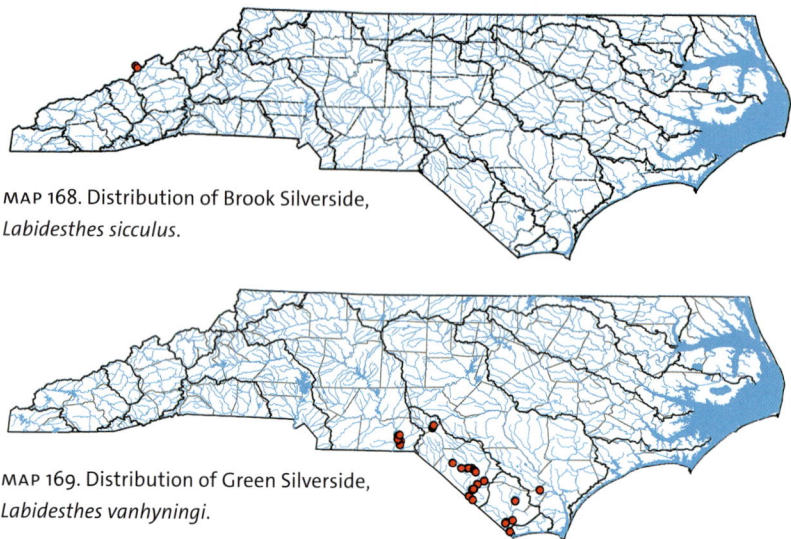

MAP 168. Distribution of Brook Silverside, *Labidesthes sicculus*.

MAP 169. Distribution of Green Silverside, *Labidesthes vanhyningi*.

3a. Posterior margins of scales finely scalloped and rough to the touch (figure 218). Dorsum peppered with melanophores (figure 218). Lateral axillary scale of pelvic fin well developed, greater than or equal to one-third length of fin. Basal third of soft dorsal and anal fins heavily scaled (figure 219) . . . . Rough Silverside, *Membras martinica* (Valenciennes, 1835)

3b. Posterior edge of scales smooth. Lateral axillary scale of pelvic fin absent or poorly developed, less than or equal to one-fourth length of fin. Base of soft dorsal and anal fins with few or no scales. . . . . *Menidia*, Couplet 4

FIGURE 218. Close-up view of the scalloped posterior margins of scales and melanophores of Rough Silverside, *Membras martinica*.

FIGURE 219. Rough Silverside, *Membras martinica*.

4a. Body very slender; depth 7.0–8.0 times SL. Lateral scales 40–50 (figure 220). Range restricted to Lake Waccamaw and Waccamaw River immediately downstream from the lake (Waccamaw basin) (map 170).........
.........Waccamaw Silverside, *Menidia extensa* Hubbs and Raney, 1946

4b. Body moderately slender; depth 4.8–6.0 times SL. Lateral scales 34–38 (40). Not found in Waccamaw basin......................................5

FIGURE 220. Waccamaw Silverside, *Menidia extensa*.

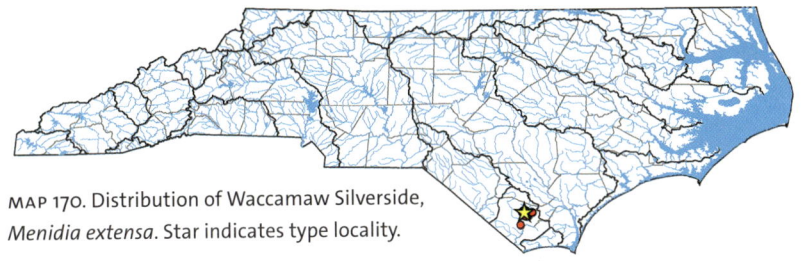

MAP 170. Distribution of Waccamaw Silverside, *Menidia extensa*. Star indicates type locality.

5a. Origin of spinous dorsal fin over or slightly posterior to origin of anal fin (figure 221). Anal fin rays (21) 22–24. Scale rows on nape (17) 18–20 ....Atlantic Silverside, *Menidia menidia* (Linnaeus, 1766)

5b. Origin of spinous dorsal fin anterior to origin of anal fin (figure 222). Anal fin rays 15–18 (19). Scale rows on nape 14–16 (17) (map 171)...............
.........................Inland Silverside, *Menidia beryllina* (Cope, 1867)

FIGURE 221. Atlantic Silverside, *Menidia menidia*, with bar indicating origin of the spinous dorsal fin over or posterior to anus.

FIGURE 222. Inland Silverside, *Menidia beryllina*, with bar indicating origin of the spinous dorsal fin anterior to anus.

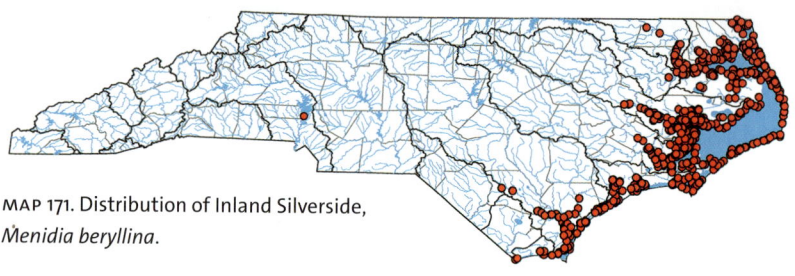

MAP 171. Distribution of Inland Silverside, *Menidia beryllina*.

17

Topminnows
Family Fundulidae

The Topminnow family in North Carolina consists of eleven scientifically described and one undescribed species occurring primarily within the eastern Coastal Plain and the estuarine marshes along the Atlantic Coast (tables 2 and 19; Menhinick, 1991; Tracy, Rohde, and Hogue, 2020; Tracy et al., 2022). Colloquial names for these species include killifishes, mud-minnows, and top minnows.

Topminnows range in size from the diminutive *Lucania* spp. at 50 mm TL to the 200 mm TL Striped Killifish. Because of their abundance and the ease by which they can be collected, they are often sold and used as baitfish along the coast. Most of our species inhabit a variety of coastal aquatic environments (table 20) and have a wide-ranging tolerance to salinities. The Bluefin Killifish, "Lake Phelps" Killifish, Speckled Killifish, and Waccamaw Killifish are known to inhabit only freshwater environments.

The Golden Topminnow is a recent, naturally occurring migrant from South Carolina; it was unknown to occur in North Carolina until 2007, when it was first discovered in Marlowe Branch, Columbus County (Waccamaw basin). The Bluefin Killifish is our state's only nonindigenous topminnow. The Speckled Killifish is suspected of being introduced in the Catawba basin. *Fundulus* specimens similar to the "Lake Phelps" Killifish have been discovered in Shearon Harris Lake (Wake and Chatham Counties, Cape Fear basin) but could be an introduction of the Banded Killifish that has rapidly assumed traits associated with a lacustrine environment (Tracy, Rohde, and Hogue, 2020). The Waccamaw Killifish is endemic to Lake Waccamaw, Columbus County (Waccamaw basin), and is found nowhere else in the world.

TABLE 19. Species of topminnows in North Carolina

Scientific Name, Common Name	Common Name	Imperilment Status
Fundulus chrysotus[a]	Golden Topminnow	
Fundulus confluentus	Marsh Killifish	
Fundulus diaphanus	Banded Killifish	
Fundulus heteroclitus	Mummichog	
Fundulus lineolatus	Lined Topminnow	
Fundulus luciae	Spotfin Killifish	
Fundulus majalis	Striped Killifish	
Fundulus rathbuni	Speckled Killifish	
Fundulus waccamensis[a]	Waccamaw Killifish	state special concern
Fundulus sp. "Lake Phelps" Killifish		state significantly rare
Lucania goodei[a]	Bluefin Killifish	
Lucania parva	Rainwater Killifish	

Note: Names in quotation marks denote a scientifically undescribed species.
[a] Species found in only one river basin.

The Lined Topminnow is our most widely distributed species, being found in eleven basins (table 2). The Cape Fear basin contains the most species, eight, whereas the Catawba and Lumber basins each have only one species. No species occurs west of the Appalachian Mountains (Tracy, Rohde, and Hogue, 2020).

Key characteristics for the proper identification of topminnows include the positioning of the dorsal fin relative to the snout and caudal fin; origin of the dorsal fin relative to the origin of the anal fin; color patterns; number of dorsal fin rays; number of gill rakers; and lateral scale count. Several species can co-occur, rendering field identification a challenge.

TABLE 20. Regions and habitats of topminnows

Scientific Name	Common Name	Physiographic Region	Habitats
Fundulus chrysotus	Golden Topminnow	Southeastern Coastal Plain	Open, sunlit, quiet, slow, shallow, warm, heavily vegetated waters of marshes, swamps, lake shores, sloughs, drainage ditches, borrow pits, and creek waters; also occurring in brackish waters
Fundulus confluentus	Marsh Killifish	Eastern Coastal Plain	Freshwater rivers and streams and brackish water tidal streams, coastal bays, marshes, channels, and over seagrass flats
Fundulus diaphanus	Banded Killifish	Primarily northeastern Coastal Plain	Calm, slow, and clear water of rivers and creeks, but also occurring from small inland streams to wide tidal rivers with low salinity, usually over a bottom of open sand
Fundulus heteroclitus	Mummichog	Eastern Coastal Plain	Tidal marshes, creeks, and ditches over mud flats and in or near vegetation, but also often occurring in fresh water
Fundulus lineolatus	Lined Topminnow	Sand Hills, Coastal Plain	Freshwater, soft-water, dystrophic, acidic, clear, or tannin-stained quiet portions of streams, sloughs, drainage ditches, borrow pits, and ponds, especially near submerged or emergent vegetation
Fundulus luciae	Spotfin Killifish	Southeastern Coastal Plain	Estuarine, typically in intertidal salt marshes
Fundulus majalis	Striped Killifish	Eastern Coastal Plain	High-saline inlets, bays, estuaries, and marshes and along beaches

TABLE 20. *(continued)*

Scientific Name	Common Name	Physiographic Region	Habitats
Fundulus rathbuni	Speckled Killifish	Central Piedmont	Fresh water, common in pools and runs of streams, usually over mud or sand bottoms
Fundulus waccamensis	Waccamaw Killifish	Coastal Plain—Lake Waccamaw	Fresh water, occurring in large schools in shallow water along the sandy to muddy shoreline, often associated with submerged or emergent vegetation
Fundulus sp. "Lake Phelps" Killifish		Coastal Plain—Lake Phelps	Fresh water, occurring in large schools in shallow water along the sandy to muddy shoreline, often associated with submerged or emergent vegetation
Lucania goodei	Bluefin Killifish	Wilmington, New Hanover Co.	Fresh water, occurring only in the lake and outfall
Lucania parva	Rainwater Killifish	Eastern Coastal Plain	Saltwater environments, but also occurring in some freshwater habitats; usually associated with dense vegetation

Note: Names in quotation marks denote a scientifically undescribed species.

Sources: Adapted from Hardy (1980); Kells and Carpenter (2011); Krabbenhoft, Rohde, and Quattro (2009); Lee (1980); Rohde et al. (2009); and Shute, Lindquist, and Shute (1983).

Identification Key to the Species of Topminnows (Family Fundulidae)

1a. Dorsal fin origin closer to preopercle than to caudal fin base (figure 223) ....*Lucania*, Couplet 2

1b. Dorsal fin origin closer to caudal fin base than to preopercle (figure 223) ....*Fundulus*, Couplet 3

FIGURE 223. Arrows and bars indicating difference in distance of the dorsal fin origin to the preopercle and the caudal fin base. A—Bluefin Killifish, *Lucania goodei*; B—Striped Killifish, *Fundulus majalis*.

2a. Lateral stripe black, extending from snout to caudal fin spot (figure 224). Dorsal fin rays 8–11. Restricted to Burnt Mill Creek and an impoundment of the creek at Anne McCrary Park in Wilmington, New Hanover County, and to an unnamed swamp tributary to the Northeast Cape Fear River in Pender County (Cape Fear basin) (map 172)...................... ....Bluefin Killifish, *Lucania goodei* Jordan, 1880

2b. Lateral stripe inconspicuous; caudal fin spot absent (figure 225). Dorsal fin rays 11 or 12 (map 173).. ....Rainwater Killifish, *Lucania parva* (Baird and Girard, 1855)

FIGURE 224. Bluefin Killifish, *Lucania goodei*. A—male; B—female.

FIGURE 225. Rainwater Killifish, *Lucania parva*. A—male; B—female.

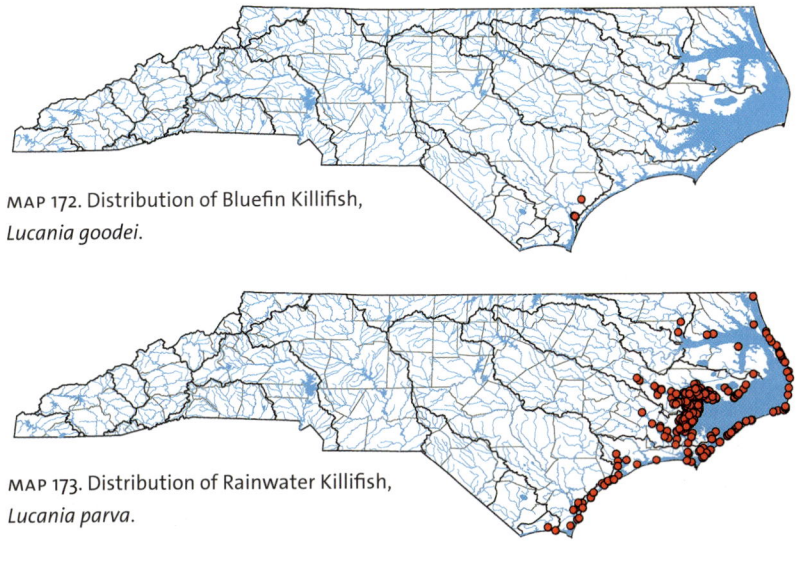

MAP 172. Distribution of Bluefin Killifish, *Lucania goodei*.

MAP 173. Distribution of Rainwater Killifish, *Lucania parva*.

3a. Dorsal fin rays 9 or fewer. Dorsal fin origin behind anal fin origin (figure 226) 4

3b. Dorsal fin rays 10 or more. Dorsal fin origin in front of or slightly behind anal fin origin (figure 226) 6

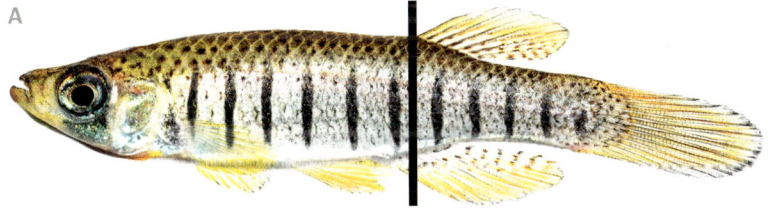

FIGURE 226. Bars indicating position of the dorsal fin relative to that of the anal fin. A— posterior to anal fin in Lined Topminnow, *Fundulus lineolatus*; B—anterior to anal fin in Waccamaw Killifish, *Fundulus waccamensis*.

4a. Predorsal median stripe distinct and dark. Male with ten to thirteen dark bars on the side and a black spot on the dorsal fin (figure 227). Female plain gray with no spot on the dorsal fin.....................................
...........................Spotfin Killifish, *Fundulus luciae* (Baird, 1855)

4b. Predorsal stripe light or absent. No black spot on the dorsal fin.........5

FIGURE 227. Spotfin Killifish, *Fundulus luciae*, with arrow indicating location of black predorsal stripe.

5a. Dark blotch below eye. Male with nine to thirteen narrow dark bars on the side; female with six to eight narrow longitudinal black stripes (figure 228) (map 174)...
..................Lined Topminnow, *Fundulus lineolatus* (Agassiz, 1854)

5b. No dark blotch below eye; many small golden or pearly (in females and juveniles) to red (in males) spots on the side (figure 229). Male usually with eight to twelve faint green, indistinct vertical bars on the side; females without vertical bars. Range currently restricted to Waccamaw basin (map 175)......Golden Topminnow, *Fundulus chrysotus* (Günther, 1866)

FIGURE 228. Lined Topminnow, *Fundulus lineolatus*. A—male; B—female.

FIGURE 229. Golden Topminnow, *Fundulus chrysotus*. A—male; B—female.

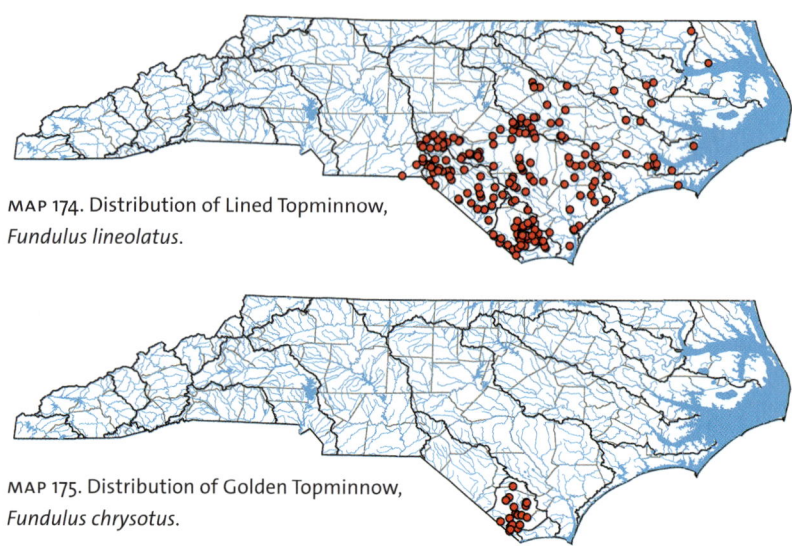

MAP 174. Distribution of Lined Topminnow, *Fundulus lineolatus.*

MAP 175. Distribution of Golden Topminnow, *Fundulus chrysotus.*

6a. Bars or distinct stripes present in one or both sexes. Range restricted to Coastal Plain...7
6b. Sides of adults with no bars (figure 230). Males without irregular black spots on sides, head spotted; females with irregular black spots on sides and head. Range restricted to Piedmont (map 176)......................
............Speckled Killifish, *Fundulus rathbuni* Jordan and Meek, 1889

FIGURE 230. Speckled Killifish, *Fundulus rathbuni.* A—male; B—female.

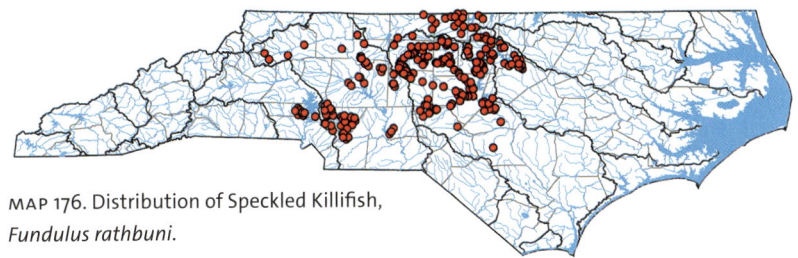

MAP 176. Distribution of Speckled Killifish, *Fundulus rathbuni*.

7a. Caudal fin truncate or emarginate. Dorsal fin with no black spot. Males with fifteen to twenty dark bars, usually wider than light interspaces; females with ca. twelve to sixteen narrow dark bars. Gill rakers 4 or 5. Lateral scales 37–64...8

7b. Caudal fin rounded or slightly rounded. One sex often with a black spot at the posterior of the dorsal fin. Bars, if present, fewer than fifteen (but up to eighteen in female Marsh Killifish, *Fundulus confluentus*). Gill rakers 5–10. Lateral scales (29) 30–36 (39)...10

8a. Lateral scales (34) 36–39 (46). In adults, depth of caudal peduncle 2.0–2.8 times length of caudal peduncle (figure 231) (map 177).....................
.....................Banded Killifish, *Fundulus diaphanus* (Lesueur, 1817)

8b. Lateral scales (50) 52–58 (64). In adults, depth of caudal peduncle 2.8–3.5 times length of caudal peduncle...9

FIGURE 231. Banded Killifish, *Fundulus diaphanus*. A—male; B—female or juvenile.

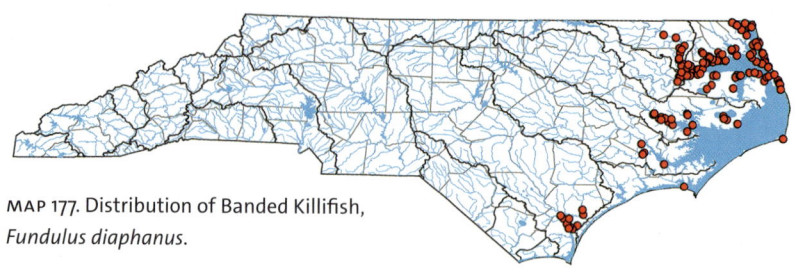

MAP 177. Distribution of Banded Killifish,
Fundulus diaphanus.

9a. Restricted to Lake Waccamaw and adjacent canals, Columbus County
(Waccamaw basin) (figure 232) (map 178)...............................
.....Waccamaw Killifish, *Fundulus waccamensis* Hubbs and Raney, 1946
9b. Restricted to Lake Phelps in Washington and Tyrell Counties (Albemarle
Sound basin), and Shearon Harris Lake in Wake and Chatham Counties
(Cape Fear basin) (figure 233) (map 179)...............................
...."Lake Phelps" Killifish, *Fundulus* sp.

FIGURE 232. Waccamaw Killifish, *Fundulus waccamensis.* A—male, B—female.

FIGURE 233. "Lake Phelps" Killifish, *Fundulus* sp. A—male; B—female or juvenile.

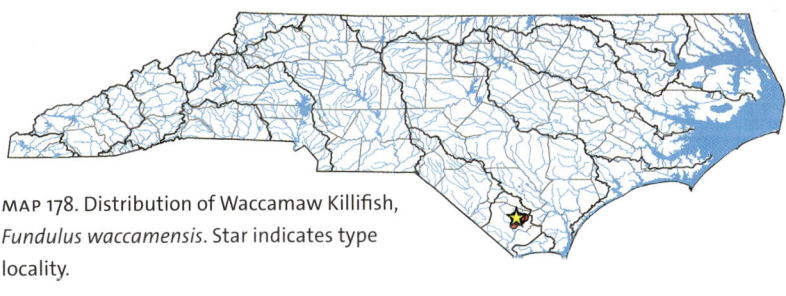

MAP 178. Distribution of Waccamaw Killifish, *Fundulus waccamensis*. Star indicates type locality.

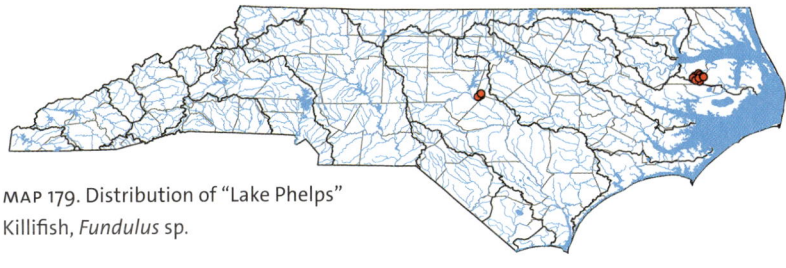

MAP 179. Distribution of "Lake Phelps" Killifish, *Fundulus* sp.

10a. Snout long; eye width ca. 1.5 times snout length. Dorsal fin rays 14 or 15 (12–16). Male with black bars on the side and a dark spot in the dorsal fin; female with black stripes (figure 234)......................................
.....................Striped Killifish, *Fundulus majalis* (Walbaum, 1792)

10b. Snout short; eye width ca. equal to snout length. Dorsal fin rays 10–12. Both sexes may have bars on the body....................................11

FIGURE 234. Striped Killifish, *Fundulus majalis*. A—male; B—female.

11a. Gill rakers 5. Males with ca. eighteen dark bars, gold spot anterior to the dorsal fin, and black spots on the dorsum (figure 235). Females with fourteen narrow black bars on sides, black spot on the posterior part of the dorsal fin, and poorly defined longitudinal rows of black dots on the dorsum (figure 235) (map 180)..
............Marsh Killifish, *Fundulus confluentus* Goode and Bean, 1879

11b. Gill rakers 9 or 10. Males with ca. fourteen narrow white bars that fade anteriorly, blotch sometimes present on last rays of dorsal fin, and black spot anterior to the dorsal fin on the dorsum; dorsal, anal, and caudal fins often dark with light spots and light margin (figure 236). Females with ca. twelve dark bars that fade anteriorly (often absent in adults) and no black spots on the dorsal, anal, or caudal fins (figure 236) (map 181).............
.....................Mummichog, *Fundulus heteroclitus* (Linnaeus, 1766)

FIGURE 235. Marsh Killifish, *Fundulus confluentus*. A—male, with arrow indicating location of gold spot anterior to dorsal fin; B—female.

FIGURE 236. Mummichog, *Fundulus heteroclitus*. A—male, with arrow indicating location of black spot anterior to dorsal fin; B—female.

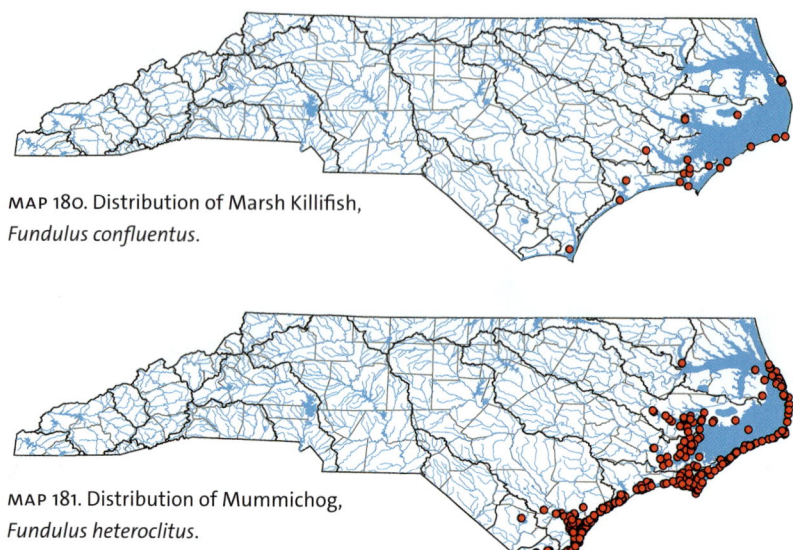

MAP 180. Distribution of Marsh Killifish,
Fundulus confluentus.

MAP 181. Distribution of Mummichog,
Fundulus heteroclitus.

Livebearers
Family Poeciliidae

North Carolina is home to four species of livebearers (tables 2 and 21). The common name of the family stems from the fact that a female gives birth to live young, rather than depositing or scattering her eggs externally. A male fish impregnates the female using specialized anal fin rays called a gonopodium. Members of this family are well known because certain species of livebearers are frequently sold in pet shops. Livebearers are usually referred to colloquially as guppies or mosquitofish.

Livebearers may be found in many freshwater Mountain and Piedmont streams and in Coastal brackish and estuarine waters in North Carolina. They can be very abundant in shallow, quiet water along the silty edges and side pools of slow-moving streams and in heavily vegetated ponds, ditches, and sloughs. They may be found from near Murphy in Cherokee County to Cape Hatteras in Dare County. The Eastern Mosquitofish, our most widely distributed and commonly collected species, is indigenous (native) in all river basins east of the Mountains. However, it is nonindigenous in the middle and lower portions of the French Broad basin in Henderson, Buncombe, and Madison Counties. The Least Killifish and Sailfin Molly are generally confined to the southeastern corner of the state in the Cape Fear, White Oak, Shallotte, and Waccamaw basins.

The largest of our livebearers is the Sailfin Molly, which reaches a maximum TL of ca. 150 mm (Rohde et al., 2009). Our smallest and one of the smallest vertebrates in the world to give birth to live young is the Least Killifish, which reaches a maximum TL of only 36 mm (Rohde et al., 2009). To put the diminutive size of the Least Killifish into proper perspective, it is only 0.8 percent as big as our largest freshwater fish species in North Carolina, the Atlantic Sturgeon.

Our only imperiled species is the Least Killifish; this status is due to its limited distribution and the anthropogenic impacts upon its habitats and water

TABLE 21. Species of livebearers in North Carolina

Scientific Name	Common Name	Imperilment Status
Gambusia affinis	Western Mosquitofish	
Gambusia holbrooki	Eastern Mosquitofish	
Heterandria formosa	Least Killifish	state special concern
Poecilia latipinna	Sailfin Molly	

quality (table 21; NCAC, 2023; NCNHP, 2022; NCWRC, 2021). None of the species are endemic to a specific river basin in North Carolina (table 2).

Key characteristics for the proper identification of livebearers include the position of the origin of the dorsal fin relative to that of the anal fin; pigmentation patterns along the side and on the caudal fin; the number of dorsal and anal fin rays; and the presence or absence of teeth on the gonopodium. The viewing of gonopodium of *Gambusia* and the counting of the dorsal fin rays are best accomplished with the aid of a dissecting microscope.

Identification Key to the Species of Livebearers (Family Poeciliidae)

1a. Dorsal fin origin located well anterior of the anal fin origin. Body with ca. seven dark stripes (figure 237) (map 182)..............................
..........................Sailfin Molly, *Poecilia latipinna* (Lesueur, 1821)

1b. Dorsal fin origin located posterior to the anal fin origin. Body with one stripe or none..2

FIGURE 237. Sailfin Molly, *Poecilia latipinna*. A—male; B—female.

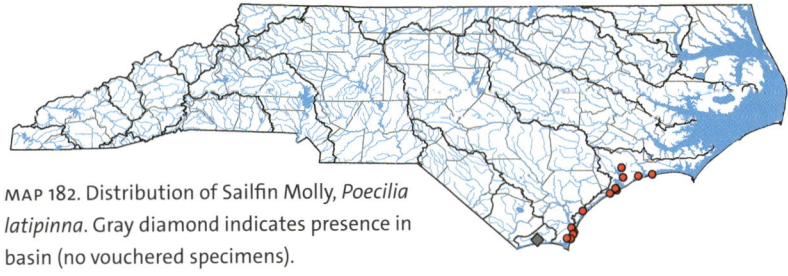

MAP 182. Distribution of Sailfin Molly, *Poecilia latipinna*. Gray diamond indicates presence in basin (no vouchered specimens).

2a. Side with a broad lateral stripe. Subocular bar absent (figure 238). Black spot on the dorsal fin base; black spot on female's anal fin. Caudal fin without vertical rows of dots. Range restricted to lower Cape Fear and Waccamaw basins (map 183)...
......................Least Killifish, *Heterandria formosa* (Girard, 1859)

2b. Lateral stripe absent. Subocular bar usually present (figure 239). No black spot on the dorsal or anal fins. Caudal fin with two or three vertical rows of dots..*Gambusia*, Couplet 3

FIGURE 238. Least Killifish, *Heterandria formosa*. A—male; B—female.

FIGURE 239. Eastern Mosquitofish, *Gambusia holbrooki*, with arrow indicating subocular bar.

MAP 183. Distribution of Least Killifish, *Heterandria formosa*.

3a. Dorsal fin rays usually 7 or 8 (figure 240). Anal fin rays usually 10. Third ray of the anal fin (the first ray of the gonopodium) of the male with a series of prominent teeth (observable only with the aid of a dissecting microscope) (figure 241). Widespread across all Atlantic slope basins, except for the Savannah basin. Melanistic populations occur in southeastern North Carolina (figure 240) (map 184).......................................
.................Eastern Mosquitofish, *Gambusia holbrooki* Girard, 1859

3b. Dorsal fin rays usually 6 (figure 242). Anal fin rays usually 9. Third ray of the anal fin of the male without teeth, smooth (observable only with the aid of a dissecting microscope) (figure 241). Range restricted to Hiwassee, Little Tennessee, and French Broad basins (map 185).....................
........Western Mosquitofish, *Gambusia affinis* (Baird and Girard, 1853)

FIGURE 240. Eastern Mosquitofish, *Gambusia holbrooki*. A— melanistic male from southeastern North Carolina; B—female.

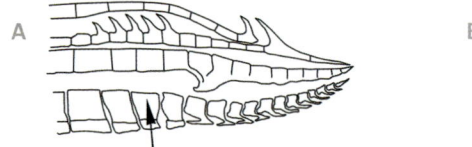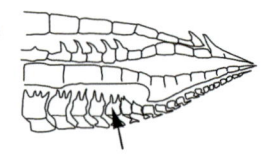

FIGURE 241. Gonopodia with arrows indicating presence or absence of prominent teeth. A—Eastern Mosquitofish, *Gambusia holbrooki*; B—Western Mosquitofish, *Gambusia affinis*. Illustration adapted from Etnier and Starnes (1993) and Rohde et al. (2009).

FIGURE 242. Western Mosquitofish, *Gambusia affinis*. A—male; B—female. Ohio.

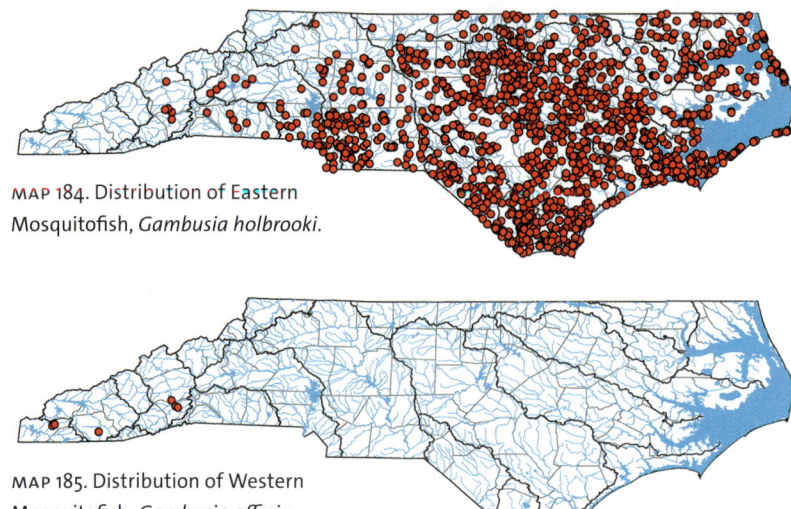

MAP 184. Distribution of Eastern Mosquitofish, *Gambusia holbrooki*.

MAP 185. Distribution of Western Mosquitofish, *Gambusia affinis*.

Mullets
Family Mugilidae

North Carolina is home to three species of mullets: the Mountain Mullet, Striped Mullet, and White Mullet (tables 2 and 22). Colloquial names for mullet include black mullet, callifaver mullet, common mullet, finger mullet, grey (gray) mullet, jumping mullet, and silver mullet. Another name for mullets that is sure to raise some eyebrows is turd wrestlers, which originates from the observation that mullets feed on plankton and detritus either via coprophagy or via suction from the sediment (Powers, 2020).

The mullets in North Carolina range in size from ca. 356 mm TL for the Mountain Mullet, to ca. 915 mm TL for the White Mullet, to ca. 1,220 mm TL for the Striped Mullet (Kells and Carpenter, 2011). All three species are catadromous, meaning they live in fresh or estuarine waters but spawn in the ocean. The Mountain Mullet may also be amphidromous, meaning the species migrates between salt and fresh water. Those migrations are not directly related to spawning but are related to other activity such as feeding (Rohde et al., 2009).

The Mountain Mullet is a rarely encountered species in North Carolina waters, where it is a seasonal inhabitant of fresh water. However, because it is difficult to capture with most collecting gear, it may be more common in US waters than is generally thought (Pezold and Edwards, 1983). The Striped Mullet is a seasonal inhabitant in all Coastal Plain basins except for the Waccamaw and Lumber basins and can be found as far upstream as near the Fall Zone near Rockingham (Anson and Richmond Counties, Yadkin basin), at Lillington on the mainstem of the Cape Fear River (Harnett County, Cape Fear basin), and at Raleigh (Wake County, Neuse basin). The White Mullet has rarely been found distant from brackish and saltwater coastal waters. Vouchered specimens at the North Carolina Museum of Natural Sciences document the species only

TABLE 22. Species of mullets in North Carolina

Scientific Name	Common Name	Management Status
Dajaus monticola	Mountain Mullet	
Mugil cephalus	Striped Mullet	Managed by NCDMF and NCWRC
Mugil curema	White Mullet	Managed by NCDMF and NCWRC

as far upstream as the Neuse River at New Bern (Craven and Pamlico Counties, Neuse basin) and in Northeast Creek near Jacksonville (Onslow County, White Oak basin). None of the species are endemic to a specific river basin in North Carolina (table 2).

Mullets are important prey items for piscivorous species such as Bluefish, *Pomatomus saltatrix*; King Mackerel, *Scomberomorus cavalla*; Red Drum; sand flounders; Spanish Mackerel, *Scomberomorus maculatus*; Spotted Seatrout, *Cynoscion nebulosus*; Striped Bass; and many other commercially and recreationally important species (Manooch, 1984). None of the mullet species are considered imperiled in North Carolina (NCAC, 2023; NCNHP, 2022; NCWRC, 2021), but the recreational and commercial harvesting (take) of both the Striped Mullet and White Mullet are regulated by the North Carolina Division of Marine Fisheries (NCDMF, 2020).

Key characteristics for the proper identification of mullets include the presence or absence of an adipose eyelid; extent of scalation of the second dorsal and anal fins; and body pigmentation.

Identification Key to the Species of Mullets (Family Mugilidae)

1a. Adipose eyelid present in individuals over 30 mm TL (figure 243). Lower jaw with a median knob (inside of mouth). Caudal peduncle without a dark spot. Fins not yellow, typically dusky.............*Mugil*, Couplet 2

1b. Adipose eyelid absent (figure 243). Lower jaw without a median knob (inside of mouth). Caudal peduncle with a black spot. All fins suffused with yellow (figure 244) (map 186)...
.....................Mountain Mullet, *Dajaus monticola* (Bancroft, 1834)

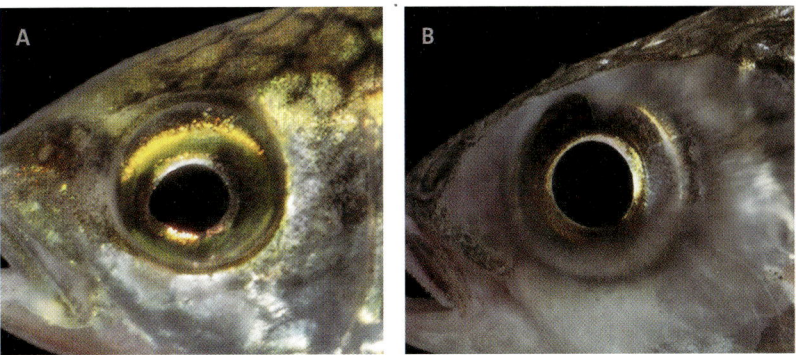

FIGURE 243. Adipose eyelid. A—absent in Mountain Mullet, *Dajaus monticola*. Mexico. B—present in Striped Mullet, *Mugil cephalus*.

FIGURE 244. Mountain Mullet, *Dajaus monticola*. Mexico.

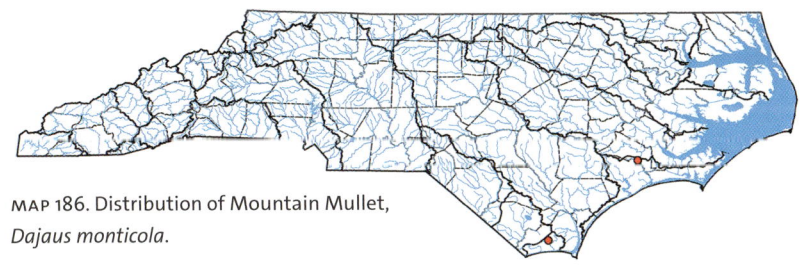

MAP 186. Distribution of Mountain Mullet, *Dajaus monticola*.

2a. Second dorsal fin and anal fin unscaled (a few scales may be present in anterior basal portion of both fins) (figure 245). Stripes present on body (figure 246) (map 187).... Striped Mullet, *Mugil cephalus* Linnaeus, 1758

2b. Second dorsal fin and anal fin well scaled (figure 245). No stripes present on body (figure 246). Occurs primarily in brackish and salt water........
...........................White Mullet, *Mugil curema* Valenciennes, 1836

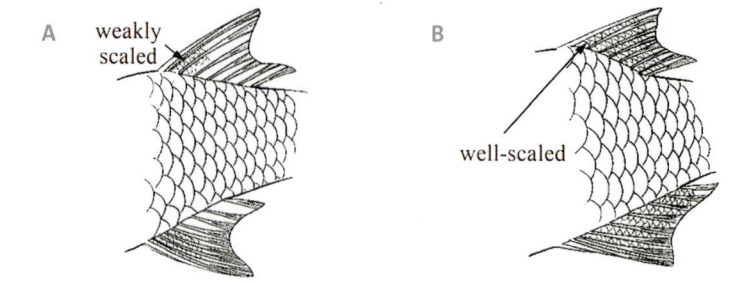

FIGURE 245. Second dorsal and anal fins. A—unscaled or weakly scaled; B—well scaled. Source: Food and Agriculture Organization of the United Nations, 2002, I. J. Harrison, "Mugilidae. Mullets", https://www.fao.org/documents/card/en/c/eb133bb9-b358-5887 -b82f-23c9f8ee94fd/. Reproduced with permission.

FIGURE 246. A—Striped Mullet, *Mugil cephalus*; B—White Mullet, *Mugil curema*.

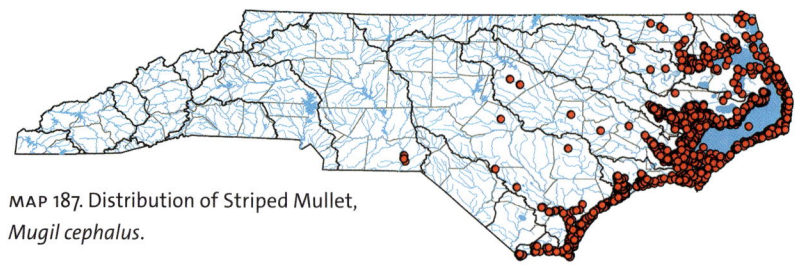

MAP 187. Distribution of Striped Mullet, *Mugil cephalus*.

20

Temperate Basses
Family Moronidae

North Carolina is home to three species of temperate basses, plus one hybrid (tables 2 and 23). Colloquial names for them include Waccamaws (perhaps because of the White Perch's abundance in Lake Waccamaw), silver perch (but not to be confused with the true Silver Perch, an estuarine and marine species), wiper, white lightning, rockfish (because of the Striped Bass's spawning habits over shoals and rocky substrates or living among rocky ledges), blue-nosed perch, gray perch, hybrids, silver bass, stripe, and striper. Hybrids of the Striped and White Bass are commonly called Bodie bass and palmetto bass.

The Striped Bass is one of our largest freshwater species, exceeded in length only by the Atlantic Sturgeon and Lake Sturgeon. Maximum TLs, reported by Rohde et al. (2009) for three of the four species, are the Striped Bass—2,000 mm TL; White Bass—450 mm TL; and White Perch—350 mm TL. Reports of the Striped Bass weighing more than forty-five kilograms in the late 1800s were noted by Smith (1907) and Jenkins and Burkhead (1994). The Striped Bass in North Carolina waters no longer reaches that size, although it can get as heavy as twenty-three kilograms or more. Recent data show that the Striped Bass along the Atlantic Coast may live as long as thirty years or more (McCargo, 2020).

Temperate basses in North Carolina may be found primarily east of the Appalachian Mountains in our large rivers such as the Catawba, Yadkin, Cape Fear, Neuse, Tar, and Roanoke; in Piedmont reservoirs and the tail races below these dams; in natural lakes such as lakes Mattamuskeet and Waccamaw; and in the Albemarle and Pamlico Sounds (table 2). Historically, there were only two indigenous (native) species of temperate basses in North Carolina—the Striped Bass and White Perch. Although populations of the Striped Bass along the Atlantic Coast have diminished over the past 300 years, both species are

TABLE 23. Species of temperate basses in North Carolina

Scientific Name	Common Name
Morone americana	White Perch
Morone chrysops	White Bass
Morone saxatilis	Striped Bass
Morone saxatilis x *Morone chrysops*	

Note: All species are managed by the NCWRC, and Striped Bass is also managed by the NCDMF.

still found in Coastal Plain and nearshore coastal waters with the anadromous Striped Bass making late winter–early spring spawning runs up the larger rivers to the Fall Zone at the eastern edge of the Piedmont. Today, the Striped Bass is found in all basins from the Catawba, where it has been introduced, eastward to the Atlantic Ocean, including the Waccamaw and Lumber basins. Landlocked populations, due to stocking, can be found in many of the larger reservoirs in the Hiwassee, Catawba, Yadkin, Cape Fear, and Roanoke basins (Tracy, Rohde, and Hogue, 2020; NCWRC, 2020).

The White Perch makes semi-anadromous spawning runs from brackish waters up the coastal rivers to fresh water, but not to the same extent as the Striped Bass. Today, the White Perch is abundant in reservoirs and rivers upstream from the Fall Zone, where it is considered an unauthorized bait-bucket introduction by anglers. Downstream, Coastal Plain populations are considered indigenous (Tracy, Rohde, and Hogue, 2020). The White Perch is now known from throughout the entire Catawba basin and Yadkin Chain-of-Lakes (Yadkin basin), from lakes Townsend, Cammack, and Shearon Harris in the Cape Fear basin, to Kerr Reservoir in the Roanoke basin. The White Bass, a nonindigenous (nonnative) species, was introduced into North Carolina waters as a new sport fish in the 1950s and today is found in our larger rivers and reservoirs from the Hiwassee to the Roanoke basins (Tracy, Rohde, and Hogue, 2020).

The Striped Bass hybrid is an aquaculture-created hybrid, which did not occur naturally in North Carolina. These hybrids, first created in the mid-1960s, can be fertile and are known to back-cross with either parent in the wild (Hodson, 1989). They are currently stocked by the North Carolina Wildlife Resources Commission (NCWRC) in Moss Lake (Broad basin), Lake Norman (Catawba basin); Lake Thom-A-Lex, Salem Lake, and W. Kerr Scott Reservoir (Yadkin basin); Oak Hollow Lake and Lake Townsend (Cape Fear basin);

and in Hyco Reservoir (Roanoke basin) (NCWRC, 2010, 2019a, 2020) and are stocked privately in farm and golf course ponds. They are also found in Lake Chatuge (Hiwassee basin), in the Neuse River at Milburnie (Wake County, Neuse basin), in the Cape Fear River (Cape Fear basin), and in the Albemarle and Pamlico Sounds (NCWRC, 2020).

In North Carolina all of the species in the family Moronidae are classified as commercially important game species and are managed and regulated with seasonal and river-basin-specific creel and landing limits by the North Carolina Division of Marine Fisheries and the NCWRC (table 23; NCDMF, 2020; NCWRC, 2022). Implementation of strict harvesting quotas, improved instream flows downstream of reservoirs, and more strict enforcement of water quality standards have helped the Striped Bass on its road to recovery, although some Striped Bass populations continue to be severely depleted due to poor recruitment. None of the species are listed as imperiled nor are endemic to a specific river basin in North Carolina (table 2; NCAC, 2023; NCNHP, 2022; NCWRC, 2021).

Key characteristics for the proper identification of temperate basses include the length and thickness of the second and third anal fin spines; the presence or absence and shape of a medial tooth patch on the tongue; whether the two dorsal fins are separate or joined; body shape; and the intactness of the lateral striping along the sides.

Identification Key to the Species of Temperate Basses (Family Moronidae)

1a. Two dorsal fins joined distinctly by a membrane (figure 247). Second anal fin spine thicker and almost as long as the third (figure 247). Median tooth patch on tongue absent (figure 248) (map 188)......................
.........................White Perch, *Morone americana* (Gmelin, 1789)

1b. Two dorsal fins slightly separate, not joined by a membrane (figure 249). Second anal fin spine shorter and thinner or equal to the third. Median tooth patch on tongue present (figure 248)...............................2

FIGURE 247. White Perch, *Morone americana*.

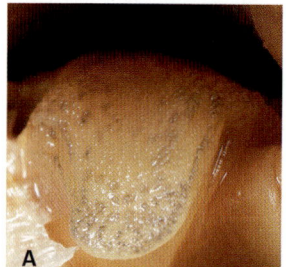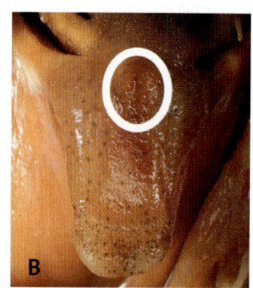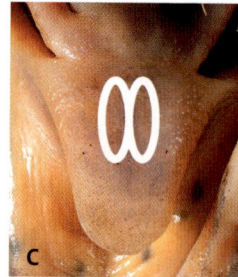

FIGURE 248. Tongues of temperate basses. A—no tooth patch in White Perch, *Morone americana*; B—single, medial, and oval tooth patch in White Bass, *Morone chrysops*; C—well-separated medial tooth patches in Striped Bass, *Morone saxatilis*.

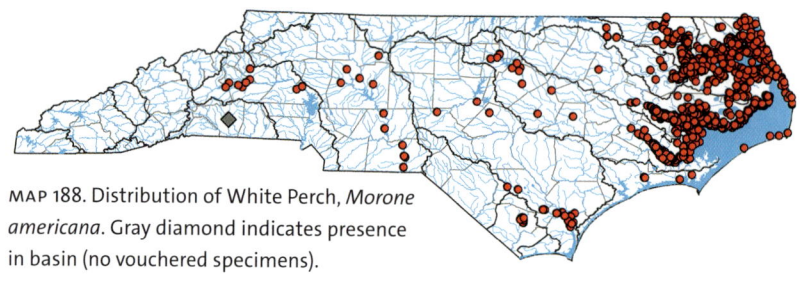

MAP 188. Distribution of White Perch, *Morone americana*. Gray diamond indicates presence in basin (no vouchered specimens).

FIGURE 249. Dorsal fins of Striped Bass, *Morone saxatilis*, slightly separate, not joined by a membrane. South Carolina.

2a. One medial oval tooth patch on tongue, occasionally divided (figures 248 and 250) (map 189)......White Bass, *Morone chrysops* (Rafinesque, 1820)
2b. Median tooth patch on tongue elongate, clearly divided (figure 249)....3

FIGURE 250. White Bass, *Morone chrysops*, juvenile. South Carolina.

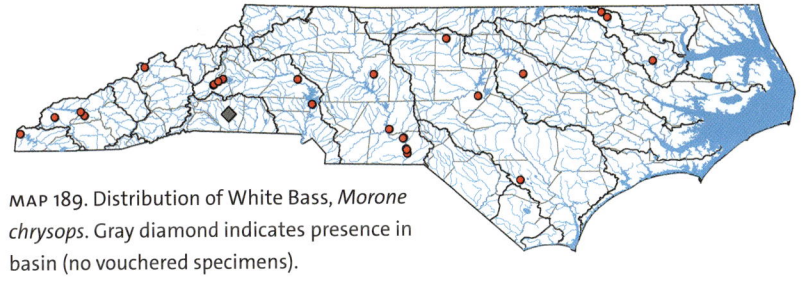

MAP 189. Distribution of White Bass, *Morone chrysops*. Gray diamond indicates presence in basin (no vouchered specimens).

3a. Body elongate; depth less than 33 percent of SL; slight arching of the back. Dorsum color more bluish than green. Bold lateral striping, occasionally broken (figure 251) (map 190)...
...........................Striped Bass, *Morone saxatilis* (Walbaum, 1792)
3b. Body not elongate; depth greater than 33 percent of SL; pronounced arching of the back. Dorsum color more greenish than blue. Bold lateral striping, often broken (figure 252) (map 191)....................................
...............Striped Bass hybrid and Bodie bass (female Striped Bass × male White Bass)

FIGURE 251. Striped Bass, *Morone saxatilis*. South Carolina.

FIGURE 252. Striped Bass hybrid (Bodie bass). Photograph courtesy of Kelsey Roberts.

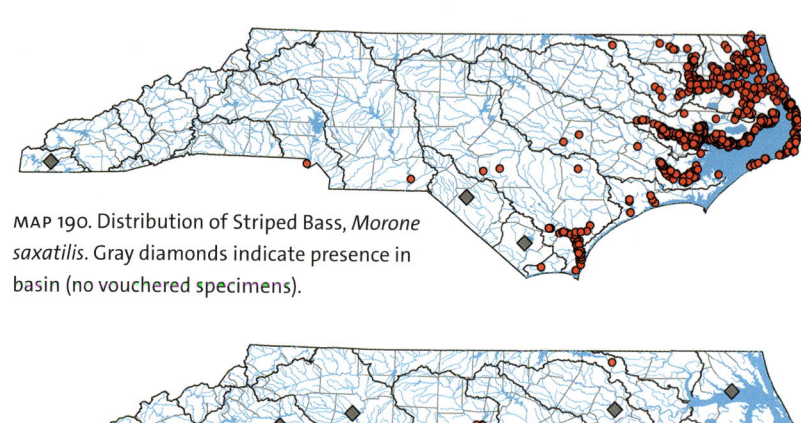

MAP 190. Distribution of Striped Bass, *Morone saxatilis*. Gray diamonds indicate presence in basin (no vouchered specimens).

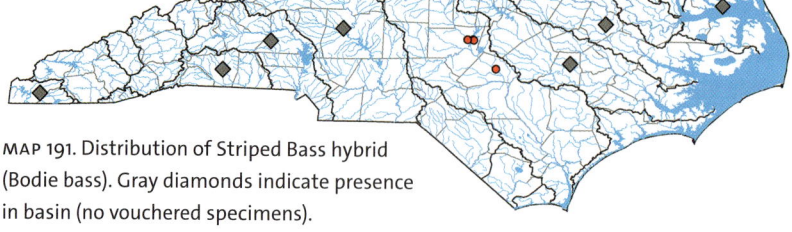

MAP 191. Distribution of Striped Bass hybrid (Bodie bass). Gray diamonds indicate presence in basin (no vouchered specimens).

Drums and Croakers
Family Sciaenidae

North Carolina, including its marine waters, is home to eighteen species within the family Sciaenidae (NCFishes.com). Of those eighteen species, four may be encountered in its coastal streams and estuarine environments and one in fresh water (tables 2 and 24).Colloquial names for the species in this family are many due to their recreational and commercial importance (table 24). Even Smith (1907) opined that "the common names of this species [referring to Weakfish, *Cynoscion regalis*] are numerous, and some of them are very improper," and "the local names applied to this species [referring to Spotted Seatrout] are indefensible, but will probably never be supplanted by appropriate ones." The five species vary greatly in size. The Silver Perch is the smallest species at ca. 300 mm TL, while the adult Red Drum may reach almost 1,600 mm TL (Kells and Carpenter, 2011).

Except for the Freshwater Drum, all of the other species of drums and croakers are found along North Carolina's coast. The Silver Perch, Atlantic Croaker, and Red Drum occasionally stray into fresh waters (Tracy, Rohde, and Hogue, 2020) but spend most of their lives in estuarine or marine waters. The Spot is primarily an estuarine species that may be found seasonally in freshwater habitats (Tracy, Rohde, and Hogue, 2020). The occupied habitats of Atlantic Croaker, Red Drum, Silver Perch, and Spot are variable and species-dependent, from shallow inlets and near river mouths to offshore deep reefs and hard bottoms (Kells and Carpenter, 2011).

The Freshwater Drum, as its names implies, is found exclusively in fresh water (table 2). It is indigenous to the western Mountain river basins but up until 2023 was found only in the lower French Broad River in Madison County, where there are some backwaters, slower currents, and deep pools. Beginning

TABLE 24. Species of drums and croakers found in or along the coast of North Carolina

Scientific Name	Common Name	Vernacular Name	Imperilment or Management Status
Aplodinotus grunniens	Freshwater Drum	Gasper-gou, sheepshead	state special concern
Bairdiella chrysoura	Silver Perch	Perch (white, sand, and yellow-finned), yellow-tail, and silver croaker	
Leiostomus xanthurus	Spot	Jimmy, chub, roach, goddy, Lafayette, Norfolk spot	Managed by NCDMF
Micropogonias undulatus	Atlantic Croaker	Croaker, crocus, hard-head	Managed by NCDMF
Sciaenops ocellatus	Red Drum	Puppy drum, drum, channel bass, red-fish, spotted bass	Managed by NCDMF, NCWRC

in March 2023 the Freshwater Drum is being reintroduced by North Carolina Wildlife Resources Commission staff back into the upper French Broad basin near Hendersonville, Henderson County (Luke Etchison, NCWRC, personal communication). The Freshwater Drum was illegally introduced into John H. Kerr Reservoir (Roanoke basin) possibly as recently as the 1980s or 1990s. Recently, it has been collected upstream from John H. Kerr Reservoir in the Dan River at Milton in Caswell County (Kelsey Roberts, NCWRC, personal communication). Anecdotal reports now document its occurrence in Lake Gaston and Roanoke Rapids Lake and within the Roanoke Rapids Dam bypass reach. More occurrences in the mainstem of the Roanoke River downstream from Roanoke Rapids Dam are to be expected in the future (Tracy, Rohde, and Hogue, 2020).

The Freshwater Drum is the only species in the family that is considered imperiled in North Carolina (table 24; NCAC, 2023; NCNHP, 2022; NCWRC, 2021). The recreational and commercial harvesting (take) of the Atlantic Croaker, Red Drum, and Spot are state-regulated by the North Carolina Division of Marine Fisheries and the NCWRC (table 24; NCDMF, 2020; NCWRC, 2022). None of the species are endemic to a specific river basin in North Carolina (table 2).

Key characteristics for the proper identification of drums and croakers include the presence (and number) or absence of chin barbels; preopercle serrations; body shape; and body coloration.

Identification Key to the Freshwater and Estuarine Species of Drums and Croakers (Family Sciaenidae)

1a. Chin or underside of lower jaw with barbels (figure 253)...................
............Atlantic Croaker, *Micropogonias undulatus* (Linnaeus, 1766)
1b. Chin without barbels..2

2a. Preopercle serrate often with one or more distinct bony spines at angle or prominent serration on posterior margin (figure 254).....................
.......................Silver Perch, *Bairdiella chrysoura* (Lacepède, 1802)
2b. Preopercle smooth or slightly denticulate or ciliate, never with a strong bony spine or serration in adults..3

FIGURE 253. Atlantic Croaker, *Micropogonias undulatus*.

FIGURE 254. Silver Perch, *Bairdiella chrysoura*.

3a. Body short and deep, dorsal profile strongly elevated or arched on nape. No spot at base of caudal fin (figure 255)..................................4

3b. Body elongate, dorsal profile not strongly elevated or arched on nape. One or more ocellated spots at base of caudal fin and more spots sometimes present on posterior portion of body (figure 255).........................
...........................Red Drum, *Sciaenops ocellatus* (Linnaeus, 1766)

FIGURE 255. A—Spot, *Leiostomus xanthurus*; B—Red Drum, *Sciaenops ocellatus*.

4a. Body with spots, bars, or stripes. Lower pharyngeal tooth plates not fused (visible upon dissection) (figure 255) (map 192)...........................
...............................Spot, *Leiostomus xanthurus* Lacepède, 1802

4b. Body uniformly silvery, darker dorsally. Lower pharyngeal tooth plates fused into a single triangular plate (visible upon dissection) (figure 256) (map 193).....Freshwater Drum, *Aplodinotus grunniens* Rafinesque, 1819

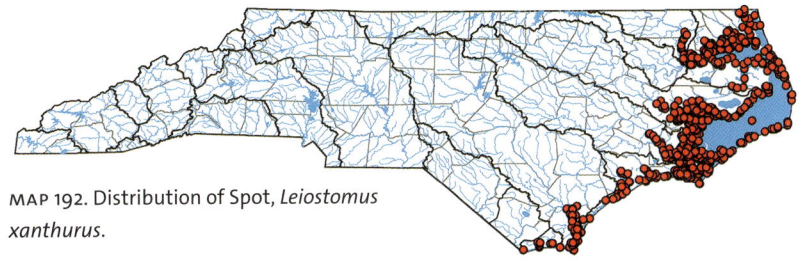

MAP 192. Distribution of Spot, *Leiostomus xanthurus*.

FIGURE 256. Freshwater Drum, *Aplodinotus grunniens*. Photograph courtesy of Luke Etchison.

MAP 193. Distribution of Freshwater Drum, *Aplodinotus grunniens*.

Perches and Darters
Family Percidae

North Carolina is home to thirty-eight brightly colored and beautiful species of perches and darters that rival their tropical counterparts (table 25). They can be found throughout our state in reservoirs, creeks, large and small rivers, swamps, and channelized streams (table 2). Colloquial names for perches and darters include jackfish, jack salmon, lake perch, pike, pike perch, raccoon perch, redfin perch, ringed perch, river slicks, walleyed pike, and many more.

Darter habitats range from turbulent and fast, cold, clear Mountain streams to warm and turbid Piedmont streams to slow-moving, tannin-colored Sand Hills and Coastal Plain streams. Darters are generally found in riffles and runs, whereas the Yellow Perch can also be found in reservoirs and ponds. The Sauger and Walleye are found in reservoirs and in pools and deep runs in low- to moderate-gradient rivers. At least two species, the Banded Darter and River-weed Darter, are closely associated with hornleaf riverweed, *Podostemum ceratophyllum*, an aquatic plant that grows attached to rocks in riffles and runs. Perches and darters range in size from ca. 30 mm TL for the Carolina and Swamp Darters to the Sauger and Walleye that can reach ca. 300 mm TL. The Sauger, Walleye, and Yellow Perch are widely sought-after game species noted for their delectability.

The family Percidae is our second most diverse family in North Carolina. Each of North Carolina's 100 counties has at least one species found within its borders. The Yellow Perch is found in eighteen of our twenty-one basins but has yet to be found in the Nolichucky, Watauga, and Shallotte basins (table 2). It has been introduced into the Hiwassee, Little Tennessee, Pigeon, and French Broad basins. Our most diverse basins are the Pigeon and French Broad, where, in each basin, there are currently thirteen indigenous (native) species

TABLE 25. Species of perches and darters in North Carolina

Scientific Name	Common Name	Imperilment or Management Status
Etheostoma blennioides	Greenside Darter	
Etheostoma brevispinum	Carolina Fantail Darter	
Etheostoma collis	Carolina Darter	state special concern
Etheostoma flabellare	Fantail Darter	
Etheostoma fusiforme	Swamp Darter	
Etheostoma gutselli	Tuckasegee Darter	
Etheostoma inscriptum[a]	Turquoise Darter	state threatened
Etheostoma kanawhae[a]	Kanawha Darter	state significantly rare
Etheostoma maculaticeps	Southern Tessellated Darter	
Etheostoma mariae[a]	Pinewoods Darter	state special concern
Etheostoma perlongum[a]	Waccamaw Darter	state threatened
Etheostoma podostemone[a]	Riverweed Darter	state significantly rare
Etheostoma serrifer	Sawcheek Darter	
Etheostoma simoterum	Snubnose Darter	state special concern
Etheostoma swannanoa	Swannanoa Darter	
Etheostoma thalassinum	Seagreen Darter	state special concern
Etheostoma vitreum	Glassy Darter	
Etheostoma zonale	Banded Darter	
Etheostoma sp. "Tessellated" Darter		
Nothonotus acuticeps[a]	Sharphead Darter	state threatened
Nothonotus chlorobranchius	Greenfin Darter	
Nothonotus rufilineatus	Redline Darter	
Nothonotus vulneratus[a]	Wounded Darter	state special concern
Perca flavescens	Yellow Perch	Managed by NCWRC
Percina aurantiaca	Tangerine Darter	
Percina burtonia	Blotchside Logperch	state endangered
Percina caprodes	Logperch	state threatened
Percina crassa	Piedmont Darter	

TABLE 25. (*continued*)

Scientific Name	Common Name	Imperilment or Management Status
Percina evides	Gilt Darter	
Percina gymnocephala[a]	Appalachia Darter	state significantly rare
Percina nevisensis	Chainback Darter	
Percina oxyrhynchus[a]	Sharpnose Darter	state endangered
Percina rex[a]	Roanoke Logperch	federally endangered
Percina roanoka	Roanoke Darter	
Percina squamata	Olive Darter	state special concern
Percina westfalli[a]	Sooty-banded Darter	state special concern
Sander canadensis	Sauger	Managed by NCWRC
Sander vitreus	Walleye	Managed by NCWRC

Note: Names in quotation marks denote a scientifically undescribed species.
[a] Species found in only one river basin.

and two nonindigenous (introduced) species (Swamp Darter and Yellow Perch) (graph 4). Our least speciose basin is the Watauga, where only the Greenfin Darter and Tangerine Darter are found (graph 4; table 2).

Four species have been extirpated from the French Broad basin—the Blotchside Logperch, Blueside Darter, Sickle Darter, and Wounded Darter with the latter two species having been extirpated from the state. The Walleye has long been extirpated from the Neuse basin. Twelve species are found in only one basin (tables 2 and 25).

Compared with other families of fishes—for example, minnows, sunfishes, and catfishes—few species of perches and darters have been introduced outside of their native ranges in North Carolina. It is suspected that bait-bucket dumps have led to the introduction of the Redline Darter into the Little Tennessee basin and Southern Tessellated Darter from the Yadkin basin into the New basin (MacGuigan, Orr, and Near, 2023). Transportation of aquatic plants may have led to the introduction of the Swamp Darter into the Pigeon and French Broad basins. The Sauger, Walleye, and Yellow Perch have been stocked because of their popularity as game fishes. There are twelve basins where no species of perches and darters have been introduced (graph 4).

Seventeen species are considered imperiled in North Carolina, many due to their endemism in specific basins (table 25; Krabbenhoft, Rohde, and Quattro,

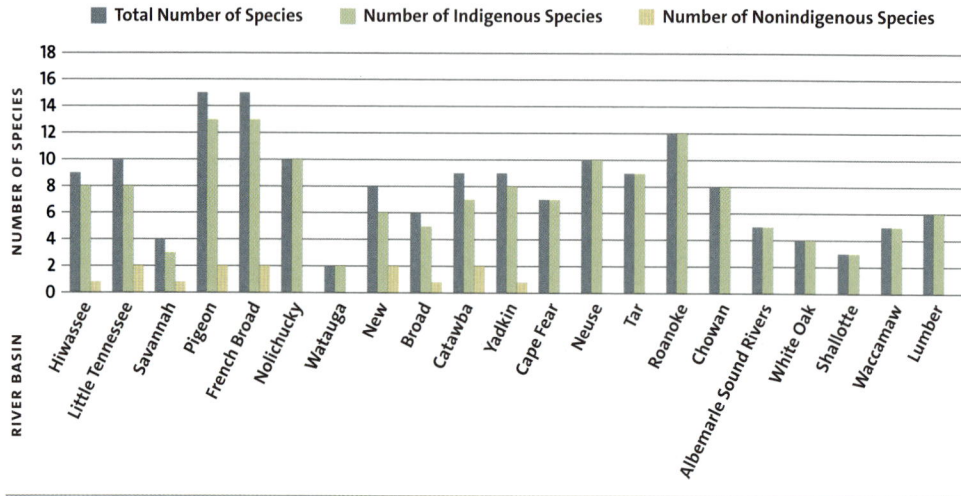

GRAPH 4. Diversity of perches and darters across North Carolina's river basins (extant species).

2006; NCAC, 2023; NCNHP, 2022; NCWRC, 2021; Roberts and Rosenberger, 2008). The Sauger, Walleye, and Yellow Perch are classified and managed as game species by the North Carolina Wildlife Resources Commission (table 25; NCWRC, 2022).

Key characteristics for the proper identification of perches and darters include the presence/absence of modified scales on the belly; the presence/absence of scales on the nape and cheek; the presence/absence of a frenum; the number and thickness of anal fin spines; lateral line shape and scale counts; the number of unpored lateral line scales; the number of spines and rays in the dorsal fins; overall color patterns; and the geographical distributions of the species. Most species can be easily distinguished from one another, with a few exceptions.

Recently, MacGuigan, Orr, and Near (2023) resurrected *Etheostoma maculaticeps* (Cope, 1870) from synonymy with *Etheostoma olmstedi* Storer, 1842. The Southern Tessellated Darter, *E. maculaticeps*, rather than *E. olmstedi*, now occupies the New, Broad, Catawba, Yadkin, Cape Fear, Shallotte, Waccamaw, and Lumber basins. Two species, *E. nigrum* Rafinesque, 1820, Johnny Darter, and *E. olmstedi*, occurring in the White Oak, Neuse, Tar, Roanoke, Chowan, and Albemarle Sound basins, are now considered by MacGuigan, Orr, and Near (2023) to be an undescribed species, the "Tessellated" Darter, *Etheostoma* sp.

Identification Key to the Species of Perches and Darters (Family Percidae)[1]

1a. Preopercle strongly serrate. Branchiostegal rays 7 2

1b. Preopercle smooth or occasionally weakly serrate. Branchiostegal rays 6, rarely 5 4

2a. Sides with six to eight prominent bars (figure 257). Teeth small. Anal fin rays 6–9 (map 194) . . . . .Yellow Perch, *Perca flavescens* (Mitchill, 1814)

2b. Sides uniform or mottled. Prominent canine teeth. Anal fin rays 11–14 . . . .*Sander*, Couplet 3

FIGURE 257. Yellow Perch, *Perca flavescens*.

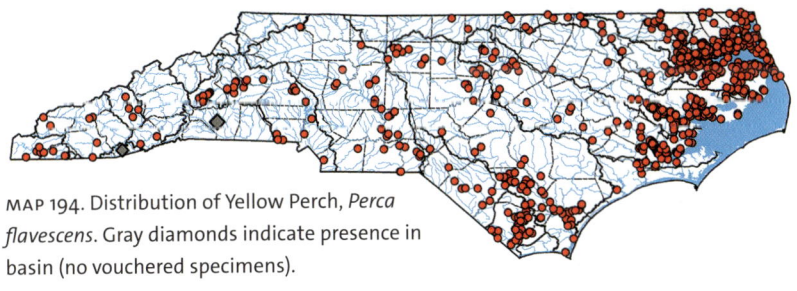

MAP 194. Distribution of Yellow Perch, *Perca flavescens*. Gray diamonds indicate presence in basin (no vouchered specimens).

1. Identification key excludes the Blueside Darter and Sickle Darter, which are extirpated from North Carolina.

3a. Last two membranes of spinous dorsal fin nearly black; other membranes usually with many small melanophores (figure 258). Cheek naked or with few scales (map 195)..............Walleye, *Sander vitreus* (Mitchill, 1818)

3b. Last two membranes of spinous dorsal fin not black but fin membranes containing several round, dark spots forming rows (figure 258). Cheek well scaled (map 196)...
......................Sauger, *Sander canadensis* (Griffith and Smith, 1834)

FIGURE 258. A—Walleye, *Sander vitreus*; B—Sauger, *Sander canadensis*. Photograph courtesy of Luke Etchison (Sauger).

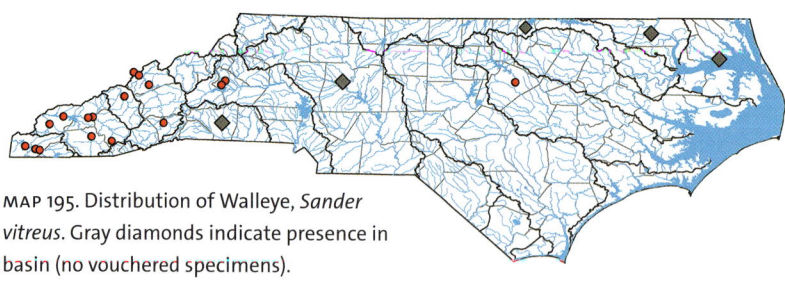

MAP 195. Distribution of Walleye, *Sander vitreus*. Gray diamonds indicate presence in basin (no vouchered specimens).

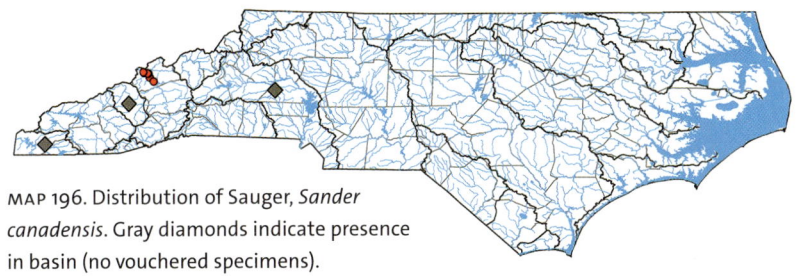

MAP 196. Distribution of Sauger, *Sander canadensis*. Gray diamonds indicate presence in basin (no vouchered specimens).

4a. One to three enlarged and often spiny scales between pelvic fins and one enlarged scale usually present just behind midbreast indentation (figure 259)..*Percina*, Couplet 5
4b. Enlarged scales absent between pelvic fins or on middle of breast (figure 259) ....*Etheostoma* and *Nothonotus*, Couplet 16

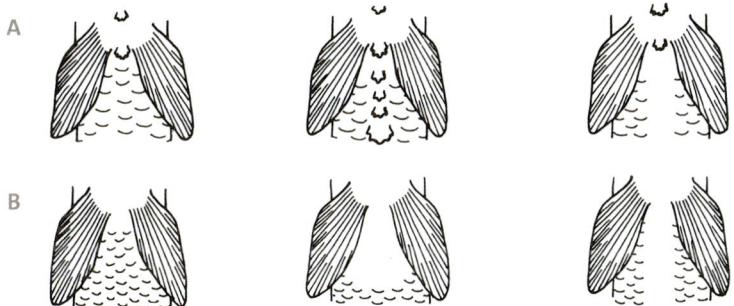

FIGURE 259. Belly scalation. A—enlarged scales present in *Percina* spp.; B—enlarged scales absent in *Etheostoma* and *Nothonotus* spp. Illustration adapted from Menhinick (1991).

5a. Mouth inferior; snout conical and fleshy, "piglike." Lateral blotches usually separate, not forming a continuous lateral stripe......................6
5b. Mouth terminal or subterminal; snout not conical and fleshy, not "piglike." Lateral blotches usually connected to form a continuous lateral stripe...8

306 | CHAPTER 22

6a. Sides with nine or ten horizontally oblong blotches (figure 260). Soft dorsal fin rays 13 or 14. Nape, cheek, and opercle naked or weakly scaled. Range restricted to Nolichucky basin (map 197)...........................
...........................Blotchside Logperch, *Percina burtoni* Fowler, 1945
6b. Sides with ten to twenty-five narrow vertical bars. Soft dorsal fin rays 15–17. Nape, cheek, and opercle well scaled..............................7

FIGURE 260. Blotchside Logperch, *Percina burtoni*.

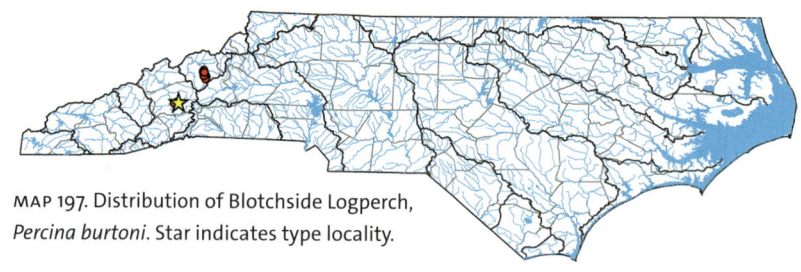

MAP 197. Distribution of Blotchside Logperch, *Percina burtoni*. Star indicates type locality.

7a. Sides with twenty to twenty-five narrow vertical bars that cross the back (figure 261). No orange stripe in the spinous dorsal fin. Range restricted to Pigeon, French Broad, and New basins (map 198).......................
.............................Logperch, *Percina caprodes* (Rafinesque, 1818)
7b. Sides with ten ovoid bars that do not cross the back (figure 261). Orange stripe in spinous dorsal fin. Range restricted to upper Roanoke basin (map 199)
............Roanoke Logperch, *Percina rex* (Jordan and Evermann, 1889)

FIGURE 261. A—Logperch, *Percina caprodes*; B—Roanoke Logperch, *Percina rex*. Photograph courtesy of Luke Etchison (Logperch).

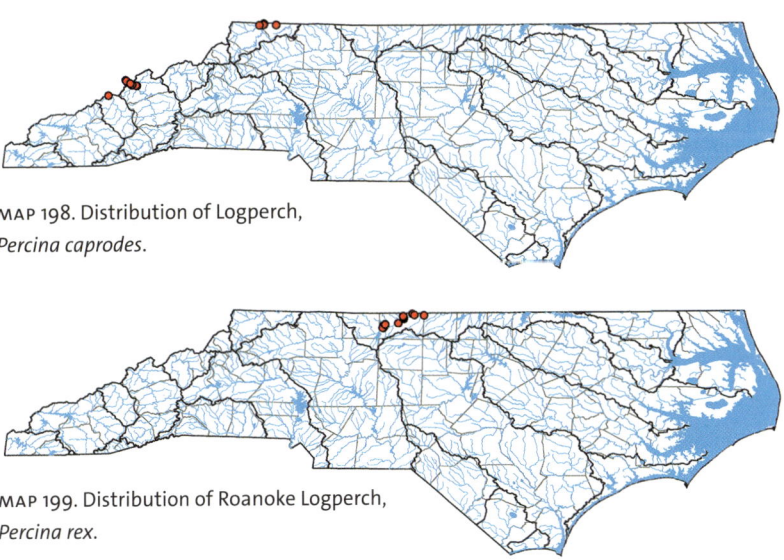

MAP 198. Distribution of Logperch, *Percina caprodes*.

MAP 199. Distribution of Roanoke Logperch, *Percina rex*.

8a. Midventral scales on belly not modified. Scales small, lateral line scales 89–100. Soft dorsal fin rays 14 or 15 (figure 262) (map 200)...............
...................Tangerine Darter, *Percina aurantiaca* (Cope, 1868)

8b. Midventral scales on belly enlarged. Scales large, lateral line scales 40–82. Soft dorsal fin rays 10–14..9

FIGURE 262. Tangerine Darter, *Percina aurantiaca.*

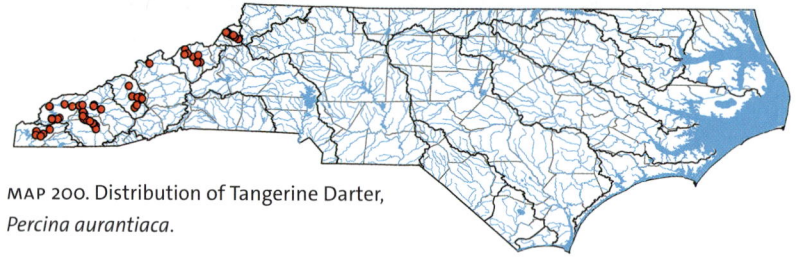

MAP 200. Distribution of Tangerine Darter, *Percina aurantiaca.*

9a. Snout pointed. Subocular bar absent. Nape, cheek, and opercle well scaled or with embedded scales..10

9b. Snout blunt. Subocular bar present. Nape, cheek, and opercle often naked or weakly scaled (except in Sooty-banded Darter, *Percina westfalli*, whose cheek and opercle are well scaled)......................................11

10a. Breast well scaled, at least laterally and posteriorly, scales small and embedded, visible without scraping. Basicaudal spot usually dark (figure 263). Range restricted to Hiwassee, Little Tennessee, Pigeon, French Broad, and Nolichucky basins (map 201)...................................
................Olive Darter, *Percina squamata* (Gilbert and Swain, 1887)

10b. Breast naked or with a few scattered embedded scales, not usually visible without scraping. Basicaudal spot usually indistinct (figure 263). Range restricted to New basin (map 202)...
.........Sharpnose Darter, *Percina oxyrhynchus* (Hubbs and Raney, 1939)

FIGURE 263. A—Olive Darter, *Percina squamata*; B—Sharpnose Darter, *Percina oxyrhynchus*. West Virginia. Photographs courtesy of Luke Etchison (Olive Darter) and David A. Neely (Sharpnose Darter).

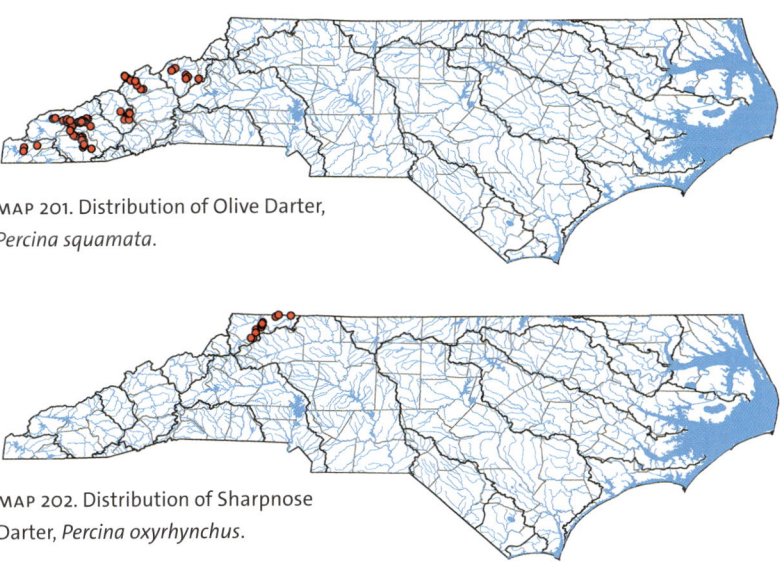

MAP 201. Distribution of Olive Darter, *Percina squamata*.

MAP 202. Distribution of Sharpnose Darter, *Percina oxyrhynchus*.

11a. Cheeks well scaled, scales often partially embedded. Breast with embedded scales. Gill membranes moderately joined, 70–80° angle. Range restricted to Savannah basin (figure 264) (map 203)........................
.....................Sooty-banded Darter, *Percina westfalli* (Fowler, 1942)

11b. Cheeks naked, occasionally weakly scaled. Breast naked except for a single enlarged scale or small central patch of scales. Gill membranes narrowly joined, 40–50° angle...12

FIGURE 264. Sooty-banded Darter, *Percina westfalli*. Photograph courtesy of Luke Etchison.

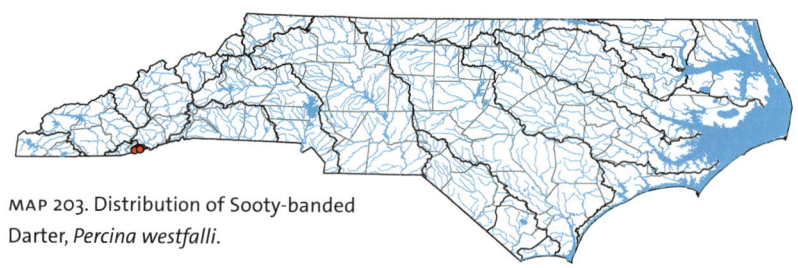

MAP 203. Distribution of Sooty-banded Darter, *Percina westfalli*.

12a. Range restricted to Hiwassee, Little Tennessee, Pigeon, French Broad, Nolichucky, and New basins...13

12b. Range restricted to Atlantic slope drainages............................14

13a. Nape sparsely scaled (figure 265). Range restricted to New basin (map 204)
................Appalachia Darter, *Percina gymnocephala* Beckham, 1980

13b. Nape usually well scaled (figure 265). Range restricted to Hiwassee, Little Tennessee, Pigeon, French Broad, and Nolichucky basins (map 205)......
.................Gilt Darter, *Percina evides* (Jordan and Copeland, 1877)

FIGURE 265. A—Appalachia Darter, *Percina gymnocephala*; B—Gilt Darter, *Percina evides*.

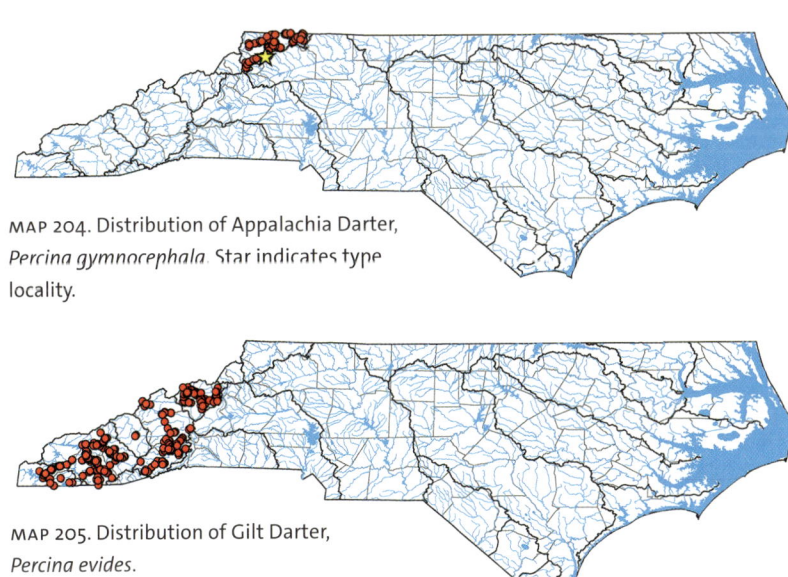

MAP 204. Distribution of Appalachia Darter, *Percina gymnocephala*. Star indicates type locality.

MAP 205. Distribution of Gilt Darter, *Percina evides*.

14a. Spinous dorsal fin with (12) 13 (15) rays. Lateral line scales (52) 54–61 (64). Dark, wavy line often present above lateral stripe (figure 266) (map 206) ....Chainback Darter, *Percina nevisensis* (Cope, 1870)

14b. Spinous dorsal fin with (10) 11 or 12 (13) rays. Lateral line scales (40) 42–52 (54). Dark, wavy line absent above lateral stripe..........................15

FIGURE 266. Chainback Darter, *Percina nevisensis*.

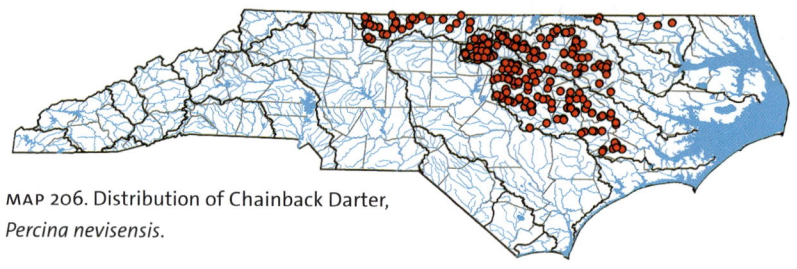

MAP 206. Distribution of Chainback Darter, *Percina nevisensis*.

15a. Chin bar black, often mottled in Piedmont forms. Spinous dorsal fin with narrow yellow band bordered above with wide black band (figure 267). Soft dorsal fin rays (11) 12 (13) (map 207).....................................
.............Piedmont Darter, *Percina crassa* (Jordan and Brayton, 1878)

15b. Chin bar absent. Spinous dorsal fin with wide orange band bordered above by narrow black band (figure 267). Soft dorsal fin rays 10 or 11 (12). Range restricted to Neuse, Tar, and Roanoke basins (map 208)...........
.............Roanoke Darter, *Percina roanoka* (Jordan and Jenkins, 1889)

FIGURE 267. A—Piedmont Darter, *Percina crassa*; B—Roanoke Darter, *Percina roanoka*.

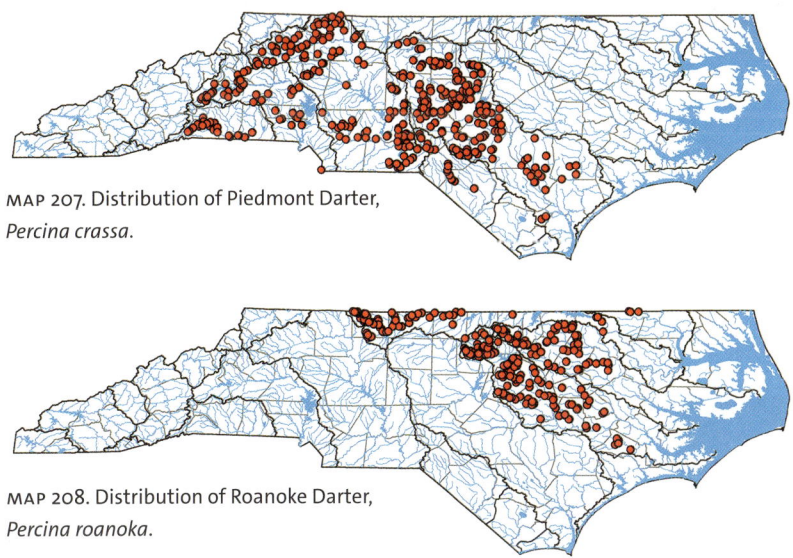

MAP 207. Distribution of Piedmont Darter, *Percina crassa*.

MAP 208. Distribution of Roanoke Darter, *Percina roanoka*.

16a. Lateral line distinctly arched anteriorly, incomplete (figure 268).......17

16b. Lateral line straight or slightly curved upward, complete or incomplete (figure 268)..19

FIGURE 268. Arrows indicating curvature of lateral line. A—distinctly arched in Sawcheek Darter, *Etheostoma serrifer*; B—slightly curved upward in Pinewoods Darter, *Etheostoma mariae*.

17a. Caudal fin base with two black basicaudal spots, often ringed with orange (figure 269). Preopercle margin serrate (map 209)........................
........Sawcheek Darter, *Etheostoma serrifer* (Hubbs and Cannon, 1935)

17b. Caudal fin base without two black basicaudal spots; usually with three vertical faint dark spots. Preopercle margin entirely smooth or partially serrate...18

FIGURE 269. Sawcheek Darter, *Etheostoma serrifer*.

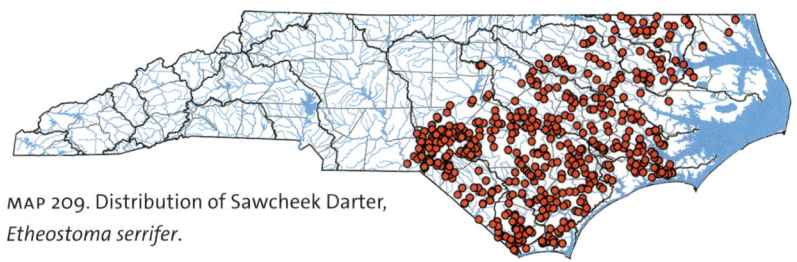

MAP 209. Distribution of Sawcheek Darter, *Etheostoma serrifer*.

18a. Anal fin spines 2 (figure 270). Breast fully scaled. Head narrow and snout pointy when viewed from above. Interorbital scales present (figure 271) (map 210)............Swamp Darter, *Etheostoma fusiforme* (Girard, 1854)

18b. Anal fin spines 1, sometimes 2 (figure 270). Breast naked to 30 percent scaled. Head broad and snout rounded when viewed from above. Interorbital scales absent (figure 271) (map 211)....................................
............Carolina Darter, *Etheostoma collis* (Hubbs and Cannon, 1935)

FIGURE 270. A—Swamp Darter, *Etheostoma fusiforme*; B—Carolina Darter, *Etheostoma collis*.

A
B

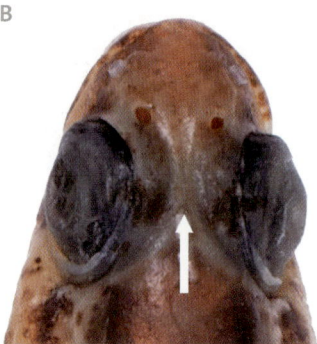

FIGURE 271.
Dorsal profiles of
heads and inter-
orbital scalation.
A—Swamp Darter,
*Etheostoma fusi-
forme*; B—Carolina
Darter, *Etheostoma
collis*.

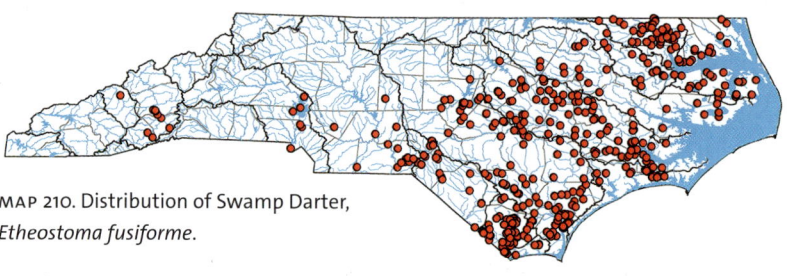

MAP 210. Distribution of Swamp Darter,
Etheostoma fusiforme.

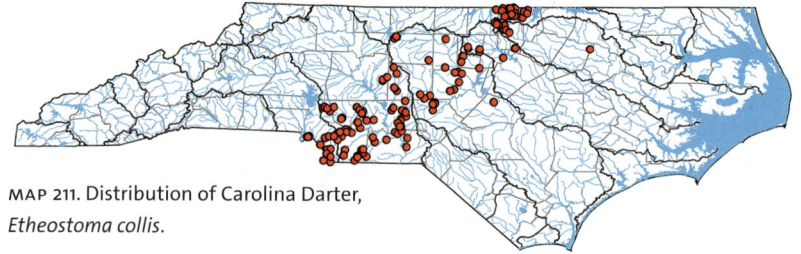

MAP 211. Distribution of Carolina Darter,
Etheostoma collis.

19a. Anus surrounded by many fleshy, conical villi (figure 272). Body yellow to
beige, translucent in life, peppered with black or brown spots (figure 272)
(map 212)...............Glassy Darter, *Etheostoma vitreum* (Cope, 1870)

19b. Anus not surrounded by many conical fleshy villi. Body color not as above
....20

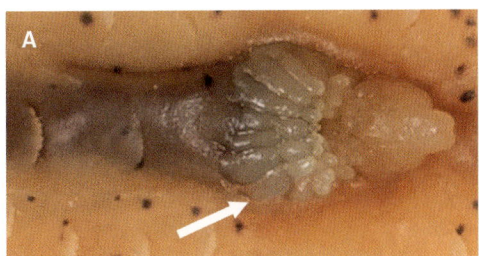

FIGURE 272. Glassy Darter,
Etheostoma vitreum.
A—arrow indicating villi;
B—full body.

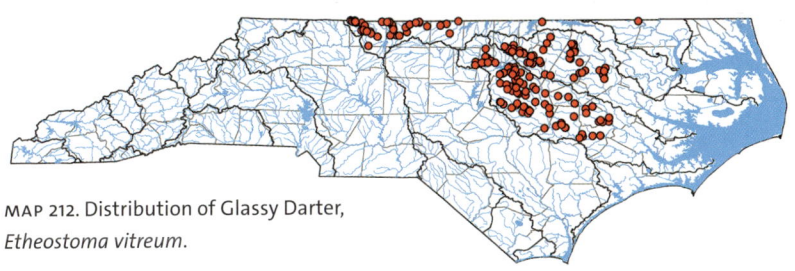

MAP 212. Distribution of Glassy Darter,
Etheostoma vitreum.

20a. Lateral line incomplete with seven to thirty-three unpored scales. Caudal
fin rounded. Black line extending from snout through eye onto opercle
....21

20b. Lateral line usually complete or if incomplete with one to four unpored
scales. Caudal fin seldom rounded. Black line usually absent from snout
to edge of opercle..22

21a. Range restricted to Broad, Catawba, and upper Yadkin basins (map 213).
Seven or fewer dark bars on side. Lateral line scales 39–57, usually 42–50
(figure 273)..
..........Carolina Fantail Darter, *Etheostoma brevispinum* (Coker, 1926)

21b. Range not restricted as above (map 214). Eight or more dark bars on side.
Lateral line scales 38–60, usually 45–55 (figure 273).......................
......................Fantail Darter, *Etheostoma flabellare* Rafinesque, 1819

FIGURE 273. A—Carolina Fantail Darter, *Etheostoma brevispinum*; B—Fantail Darter, *Etheostoma flabellare*.

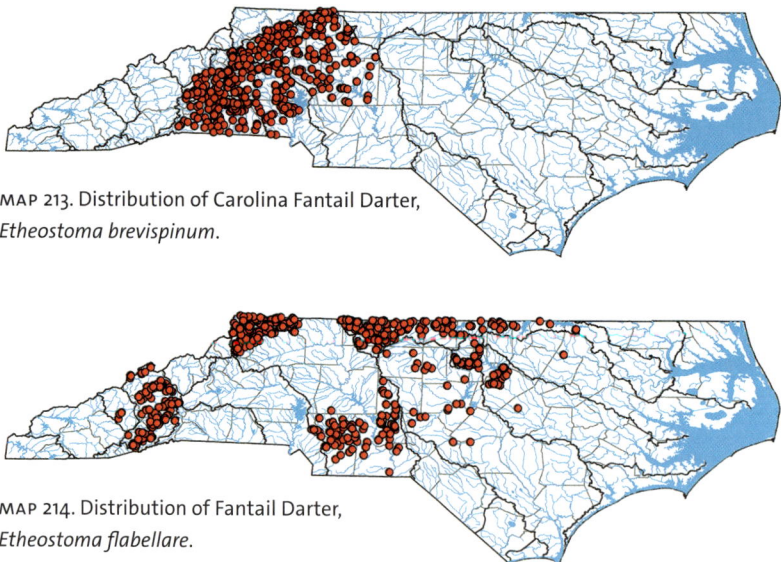

MAP 213. Distribution of Carolina Fantail Darter, *Etheostoma brevispinum*.

MAP 214. Distribution of Fantail Darter, *Etheostoma flabellare*.

22a. First anal spine stiff, as thick as or thicker than second spine. Anterior belly usually well scaled...23

22b. First anal spine flexible, as thin as second spine (except in "Tessellated" Darter, *Etheostoma* sp., which only has one anal spine). Anterior belly scalation variable, naked in the Southern Tessellated Darter, *E. maculaticeps*, and Riverweed Darter, *E. podostemone*; and scaled in the Waccamaw Darter, *E. perlongum*, and "Tessellated" Darter, *E.* sp...................35

23a. Lateral line scales 35–38 (figure 274). Range restricted to headwaters of the Lumber basin (map 215)..
...................Pinewoods Darter, *Etheostoma mariae* (Fowler, 1947)

23b. Lateral line scales 39 or more...24

FIGURE 274. Pinewoods Darter, *Etheostoma mariae*.

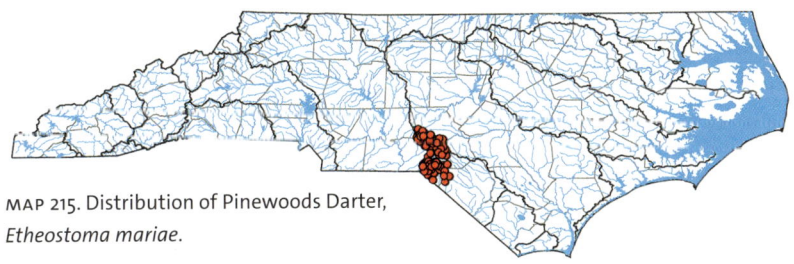

MAP 215. Distribution of Pinewoods Darter, *Etheostoma mariae*.

24a. Sides with longitudinal streaks along each scale row. Nape naked.........
....*Nothonotus*, Couplet 25

24b. Sides with no longitudinal streaks along each scale row. Nape usually scaled; scales may be embedded..28

25a. Snout short and blunt to moderately pointed. Opercle scaled. Soft dorsal, caudal, and anal fins usually edged with black, often with light reddish submarginal band, often spotted in females. Sides without complete bars. Red spots may be present on sides and caudal and soft dorsal fins......26

25b. Snout very long and pointed (figure 275). Opercle naked. Soft dorsal, caudal, and anal fins not edged with black, light submarginal band absent. Fins not spotted with black. Eleven to fifteen narrow black bars cross entire sides. No red spots on body or fins. Range restricted to Nolichucky basin (map 216)...
......................Sharphead Darter, *Nothonotus acuticeps* (Bailey, 1959)

FIGURE 275. Sharphead Darter, *Nothonotus acuticeps*.

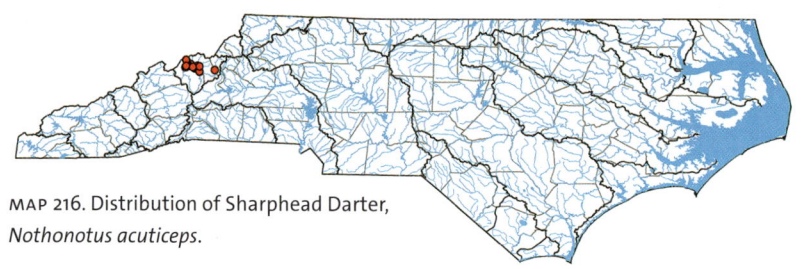

MAP 216. Distribution of Sharphead Darter, *Nothonotus acuticeps*.

26a. Cheek with three to seven dark dashes or spots, one or two of these spots in the subocular area. Basicaudal region with cream-colored blotches shaped like an hourglass (figure 276) (map 217)...........................
........................Redline Darter, *Nothonotus rufilineatus* (Cope, 1870)

26b. Cheek with no or one dark dash or spot, subocular bar present. Basicaudal region without cream-colored blotches shaped like an hourglass.......27

FIGURE 276. Redline Darter, *Nothonotus rufilineatus*. A—male; B—female.

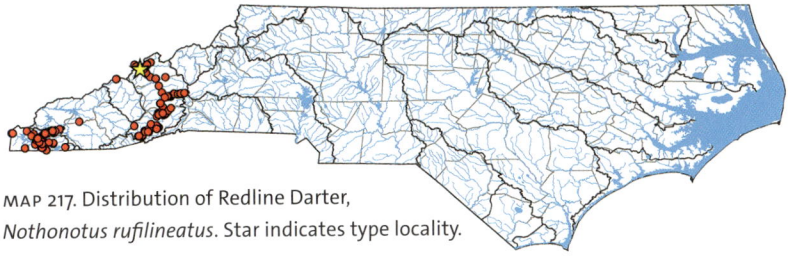

MAP 217. Distribution of Redline Darter, *Nothonotus rufilineatus*. Star indicates type locality.

27a. Snout pointed dorsally and laterally; upper lip protrudes well beyond snout. Lateral line scales (51) 54–62 (66). Adults with no light submarginal band on anal, caudal, and soft dorsal fins, margin black without thin light edge. Adults with irregularly placed red spots in center of scales on sides, red spot at tip of first interradial membrane of spinous dorsal fin, and red blotches on caudal fin base (figure 277) (map 218)................ ....Wounded Darter, *Nothonotus vulneratus* (Cope, 1870)

27b. Snout rounded; upper lip protrudes slightly beyond snout. Lateral line scales (61) 63–70 (72). Adults with light submarginal band on anal, caudal, and soft dorsal fins, margin black usually with thin light edge. Adults with no red spots on sides, with no red on spinous dorsal fin or on caudal fin bases (figure 277) (map 219)... ....Greenfin Darter, *Nothonotus chlorobranchius* (Zorach, 1972)

FIGURE 277. A—Wounded Darter, *Nothonotus vulneratus*; B—Greenfin Darter, *Nothonotus chlorobranchius*.

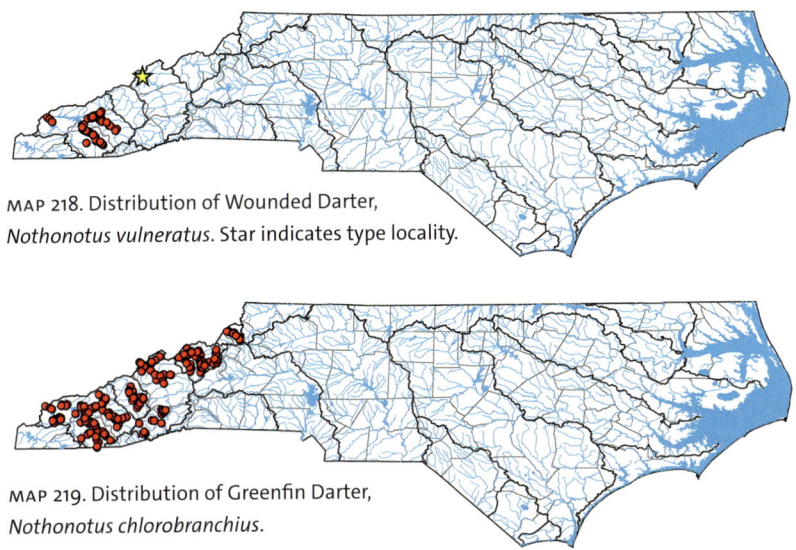

MAP 218. Distribution of Wounded Darter,
Nothonotus vulneratus. Star indicates type locality.

MAP 219. Distribution of Greenfin Darter,
Nothonotus chlorobranchius.

28a. Frenum recessed and greatly reduced (figure 278) (except in Tuckasegee Darter, *Etheostoma gutselli*). Upper jaw fits into deep groove under snout, groove as deep as width of upper jaw.....................................29

28b. Frenum broad, not recessed (figure 278) (except in Snubnose Darter, *Etheostoma simoterum*, which has a short snout and the frenum is narrow, sometimes obscured by a crease). Maxillary groove shallow............30

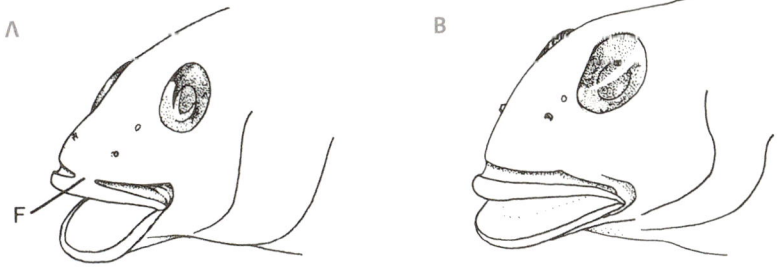

FIGURE 278. A—Frenum (*F*) broad; B—frenum recessed and greatly reduced. Illustration adapted from Jenkins and Burkhead (1994).

29a. Long nipple-like formation on upper lip (figure 279), occasionally with an overhung snout. Frenum absent. Belly completely scaled; opercle scaled (figure 280). Range restricted to Savannah, lower Pigeon, French Broad, Nolichucky, and New basins (map 220)......................................
...............Greenside Darter, *Etheostoma blennioides* Rafinesque, 1819

29b. Nipple-like formation on the upper lip and overhung snout absent (figure 279). Frenum well developed. Belly naked anteriorly in most specimens; opercle usually naked (figure 280). Range restricted to Little Tennessee and Pigeon basins (map 221)...
...............Tuckasegee Darter, *Etheostoma gutselli* (Hildebrand, 1932)

29c. Range restricted to Hiwassee basin (figure 280) (Map 220)...............
.......Intergrades between the Greenside Darter and Tuckasegee Darter, *Etheostoma* sp. *cf. blennioides*

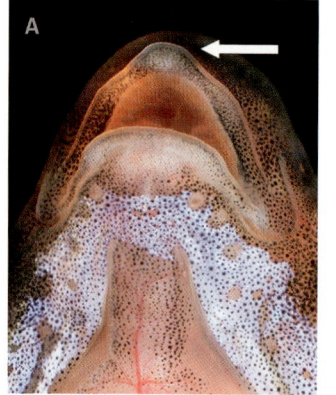 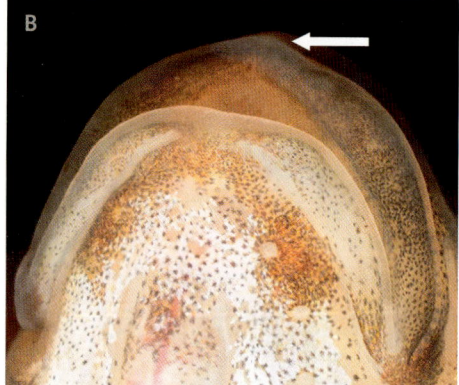

FIGURE 279. Arrows indicating nipple-like formation on the upper lip. A—present in Greenside Darter, *Etheostoma blennioides*; B—absent in Tuckasegee Darter, *Etheostoma gutselli*.

FIGURE 280. Three "Greenside" Darters. A—Greenside Darter, *Etheostoma blennioides*; B—Tuckasegee Darter, *Etheostoma gutselli*; C—intergrade between Greenside Darter and Tuckasegee Darter.

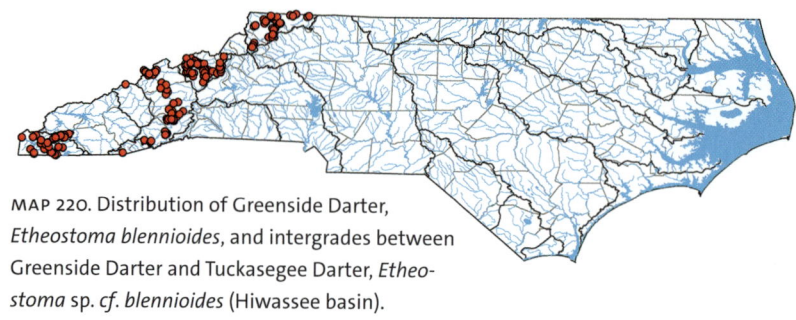

MAP 220. Distribution of Greenside Darter, *Etheostoma blennioides*, and intergrades between Greenside Darter and Tuckasegee Darter, *Etheostoma* sp. cf. *blennioides* (Hiwassee basin).

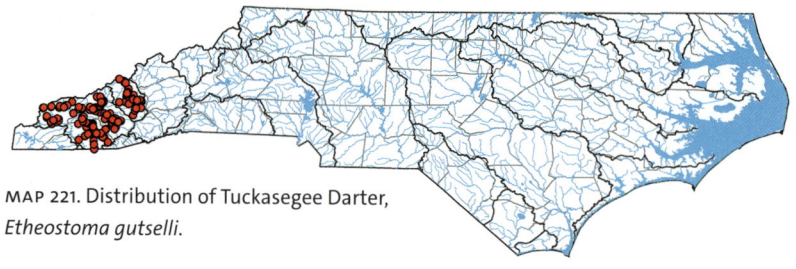

MAP 221. Distribution of Tuckasegee Darter, *Etheostoma gutselli*.

30a. Snout very blunt (figure 281). Frenum narrow, sometimes obscured by a crease. Opercle and anterior belly well scaled, cheek at least partially scaled. Range restricted to lower French Broad and Nolichucky basins (map 222)........Snubnose Darter, *Etheostoma simoterum* (Cope, 1868)

30b. Snout not blunt. Frenum broad. Opercle and cheek naked (except in Banded Darter, *Etheostoma zonale*).....................................31

FIGURE 281. Snubnose Darter, *Etheostoma simoterum*. Tennessee. Photograph courtesy of David A. Neely.

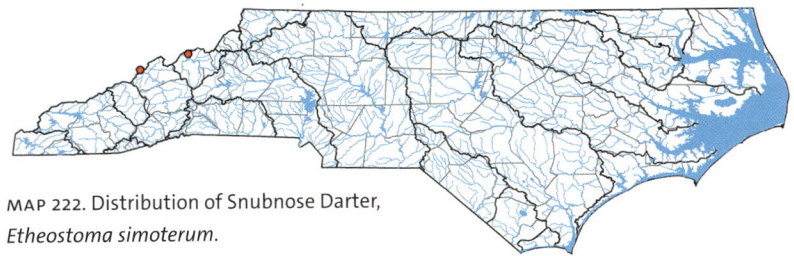

MAP 222. Distribution of Snubnose Darter, *Etheostoma simoterum*.

31a. Sides with narrow greenish bars with posterior bars encircling the caudal peduncle (figure 282). Opercle and cheeks at least partially scaled (map 223) ....Banded Darter, *Etheostoma zonale* (Cope, 1868)

31b. Posterior bars not usually encircling caudal peduncle. Opercle and cheeks naked..32

FIGURE 282. Banded Darter, *Etheostoma zonale*. A—male; B—female.

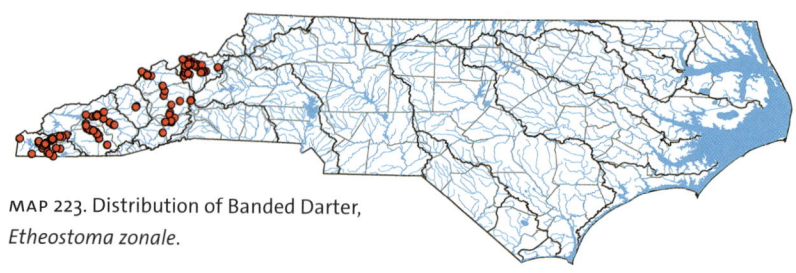

MAP 223. Distribution of Banded Darter,
Etheostoma zonale.

32a. Nape usually well scaled. Spinous dorsal fin 12–14 (figure 283). Range restricted to New basin (map 224)...
.....................Kanawha Darter, *Etheostoma kanawhae* (Raney, 1941)

32b. Nape with embedded scales near head. Spinous dorsal (9) 10–13. Not
found in the New basin...33

FIGURE 283. Kanawha Darter, *Etheostoma kanawhae.* A—male; B—female.

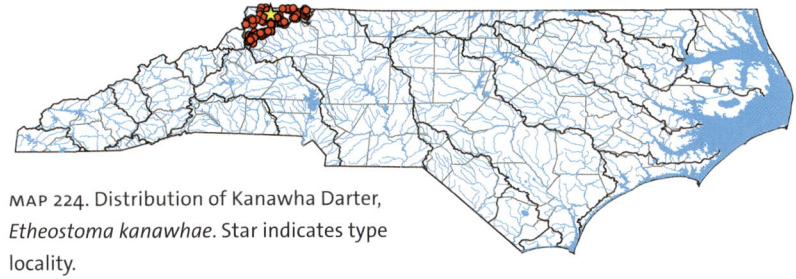

MAP 224. Distribution of Kanawha Darter, *Etheostoma kanawhae*. Star indicates type locality.

33a. Range restricted to Pigeon, French Broad, and Nolichucky basins (map 225). Dorsal spines usually 12 (figure 284)............Swannanoa Darter, *Etheostoma swannanoa* Jordan and Evermann, 1889

33b. Range restricted to Savannah, Broad, or Catawba basins. Dorsal spines usually 9 or 10..34

FIGURE 284. Swannanoa Darter, *Etheostoma swannanoa*. A—male; B—female.

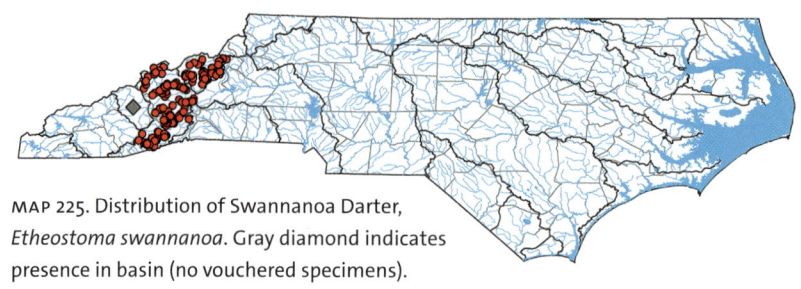

MAP 225. Distribution of Swannanoa Darter, *Etheostoma swannanoa*. Gray diamond indicates presence in basin (no vouchered specimens).

34a. Range restricted to Broad and Catawba basins (map 226). Dark dorsal saddles (6) 7; posterior three or four blotches forming vertical bars (figure 285)..

....Seagreen Darter, *Etheostoma thalassinum* (Jordan and Brayton, 1878)

34b. Range restricted to Savannah basin (map 227). Dark dorsal saddles 5 (6); posterior three or four blotches not forming vertical bars (figure 285).....

....Turquoise Darter, *Etheostoma inscriptum* (Jordan and Brayton, 1878)

FIGURE 285. A—Seagreen Darter, *Etheostoma thalassinum*; B—Turquoise Darter, *Etheostoma inscriptum*.

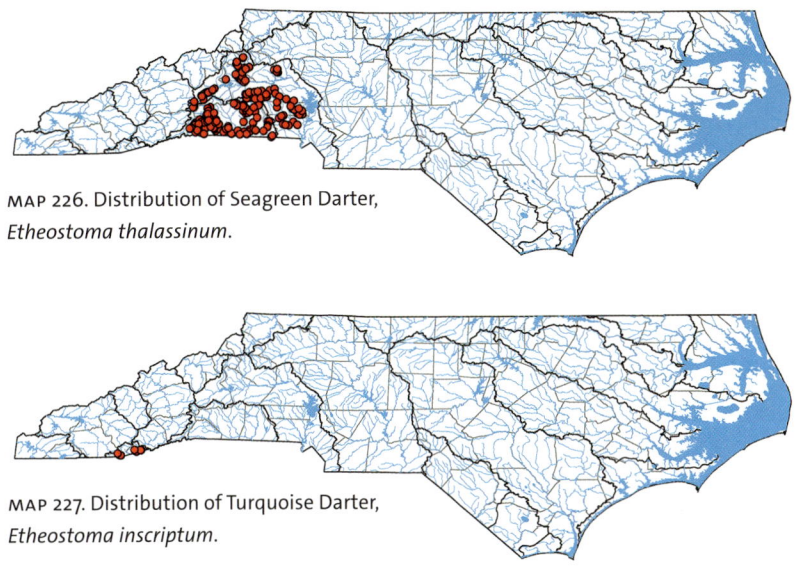

MAP 226. Distribution of Seagreen Darter, *Etheostoma thalassinum.*

MAP 227. Distribution of Turquoise Darter, *Etheostoma inscriptum.*

35a. Gill membranes broadly joined, 100–110° angle (figure 286). Caudal fin rounded (figure 287). Range restricted to Roanoke basin (map 228).......
........Riverweed Darter, *Etheostoma podostemone* Jordan and Jenkins, 1889

35b. Gill membranes narrowly joined, 55–75° angle (figure 286). Caudal fin emarginate to truncate...36

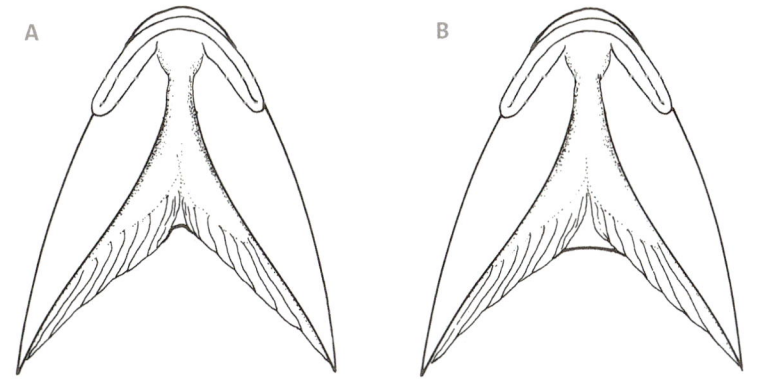

FIGURE 286. Gill membranes. A—narrowly joined; B—broadly joined. Illustration adapted from Jenkins and Burkhead (1994).

FIGURE 287. Riverweed Darter, *Etheostoma podostemone*.

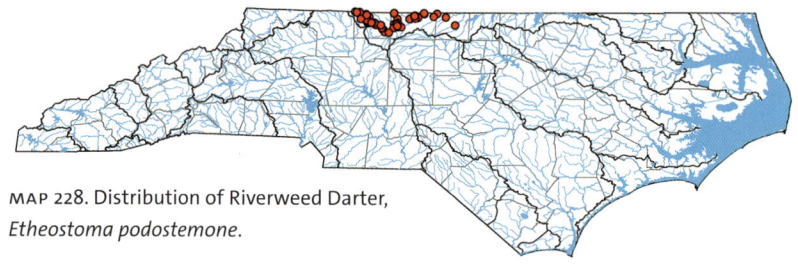

MAP 228. Distribution of Riverweed Darter,
Etheostoma podostemone.

36a. Lateral line scales (58) 60–62 (66). Anal rays (7) 8 or 9 (10) (figure 288).
Range restricted to Lake Waccamaw and Waccamaw River downstream
of the lake (Waccamaw basin) (map 229)....................................
......Waccamaw Darter, *Etheostoma perlongum* (Hubbs and Raney, 1946)
36b. Lateral line scales 39–54. Anal rays 7 or 8 (10)...........................37

FIGURE 288. Waccamaw Darter, *Etheostoma perlongum*.

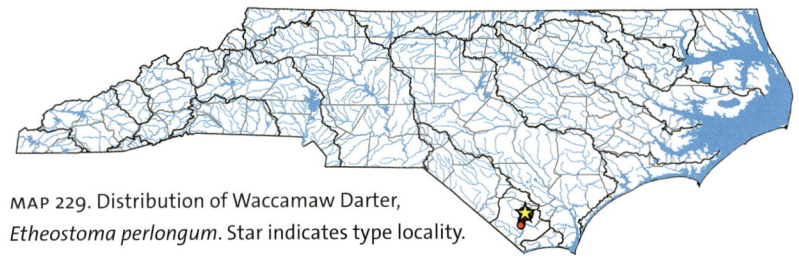

MAP 229. Distribution of Waccamaw Darter, *Etheostoma perlongum*. Star indicates type locality.

37a. One anal spine. Infraorbital canal incomplete (figures 289 and 290) (map 230)...................."Tessellated" Darter, *Etheostoma* sp. *cf. olmstedi*
37b. Two anal spines. Infraorbital canal normally complete (figure 289), often incomplete in specimens from Cape Fear basin (figure 290) (map 231)....
......Southern Tessellated Darter, *Etheostoma maculaticeps* (Cope, 1870)

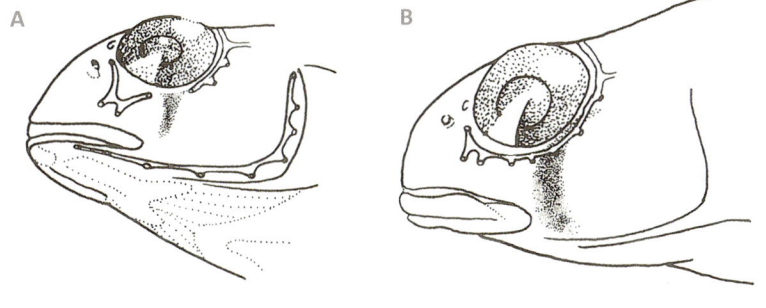

FIGURE 289 Infraorbital canals. A—incomplete in "Tessellated" Darter, *Etheostoma* sp.; B—complete in Southern Tessellated Darter, *Etheostoma maculaticeps*. Illustration adapted from Jenkins and Burkhead (1994).

FIGURE 290. A—"Tessellated" Darter, *Etheostoma* sp.; B—Southern Tessellated Darter, *Etheostoma maculaticeps*.

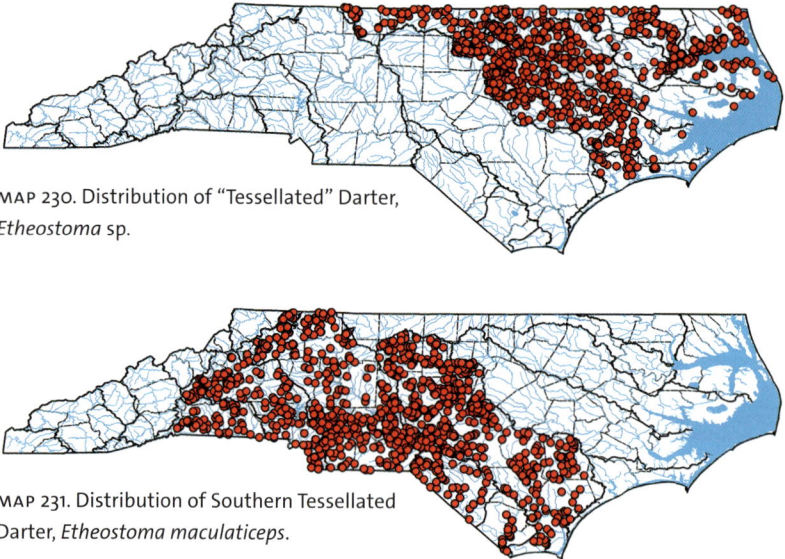

MAP 230. Distribution of "Tessellated" Darter, *Etheostoma* sp.

MAP 231. Distribution of Southern Tessellated Darter, *Etheostoma maculaticeps*.

23

Sculpins
Family Cottidae

Sculpins are unusual-looking small brown fish with big eyes atop a large, wide head; a large mouth; what appears to be a scaleless body; and big pointy fins. They often go unnoticed because they are cryptically colored to blend in with coarse substrates (that is, large gravel, cobble, and boulder) under which they often hide. North Carolina is home to three species of sculpins (tables 2 and 26). Colloquial names for sculpins include miller's thumb (in reference to the shape of a miller's thumb), molly crawlbottom, muddler, and wartfish. Sculpins range in size from ca. 95 mm TL for the Blue Ridge Sculpin to ca. 150 mm TL for the Mottled Sculpin and ca. 180 mm TL for the Banded Sculpin (Page and Burr, 2011).

Sculpins may be found in riffles and runs in fast-flowing, clear, cool-to-cold-water trout-type streams in our Mountain river basins. Historically, no species were found east of the Appalachian Mountains, except for the Blue Ridge Sculpin in the upper Roanoke basin and the Mottled Sculpin in the upper Savannah basin (table 2). The Mottled Sculpin, our most widely distributed species, is indigenous (native) in all river basins west of the Mountains, including the New and the Savannah. However, it is nonindigenous in the upper Broad basin in Polk, Henderson, and Buncombe Counties. The Blue Ridge Sculpin is the only species of sculpin found in the Roanoke basin and is our most restricted species. It is found only in the Little Dan River and in the mainstem of the Dan River upstream from the Little Dan River to the North Carolina–Virginia state line in Stokes County (Tracy, Rohde, and Hogue, 2020). The Banded Sculpin also has a limited distribution. Currently, it is found only in the Pigeon River downstream from Walters Lake into the gorge and in Big Creek (Haywood County, Pigeon basin) and the lower French Broad River and

TABLE 26. Species of sculpins in North Carolina

Scientific Name	Common Name	Imperilment Status
Cottus bairdii	Mottled Sculpin	
Cottus caeruleomentum[a]	Blue Ridge Sculpin	state special concern
Cottus carolinae	Banded Sculpin	state special concern

[a] Species found in only one river basin.

Shut-in Creek near Hot Springs, and in the French Broad River at Marshall (Madison County, French Broad basin) (Tracy, Rohde, and Hogue, 2020). The Banded Sculpin and Mottled Sculpin do not appear to be sympatric in North Carolina. In 2022, the Banded Sculpin was introduced by the North Carolina Wildlife Resources Commission into the Cheoah River, Graham County, Little Tennessee basin, to serve as an additional host for the glochidia of imperiled mussels, which are also being reintroduced into the river. The Mottled Sculpin is not known to occur naturally in the Cheoah River (Luke Etchison, NCWRC, personal communication).

The Banded Sculpin and Blue Ridge Sculpin are listed as state Special Concern species because of their limited distributions and the anthropogenic impacts upon their habitats and stream water quality (table 26; NCAC, 2023; NCNHP, 2022; NCWRC, 2021).

At the present time, sculpin identification is relatively straightforward. However, the taxonomic status of the Mottled Sculpin in North Carolina has been unsettled for a long time (see, for example, Jenkins and Burkhead, 1994), and some researchers and museums record the species as *Cottus bairdii* complex or *Cottus* sp. (Gabriela M. Hogue, North Carolina Museum of Natural Sciences, personal communication). Until sculpin taxonomy is resolved, any sculpin, except for the Banded Sculpin, found in our western Mountain streams is being called the Mottled Sculpin (Tracy, Rohde, and Hogue, 2020).

Key characteristics for the proper identification of sculpins include the pigmentation pattern of the caudal band and lateral bars; the number of pectoral fin rays; the number of preopercular spines; the size and shape of the uppermost preopercular spine; and the size of the palatine tooth patch in relation to the size of the vomerine tooth patch. Preopercular spines are best observed by examining the preserved specimen with the aid of a dissecting microscope and cutting away the overlying tissue. Palatine and vomerine teeth are also best observed on the preserved specimen with the aid of a dissecting microscope.

Identification Key to the Species of Sculpins (Family Cottidae)

1a. Range restricted to upper Dan River system in Stokes County (Roanoke basin) (map 232). Pectoral fin rays modally 14. Caudal base band at mid-height unnotched on at least one side (figures 291 and 292)...............
........................Blue Ridge Sculpin, *Cottus caeruleomentum* (Kinziger, Raesly, and Neely, 2000)

1b. Range restricted to Hiwassee, Little Tennessee, Savannah, Pigeon, French Broad, Nolichucky, Watauga, New, and Broad basins. Pectoral fin rays modally 15 or 16. Caudal base band at mid-height notched on both sides (figure 291)..2

FIGURE 291. Arrows indicating caudal base band. A—unnotched in Blue Ridge Sculpin, *Cottus caeruleomentum*; B—notched in Mottled Sculpin, *Cottus bairdii*.

FIGURE 292. Blue Ridge Sculpin, *Cottus caeruleomentum*.

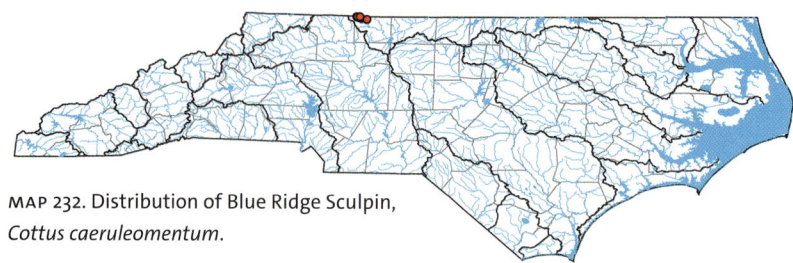

MAP 232. Distribution of Blue Ridge Sculpin, *Cottus caeruleomentum*.

2a. Preopercular spines 1 or 2, upper spine smaller than pupil width. Pectoral fin rays modally 15. Lateral bars usually diffuse, not darker on edges than elsewhere. Palatine tooth patch shorter than vomerine tooth patch (figure 293) (map 233) ....Mottled Sculpin, *Cottus bairdii* (Girard, 1850)

2b. Preopercular spines 3, upper spine usually longer than pupil width. Pectoral fin rays modally 16 or 17. Lateral bars distinct, darker on edges. Palatine tooth patch longer than vomerine tooth patch (figure 294) (map 234 ....Banded Sculpin, *Cottus carolinae* (Gill, 1861)

FIGURE 293. Mottled Sculpin, *Cottus bairdii*. A—Burningtown Creek, Macon County, Little Tennessee River basin; B—Obids Creek, Ashe County, New River basin.

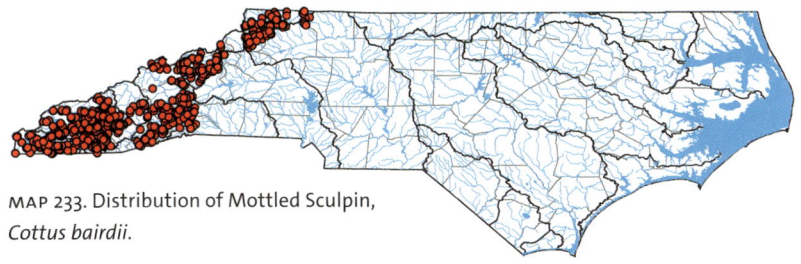

MAP 233. Distribution of Mottled Sculpin,
Cottus bairdii.

FIGURE 294. Banded Sculpin, *Cottus carolinae.*

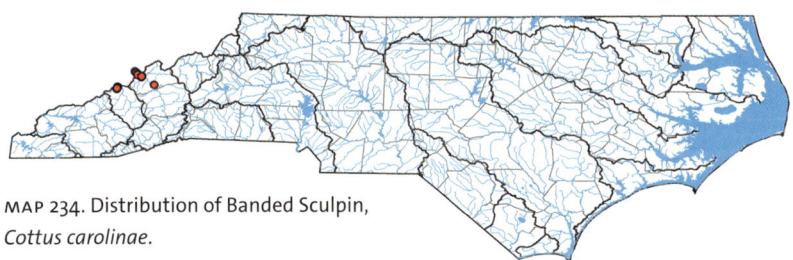

MAP 234. Distribution of Banded Sculpin,
Cottus carolinae.

Sunfishes

Family Centrarchidae

There are twenty-three species of sunfishes in North Carolina, including one undescribed species, "Bartram's" Bass, *Micropterus* sp. (tables 2 and 27). Colloquial names for the species in this family include bream, bronzeback, goggle-eye, Kentucky bass, perch, robin, sac-a-lait, shellcracker, stumpknocker, tin mouth, welshman, and many others. Sunfishes range in size from the petite *Enneacanthus* species, measuring only ca. 100 mm TL, to the black basses, *Micropterus* spp., which approach 762 mm TL.

Sunfishes are found throughout our state (table 2) in reservoirs, creeks, large and small rivers, swamps, channelized streams, and permanent wetlands. They can be found in cool, clear Mountain streams, in warm and turbid Piedmont streams, and in low-pH, tannin-colored Sand Hills and Coastal Plain streams. Sunfishes are generally found in deep pools, in slow-moving runs and snags, or beneath undercut banks. In habitats where aquatic vegetation is plentiful, they tend to blend in with their surroundings. Certain species thrive even in some of our most degraded urban streams.

Historically, nineteen species have called North Carolina home with the Cape Fear, Shallotte, Waccamaw, and Lumber basins having the most diverse faunas, each having thirteen of the nineteen indigenous (native) species. Some of our indigenous species were historically restricted to very specific river basins or physiographic regions (table 2). For example, the Roanoke Bass was endemic to the Neuse, Tar, and Roanoke basins, and the "Bartram's" Bass was restricted to the Savannah basin. Four species were found only west of the Appalachian Mountains—the Smallmouth Bass, Spotted Bass, Rock Bass, and White Crappie; other species were found only east of these mountains—the Redbreast Sunfish, Pumpkinseed, and Warmouth. The Bluegill was not found north of the White Oak basin; it was not indigenous to the New, Neuse, Tar, Roanoke,

TABLE 27. Species of sunfishes in North Carolina

Scientific Name	Common Name	Imperilment Status
Acantharchus pomotis	Mud Sunfish	
Ambloplites cavifrons	Roanoke Bass	state significantly rare
Ambloplites rupestris	Rock Bass	
Centrarchus macropterus	Flier	
Enneacanthus chaetodon	Blackbanded Sunfish	state significantly rare
Enneacanthus gloriosus	Bluespotted Sunfish	
Enneacanthus obesus	Banded Sunfish	state significantly rare
Lepomis auritus	Redbreast Sunfish	
Lepomis cyanellus	Green Sunfish	
Lepomis gibbosus	Pumpkinseed	
Lepomis gulosus	Warmouth	
Lepomis macrochirus	Bluegill	
Lepomis marginatus	Dollar Sunfish	
Lepomis microlophus	Redear Sunfish	
Lepomis punctatus	Spotted Sunfish	
Micropterus coosae[a]	Redeye Bass	
Micropterus dolomieu	Smallmouth Bass	
Micropterus henshalli	Alabama Bass	
Micropterus punctulatus	Spotted Bass	
Micropterus salmoides	Largemouth Bass	
Micropterus sp. "Bartram's" Bass		state significantly rare (noted as Redeye Bass)
Pomoxis annularis	White Crappie	
Pomoxis nigromaculatus	Black Crappie	

Note: All species are managed by the NCWRC.
[a] Species found in only one river basin.

Chowan, or Albemarle Sound basins. Even today, the Banded Sunfish, Blackbanded Sunfish, Bluespotted Sunfish, Dollar Sunfish, Flier, Mud Sunfish, and Spotted Sunfish are found almost exclusively in the Coastal Plain and Sand Hills. Historically, it is believed that the Largemouth Bass was found in every basin, except for the New. However, today, no indigenous species is endemic to a specific river basin in North Carolina (table 2).

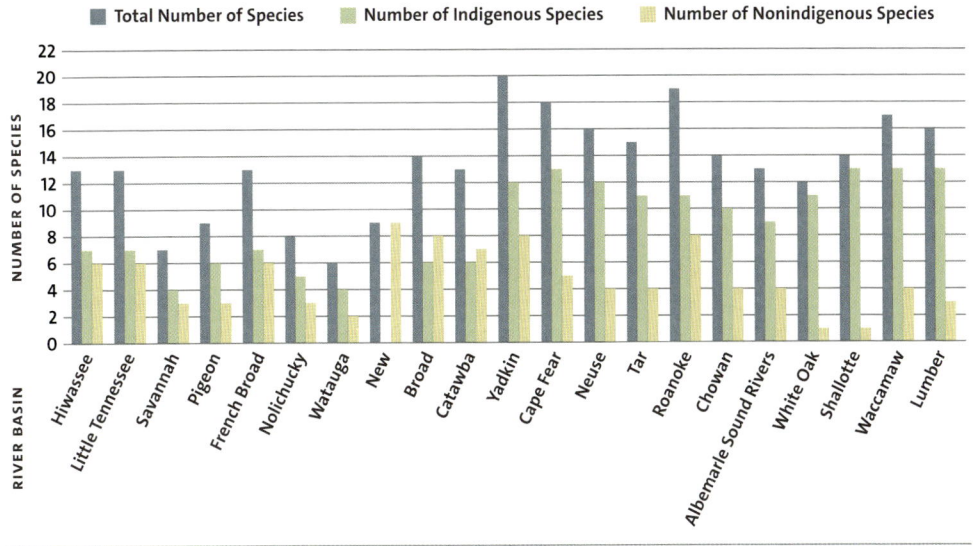

GRAPH 5. Diversity of sunfishes across North Carolina's river basins

Each of North Carolina's 100 counties has at least one species of sunfishes and black basses found within its borders, and there are three species—the Red-breast Sunfish, Bluegill, and Largemouth Bass—that can be found in each of North Carolina's twenty-one river basins (table 2; Tracy, Rohde, and Hogue, 2020). Our most diverse basin is the Yadkin, where twenty species are found, but eight are nonindigenous (graph 5). Our least diverse basin is the small head-water basin of the Watauga, where just six species are found, with two of them, the Redbreast Sunfish and Green Sunfish, being nonindigenous. Despite wide-spread introductions and stockings, the White Oak and Shallotte basins continue to have more than 90 percent of their original faunas intact (graph 5). Due to their small size and unimportance as sport fish, seven species have never been introduced into other river basins—the Banded Sunfish, Blackbanded Sunfish, Bluespotted Sunfish, Dollar Sunfish, Flier, Mud Sunfish, and Spotted Sunfish (table 2).

Because of the popularity of several species of sport fish, they have been introduced, legally or illegally, outside of their historical North Carolina ranges or have been imported into our state. In the New basin no species of sunfishes or black basses was historically found, but today nine introduced species are found there—the Black Crappie, Bluegill, Green Sunfish, Largemouth Bass, Pumpkinseed, Redbreast Sunfish, Redear Sunfish, Rock Bass, and Smallmouth

Bass (table 2). The Alabama Bass, Green Sunfish, Redear Sunfish, and Redeye Bass were not indigenous to North Carolina but are now part of our fauna. The Alabama Bass is North Carolina's most recent unauthorized introduced species and is now found in several of our Mountain and Piedmont reservoirs and large rivers (Goodfred, Hodges, and Loftis, 2023; Sammons et al., 2023). One of the deleterious consequences of these widespread introductions is that several of the genera, for example, *Lepomis*, *Ambloplites*, and *Micropterus*, readily hybridize within their genus. Hybrids of the black basses (Spotted, Largemouth, Smallmouth, and Alabama) are now common where the species co-occur. Hybrids of the endemic Roanoke Bass and introduced Rock Bass can now be found in the upper Roanoke basin and have displaced and genetically swamped the endemic Roanoke Bass.

All of the species in the family Centrarchidae are classified and managed as game species by the North Carolina Wildlife Resources Commission (NCWRC, 2022) and therefore are not considered imperiled in North Carolina (table 27; NCAC, 2023; NCNHP, 2022; NCWRC, 2021). However, the North Carolina Natural Heritage Program has listed four species as Significantly Rare (table 27; NCNHP, 2022).

Key characteristics for the proper identification of sunfishes include the shape of the caudal and pectoral fins; the pattern of blotches along the lateral line; counts of anal and dorsal fin spines and lateral line scales; size of mouth; and overall color patterns. Hybrids and juveniles of certain species can be difficult to identify.

Note on *Micropterus* spp.—Black Basses

As mentioned previously, black basses are a group of well sought-after game fishes, which include five described and one undescribed species in North Carolina (table 27). They have been widely and legally stocked by resource agencies and, in many instances, introduced illegally by fishermen. As a result, the true historical distribution of this group is challenging to unravel. Compounding the problem is that, genetically, many of these fish are appearing to be hybrids and are rapidly dispersing into new watersheds.

For several years, the NCWRC has been conducting a statewide investigation of the genetic composition of black bass populations (Kevin Dockendorf and Scott Loftis, NCWRC, personal communication). Results from the study have yet to be published.

Coinciding with the NCWRC study, Silliman et al. (2021) and Kim, Taylor, and Near (2022) provided insight into the black bass complex. Kim, Taylor, and Near (2022) determined that what was historically called the Largemouth Bass, *Micropterus salmoides* (Lacepède, 1802), was *M. nigricans* (Cuvier, 1828) and that the Florida Bass, *M. floridanus* (Lesueur, 1822), was *M. salmoides*; *M. nigricans* was elevated from synonymy for the Largemouth Bass. According to Kim, Taylor, and Near (2022), in North Carolina *M. salmoides* occurs along the southeastern Atlantic slope within the Coastal Plain north to the Cape Fear basin; hybrids of both species (*M. salmoides* × *M. nigricans*) occur in several basins throughout the Piedmont; and *M. nigricans* occurs west of the Appalachian Mountains.

Kassler et al. (2002) provided data for several meristic characters with which to separate the Largemouth Bass from Florida Bass. The most diagnostic of those characters was the number of lateral line scales:

- Largemouth Bass: 62.5 ± 2.8 (mean ± standard deviation), range = 55–66
- Florida Bass: 71.3 ± 3.9, range = 66–76

Until more widespread sampling and studies on their distribution and genetic makeup across all of North Carolina are published and consensus is reached, we will continue to consider all *M. salmoides* and *M. nigricans* in North Carolina to be the Largemouth Bass, *M. salmoides*.

Identification Key to the Species of Sunfishes (Family Centrarchidae)

  1a. Anal spines 3 (figure 295)...2
  1b. Anal spines 5 or more (figure 295).......................................18

FIGURE 295. Arrows indicating anal fin spines. A—three in Banded Sunfish, *Enneacanthus obesus*; B—six in Roanoke Bass, *Ambloplites cavifrons*.

2a. Body short and deep, not elongate. Dorsal fin not notched (appearing very slightly notched in Blackbanded Sunfish, *Enneacanthus chaetodon*).....3

2b. Body elongate. Dorsal fin moderately or deeply notched...................
....*Micropterus*, Couplet 13

3a. Caudal fin rounded.............................*Enneacanthus*, Couplet 4

3b. Caudal fin forked or emarginate.....................*Lepomis*, Couplet 6

4a. Dorsal fin with first two or three membranes black. Five or six narrow black bars extending around the body with the first bar passing through the eye (figure 296) (map 235)...
..............Blackbanded Sunfish, *Enneacanthus chaetodon* (Baird, 1855)

4b. Dorsal fin without first two or three membranes black; all dorsal membranes uniformly pigmented. Bars on sides brown and variable..........5

FIGURE 296. Blackbanded Sunfish, *Enneacanthus chaetodon*.

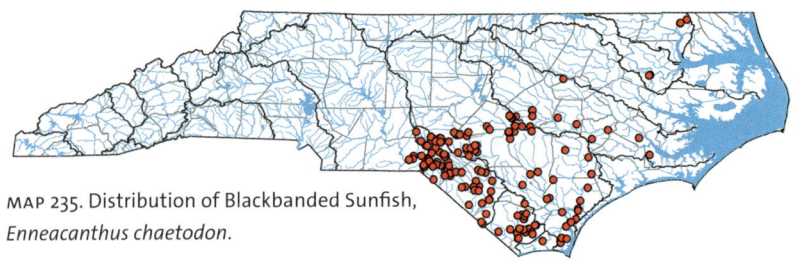

MAP 235. Distribution of Blackbanded Sunfish, *Enneacanthus chaetodon*.

5a. Circumpeduncle scales (17) 19–22 (24). Body side pattern of adult male dominated by broad, dark vertical bars. Opercular spot dark, large, and equal to or larger than eye. Iridescent color under eyes more or less a continuous crescent (figure 297) (map 236)......................................
...................Banded Sunfish, *Enneacanthus obesus* (Girard, 1854)

5b. Circumpeduncle scales (14) 16–18 (20). Body side pattern of adult male dominated by many pale spots. Dark bars on side, if present, narrow and indistinct. Opercular spot dark, smaller in diameter than eye. Iridescent color under eyes tends to be broken up into spots or short dashes (figure 297) (map 237)..
...........Bluespotted Sunfish, *Enneacanthus gloriosus* (Holbrook, 1855)

A

B

FIGURE 297. A—Banded Sunfish, *Enneacanthus obesus*; B—Bluespotted Sunfish, *Enneacanthus gloriosus*.

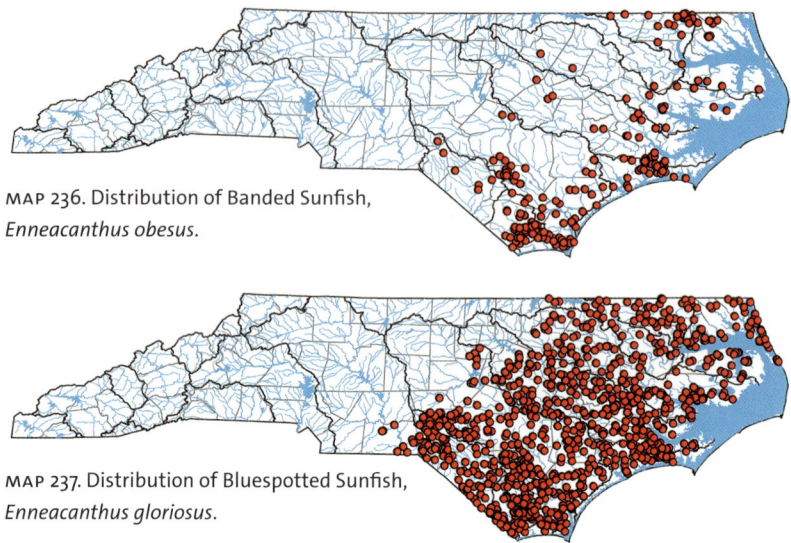

MAP 236. Distribution of Banded Sunfish, *Enneacanthus obesus.*

MAP 237. Distribution of Bluespotted Sunfish, *Enneacanthus gloriosus.*

6a. Gill rakers short, reduced to knobs (figure 298). Mouth small; upper jaw does not extend to pupil. Cheeks with numerous blue streaks (figure 299) (map 238).........Dollar Sunfish, *Lepomis marginatus* (Holbrook, 1855)

6b. Gill rakers medium to long, their length two to ten times their width (figure 298). Mouth and cheek color variable.............................7

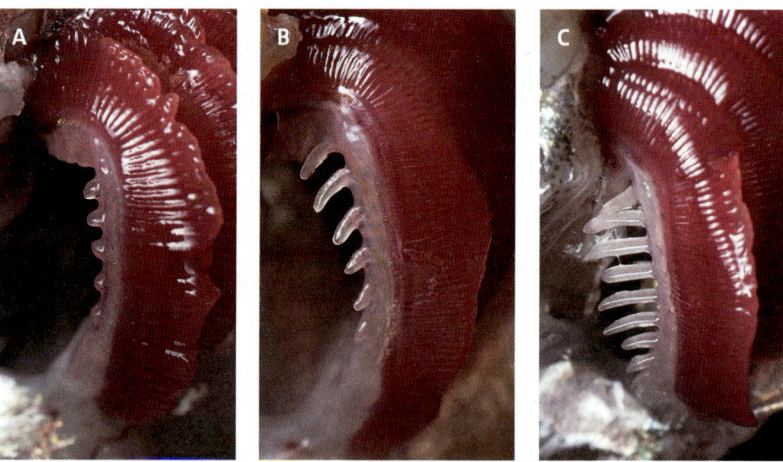

FIGURE 298. Gill rakers. A—short in Dollar Sunfish, *Lepomis marginatus*; B—medium in Redbreast Sunfish, *Lepomis auritus*; C—long in Bluegill, *Lepomis macrochirus.*

FIGURE 299. Dollar Sunfish, *Lepomis marginatus*.

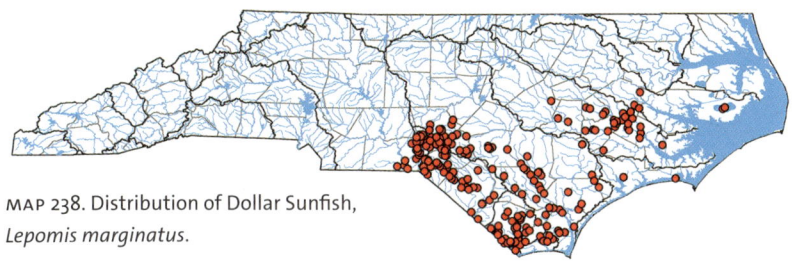

MAP 238. Distribution of Dollar Sunfish,
Lepomis marginatus.

7a. Tongue with a small median patch of teeth. Mouth large, upper jaw reaching to a point below middle of eye. Dark lines radiating back from eye usually wide and straight (figure 300) (map 239)..........................
.............................Warmouth, *Lepomis gulosus* (Cuvier, 1829)

7b. Tongue without a small median patch of teeth. Mouth small (except in Green Sunfish, *Lepomis cyanellus*). Upper jaw not reaching to a point below middle of eye. Lines radiating back from eye absent or narrow and irregular..8

FIGURE 300. Warmouth, *Lepomis gulosus*.

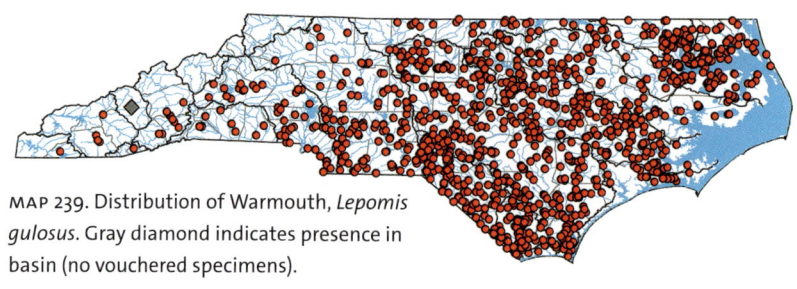

MAP 239. Distribution of Warmouth, *Lepomis gulosus*. Gray diamond indicates presence in basin (no vouchered specimens).

8a. Dark basal spot or blotch present in posterior portion of soft dorsal fin 9
8b. Dark spot or blotch absent in posterior portion of soft dorsal fin.......10

9a. Mouth large, upper jaw length greater than eye diameter. Pectoral fin rounded and short (figure 301) (map 240)..................................
......................Green Sunfish, *Lepomis cyanellus* Rafinesque, 1819
9b. Mouth small, upper jaw length equal to eye diameter. Pectoral fin pointed and long (figure 301) (map 241)...
...........................Bluegill, *Lepomis macrochirus* Rafinesque, 1819

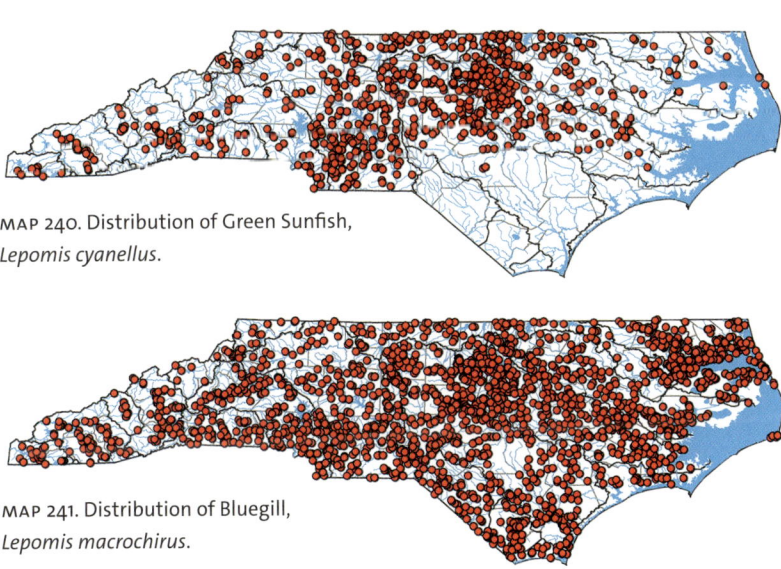

A

B

FIGURE 301. A—Green Sunfish, *Lepomis cyanellus*; B—Bluegill, *Lepomis macrochirus*.

MAP 240. Distribution of Green Sunfish, *Lepomis cyanellus*.

MAP 241. Distribution of Bluegill, *Lepomis macrochirus*.

10a. Pectoral fin long and pointed, tip extending past front of eye when fin is bent forward and pressed against head...................................11

10b. Pectoral fin short and rounded, tip not extending past eye when fin is bent forward and pressed against head......................................12

11a. Body often with distinct pale spots encircled with dusky marks. Dorsal and anal fins mottled or spotted. Cheek and opercle with dusky, wavy lines (figure 302) (map 242)...
..........................Pumpkinseed, *Lepomis gibbosus* (Linnaeus, 1758)

11b. Body lacking distinct pale spots encircled with dusky marks. Dorsal and anal fins not mottled. Cheek and opercle lacking wavy lines (figure 302) (map 243).........Redear Sunfish, *Lepomis microlophus* (Günther, 1859)

FIGURE 302. A—Pumpkinseed, *Lepomis gibbosus*; B—Redear Sunfish, *Lepomis microlophus*.

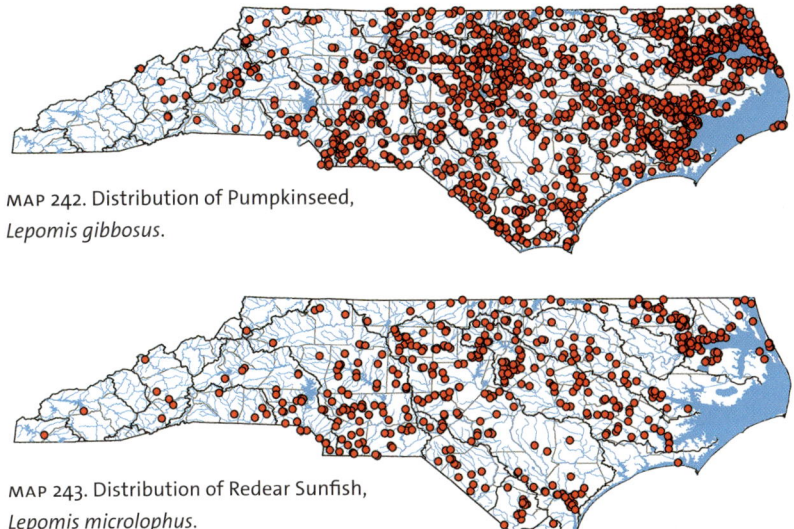

MAP 242. Distribution of Pumpkinseed,
Lepomis gibbosus.

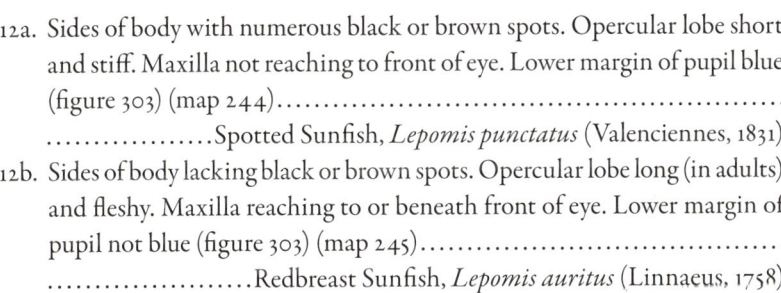

MAP 243. Distribution of Redear Sunfish,
Lepomis microlophus.

12a. Sides of body with numerous black or brown spots. Opercular lobe short and stiff. Maxilla not reaching to front of eye. Lower margin of pupil blue (figure 303) (map 244)...
.................Spotted Sunfish, *Lepomis punctatus* (Valenciennes, 1831)

12b. Sides of body lacking black or brown spots. Opercular lobe long (in adults) and fleshy. Maxilla reaching to or beneath front of eye. Lower margin of pupil not blue (figure 303) (map 245).......................................
....................Redbreast Sunfish, *Lepomis auritus* (Linnaeus, 1758)

A

B

FIGURE 303.
A—Spotted Sunfish,
Lepomis punctatus;
B—Redbreast Sunfish,
Lepomis auritus.

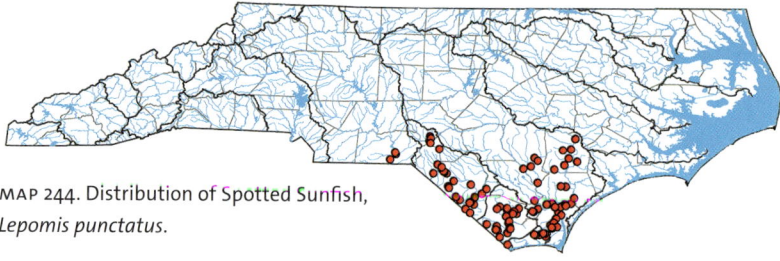

MAP 244. Distribution of Spotted Sunfish,
Lepomis punctatus.

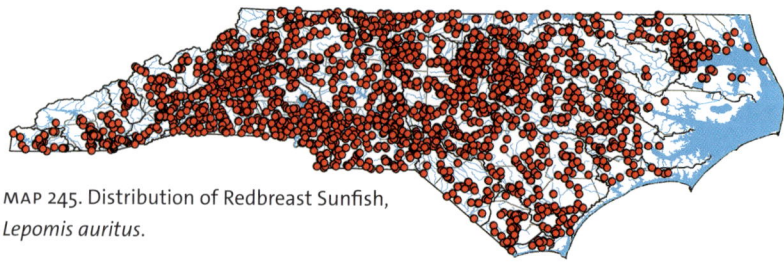

MAP 245. Distribution of Redbreast Sunfish,
Lepomis auritus.

13a. Dorsal fin deeply notched. Maxilla extending behind eye in fish greater than or equal to ca. 135 mm TL (figure 304). Base of anal and soft dorsal fins with no scales or with very few embedded scales. Tooth patch absent on tongue (map 246)..
...............Largemouth Bass, *Micropterus salmoides* (Lacepède, 1802)

13b. Dorsal fin shallowly notched. Maxilla not usually extending behind eye (figure 304). Base of anal and soft dorsal fins with scales. Tooth patch present on tongue...14

FIGURE 304. A—Largemouth Bass, *Micropterus salmoides*, juvenile, with bar indicating the maxilla extending almost to behind eye. B—Smallmouth Bass, *Micropterus dolomieu*, juvenile, with bar indicating the maxilla extending only to the middle of eye.

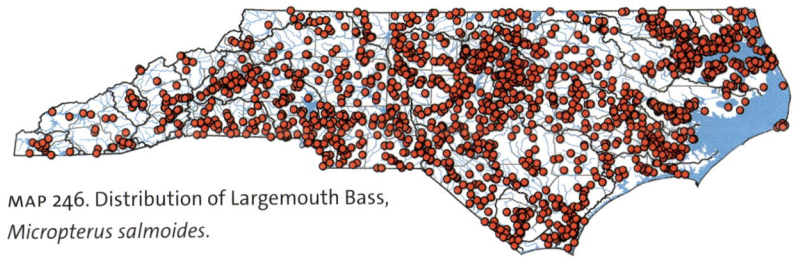

MAP 246. Distribution of Largemouth Bass, *Micropterus salmoides*.

14a. Dorsal soft rays (12) 13–15. Anal soft rays (10) 11 (12). Pectoral rays (15) 16–18. Scales above lateral line (11) 12 or 13 (14) (figure 305) (map 247)........ ....Smallmouth Bass, *Micropterus dolomieu* Lacepède, 1802

14b. Dorsal soft rays (11) 12 (13). Anal soft rays (9) 10 (11). Pectoral rays (14) 15 or 16 (17). Scales above lateral line 7–10....................................15

FIGURE 305. Smallmouth Bass, *Micropterus dolomieu*, Congaree River, South Carolina.

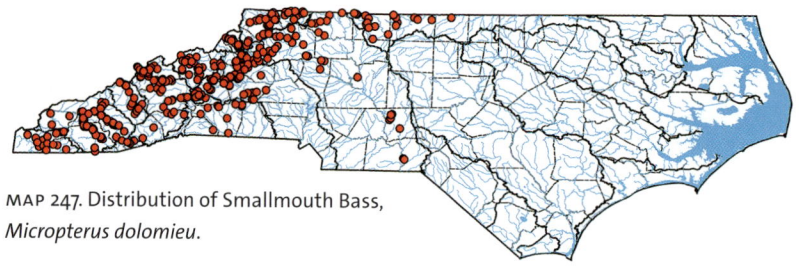

MAP 247. Distribution of Smallmouth Bass, *Micropterus dolomieu*.

15a. Color pattern consisting of a lateral series of dark blotches, which tend to be confluent and form an irregular lateral stripe. Basal caudal spot and opercular spot prominent (figure 306)...................................16

15b. Color pattern consisting of elongated vertical dark bars, which are frequently faint and obscured with age, and on the caudal peduncle spots are often modified into light-centered rhomboids. Basal caudal spot not prominent (figure 306)...17

FIGURE 306. A—Spotted Bass, *Micropterus punctulatus*. Mississippi. B—"Bartram's" Bass, *Micropterus* sp. Photograph courtesy of Luke Etchison ("Bartram's" Bass).

16a. Lateral line scales (68) 75 (84). Scales above lateral line (7) 8 (9). Circumpeduncle scales (26) 29 (32). A series of blotches on caudal peduncle (figure
307) (map 248)..
............Alabama Bass, *Micropterus henshalli* Hubbs and Bailey, 1940
16b. Lateral line scales (60) 65 (71). Scales above lateral line (5) 7 (9). Circumpeduncle scales (21) 25 (28). Solid dark line on caudal peduncle (figure 307)
(map 249)......Spotted Bass, *Micropterus punctulatus* (Rafinesque, 1819)

FIGURE 307. A—Alabama Bass, *Micropterus henshalli*. Alabama. B—Spotted Bass, *Micropterus punctulatus*. Mississippi. Photograph courtesy of Zach Alley (Alabama Bass).

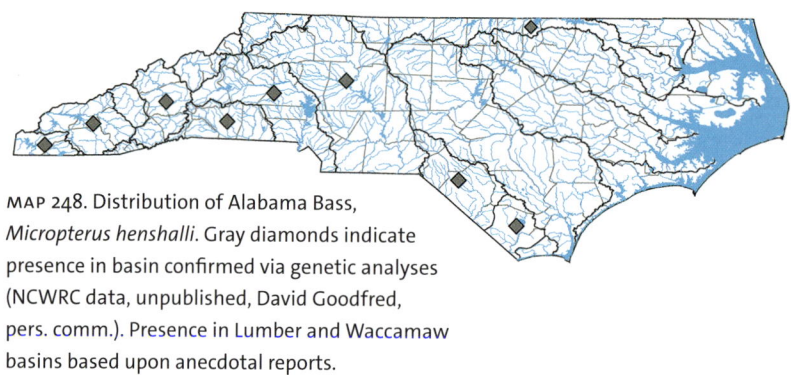

MAP 248. Distribution of Alabama Bass, *Micropterus henshalli*. Gray diamonds indicate presence in basin confirmed via genetic analyses (NCWRC data, unpublished, David Goodfred, pers. comm.). Presence in Lumber and Waccamaw basins based upon anecdotal reports.

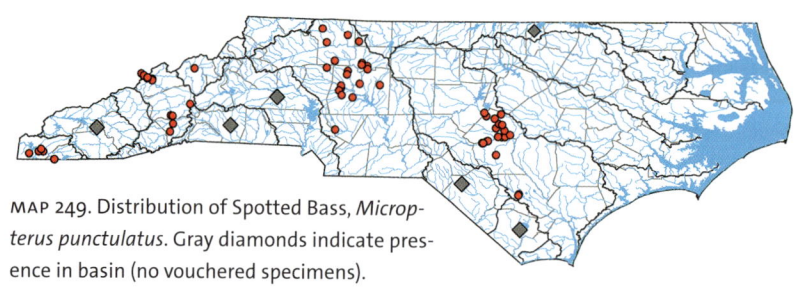

MAP 249. Distribution of Spotted Bass, *Micropterus punctulatus*. Gray diamonds indicate presence in basin (no vouchered specimens).

17a. Blotches on posterior half of lateral side of body faint or diffuse. Fin rays of posterior portions of dorsal, caudal, anal, and pectoral fins red. Known only from Hiwassee basin (figure 308) (map 250)..........................
................Redeye Bass, *Micropterus coosae* Hubbs and Bailey, 1940

17b. Blotches on entire lateral side distinct. Fin rays of posterior portions of dorsal, caudal, anal, and pectoral fins green or yellow-green, not red. Known only from Savannah basin (indigenous) and Green River in Broad basin (introduced) (figure 308) (map 251).................................
...."Bartram's" Bass, *Micropterus* sp.

FIGURE 308. A—Redeye Bass, *Micropterus coosae*. Georgia. B—"Bartram's" Bass, *Micropterus* sp. Photographs courtesy of Brian Zimmerman (Redeye Bass) and Luke Etchison ("Bartram's" Bass).

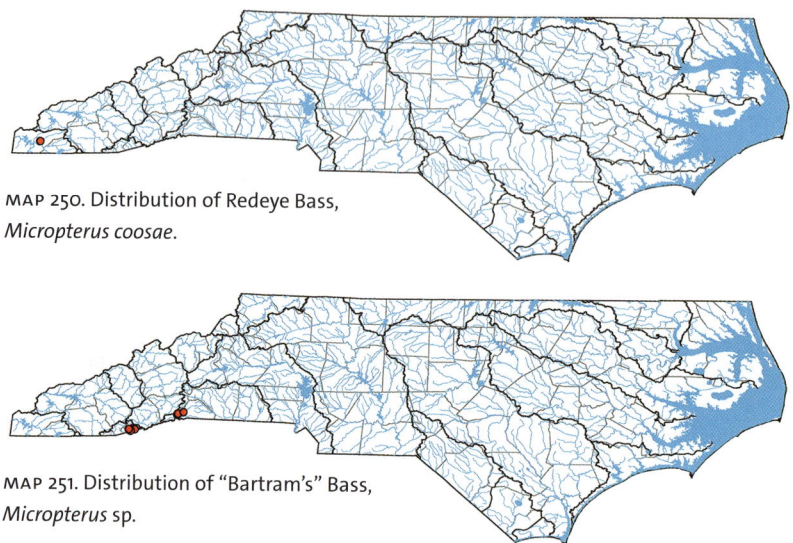

MAP 250. Distribution of Redeye Bass, *Micropterus coosae.*

MAP 251. Distribution of "Bartram's" Bass, *Micropterus* sp.

18a. Gill rakers 15 or fewer. Anal base short, 1.5–2 times in dorsal fin base...19
18b. Gill rakers 20 or more. Anal base long, 1–1.2 times in dorsal fin base...21

19a. Caudal fin rounded. Anal spines 5. Body dark green, with two to three dark stripes radiating posteriorly from eye (figure 309). Scales cycloid (map 252)...............Mud Sunfish, *Acantharchus pomotis* (Baird, 1855)
19b. Caudal fin emarginate. Anal spines (5) 6. Body olivaceous, with a dark spot on each scale forming rows of dark dots between light stripes. Scales ctenoid...*Ambloplites*, Couplet 20

FIGURE 309. Mud Sunfish, *Acantharchus pomotis.*

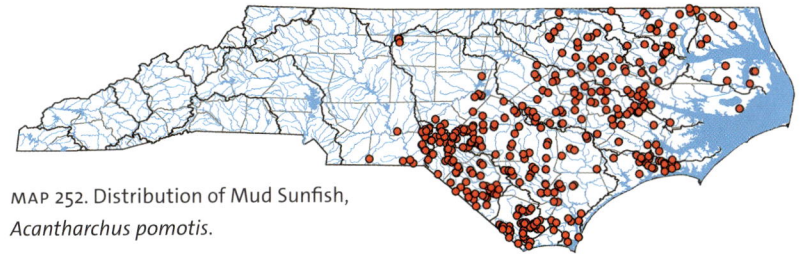

MAP 252. Distribution of Mud Sunfish, *Acantharchus pomotis*.

20a. Cheek fully scaled, the scales medium-sized and slightly or moderately embedded, the posterior margins obvious. Body lacking distinct round pale spots, although vague pale areas may be present. Anal fin of adult dark-margined (figure 310) (map 253) . . . . Rock Bass, *Ambloplites rupestris* (Rafinesque, 1817)

20b. Cheek naked or partly scaled, the scales tiny and embedded. Body often with distinct round pale spots. Anal fin of adult not dark-margined (figure 310) (map 254) . . . . Roanoke Bass, *Ambloplites cavifrons* Cope, 1868

FIGURE 310. A—Rock Bass, *Ambloplites rupestris*; B—Roanoke Bass, *Ambloplites cavifrons*.

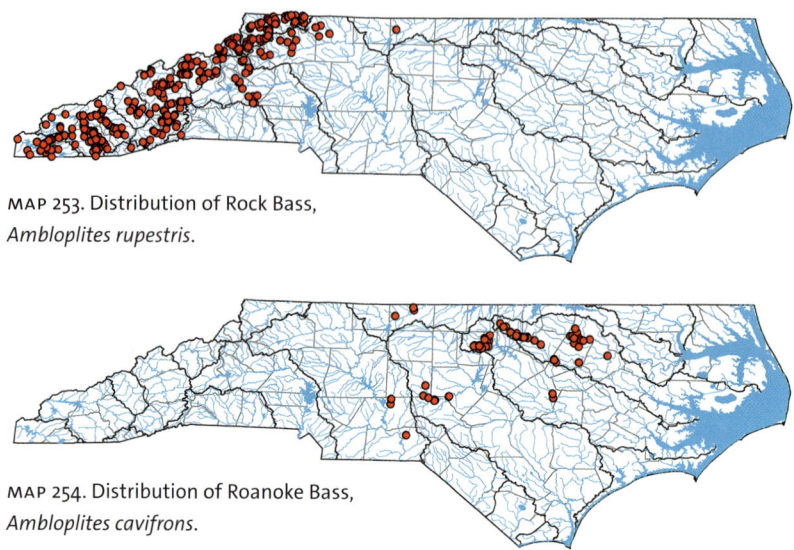

MAP 253. Distribution of Rock Bass,
Ambloplites rupestris.

MAP 254. Distribution of Roanoke Bass,
Ambloplites cavifrons.

21a. Dorsal spines 11–13. Anal spines 7 or 8. Body often greenish, each scale
with black or brown spot forming horizontal rows of dark dots on sides.
Young with a dark blotch on posterior base of dorsal fin (figure 311) (map
255).....................Flier, *Centrarchus macropterus* (Lacepède, 1801)

21b. Dorsal spines 5–8. Anal spines 5 or 6. Body light with dark spots, spots on
sides not in horizontal rows. Young without a dark blotch on posterior
base of dorsal fin......................................*Pomoxis*, Couplet 22

FIGURE 311. Flier, *Centrarchus macropterus.*

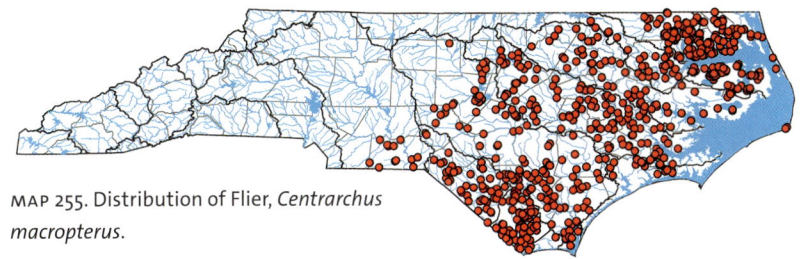

MAP 255. Distribution of Flier, *Centrarchus macropterus*.

22a. Dorsal spines (6) 7 or 8. Length of dorsal fin base greater than or equal to distance from first dorsal spine to center of eye. Sides with irregularly arranged dark flecks or small blotches (figure 312) (Map 256)..............
...................Black Crappie, *Pomoxis nigromaculatus* (Lesueur, 1829)

22b. Dorsal spines 5 or 6 (7). Length of dorsal fin base less than distance from first dorsal spine to center of eye. Sides with dark flecks tending to form vertical bars (figure 312) (map 257)...
.......................White Crappie, *Pomoxis annularis* Rafinesque, 1818

FIGURE 312. Arrows indicating length of dorsal fin base relative to distance from first dorsal spine to center of eye. A—Black Crappie, *Pomoxis nigromaculatus*; B—White Crappie, *Pomoxis annularis*.

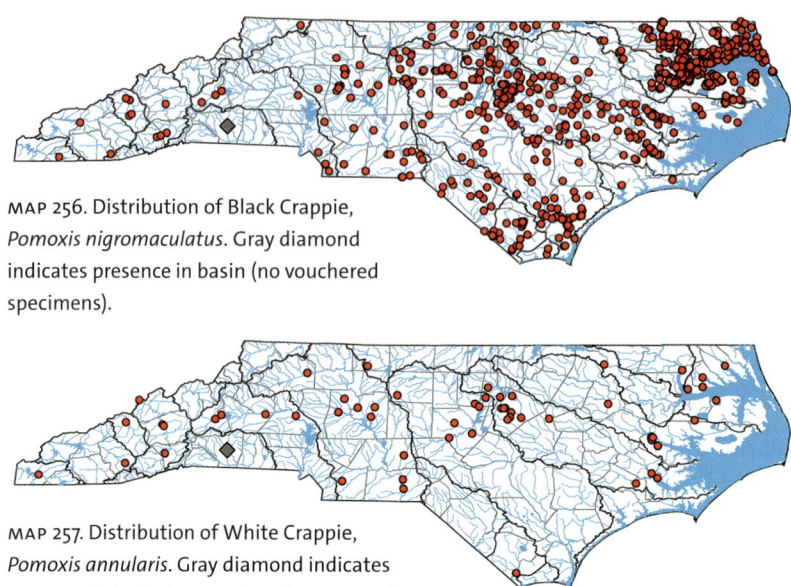

MAP 256. Distribution of Black Crappie, *Pomoxis nigromaculatus*. Gray diamond indicates presence in basin (no vouchered specimens).

MAP 257. Distribution of White Crappie, *Pomoxis annularis*. Gray diamond indicates presence in basin (no vouchered specimens).

Pygmy Sunfishes
Family Elassomatidae

North Carolina is home to three of the seven species of pygmy sunfishes known to occur in North America (table 28; Gilbert, 2004; Snelson, Krabbenhoft, and Quattro, 2009; Tracy, Rohde, and Hogue, 2020). Collectively, these seven species have been placed in the family Elassomatidae (Fricke, Eschmeyer, and van der Laan, 2002; Gilbert, 2004). However, some advocate their placement within the family Centrarchidae (Near et al., 2012). As their common name implies, pygmy sunfishes are just that—small fishes that resemble miniature sunfishes. The Carolina Pygmy Sunfish and Everglades Pygmy Sunfish range in size from ca. 20 to 32 mm TL, while the Banded Pygmy Sunfish reaches a length of ca. 50 mm TL.

All three species prefer shallow and quiet or slow-moving waters that are darkly stained (tea-colored) and of low productivity and that are acidic (low pH). Their habitats include ponds, creeks, sloughs, and roadside ditches with luxuriant submerged and emergent aquatic vegetation that can be used as shelter (Rohde and Arndt, 1987; Rohde et al., 2009).

Pygmy sunfishes are found only in the Sand Hills and throughout the Coastal Plain (table 2). The Everglades Pygmy Sunfish is found in the southeastern corner of the state. However, these populations may represent an undescribed species (Michael Sandel, University of West Alabama, personal communication). The Banded Pygmy Sunfish is our most widely distributed species, found across the Sand Hills and Coastal Plain, and is common in the southeast corner of the state.

The Carolina Pygmy Sunfish, described by Rohde and Arndt in 1987 as from the shallow edges of deep and dark Juniper Creek, is restricted to the Waccamaw basin in Brunswick and Columbus Counties. It is endemic to

TABLE 28. Species of pygmy sunfishes in North Carolina

Scientific Name	Common Name	Imperilment Status
Elassoma boehlkei[a]	Carolina Pygmy Sunfish	state threatened
Elassoma evergladei	Everglades Pygmy Sunfish	
Elassoma zonatum	Banded Pygmy Sunfish	

[a] Species found in only one river basin.

southeastern North Carolina and northeastern South Carolina and is found nowhere else in the world (Jones and Ewing, 2019; Quattro, Jones, and Rohde, 2001; Rohde et al., 2009; Sandel and Harris, 2007). Due to its limited distribution and anthropogenic impacts upon its habitat and water quality, the Carolina Pygmy Sunfish is listed as state Threatened (table 28; NCAC, 2023; NCNHP, 2022; NCWRC, 2021; Sandel and Harris, 2007).

Key characteristics for the proper identification of pygmy sunfishes include the presence or absence of scales atop the head; the presence or absence of vertical bars along the side; and the presence or absence of a postocular stripe. Several species can co-occur, especially in the Waccamaw basin, where all three species are found, rendering field identification a challenge.

Identification Key to the Species of Pygmy Sunfishes (Family Elassomatidae)

1a. Scales present on top of head. Side of body with light streaks or mottled (figure 313) (map 258)..
..............Everglades Pygmy Sunfish, *Elassoma evergladei* Jordan, 1884
1b. Scales absent on top of head. Side of body with distinct bars.............2

FIGURE 313. Everglades Pygmy Sunfish, *Elassoma evergladei*. A—male; B—female.

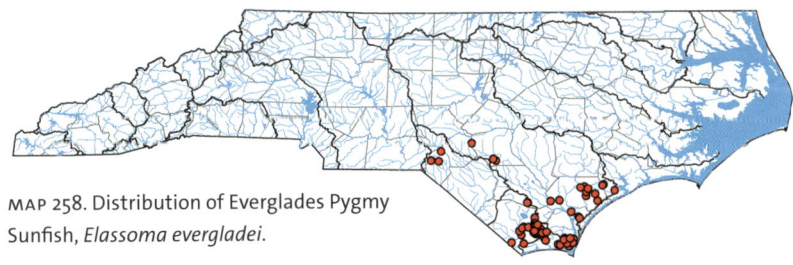

MAP 258. Distribution of Everglades Pygmy
Sunfish, *Elassoma evergladei*.

2a. Postocular stripe present. One to three dark humeral (shoulder) spot(s) present, often faint in life. Dark bars on side usually number (7) 9 (12) (figure 314) (map 259)........Banded Pygmy Sunfish, *Elassoma zonatum* Jordan, 1877[1]

2b. Postocular stripe absent. No humeral (shoulder) spot(s) present. Dark bars on side usually number ten or more (figure 315). Range restricted to Waccamaw basin (map 260)......Carolina Pygmy Sunfish, *Elassoma boehlkei* Rohde and Arndt, 1987

FIGURE 314. Banded Pygmy Sunfish, *Elassoma zonatum*. A—male. South Carolina. B—female. Arrows indicating the postocular stripe. Photograph courtesy of Zach Alley (male).

1. The Banded Pygmy Sunfish may be confused with the Fat Sleeper, *Dormitator maculatus*, (family Eleotridae) where they co-occur. However, the Banded Pygmy Sunfish has a single dorsal fin and an unscaled head, whereas the Fat Sleeper has two separated dorsal fins and a scaled head (figures 314 and 316) (Rohde et al., 2009).

FIGURE 315. Carolina Pygmy Sunfish, *Elassoma boehlkei*. A—male; B—female.

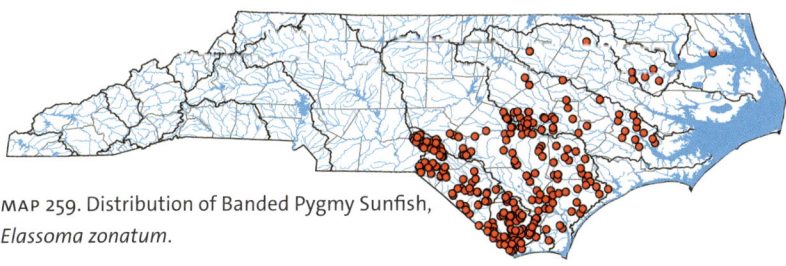

MAP 259. Distribution of Banded Pygmy Sunfish, *Elassoma zonatum*.

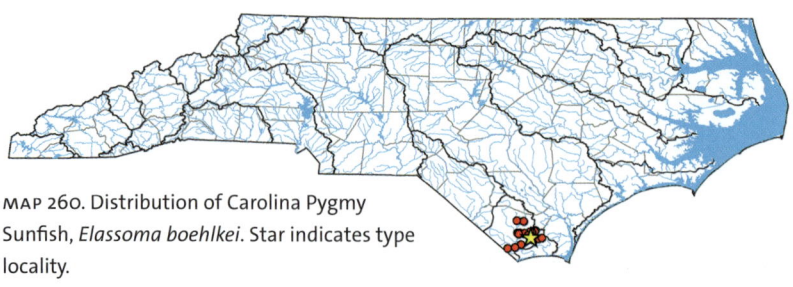

MAP 260. Distribution of Carolina Pygmy Sunfish, *Elassoma boehlkei*. Star indicates type locality.

FIGURE 316. Fat Sleeper juvenile, *Dormitator maculatus*.

ACKNOWLEDGMENTS

Scott A. Smith, Jesse L. Bissette, and Fred C. (Fritz) Rohde photographed most of the fishes within this book. We are very appreciative of the following photographers who provided additional images:

Tim Aldridge—Blue Tilapia
Zach Alley—Alabama Bass, Banded Pygmy Sunfish (male)
Ben Cantrell—Redbelly Tilapia
Robert Criswell—Blacknose Dace
Luke Etchison—"Bartram's" Bass, Black Buffalo, Freshwater Drum, Logperch, Muskellunge, Olive Darter, Quillback, River Redhorse (lips), Sauger, Sooty-banded Darter, Spotfin Chub (female), Whitetail Shiner
Dan Marotta—Bay Anchovy
David A. Neely—Kokanee (Sockeye Salmon), Paddlefish, Sharpnose Darter, Snubnose Darter
Kelsey Roberts—Bodie bass
Derek Wheaton—Spotfin Chub (male)
Brian Zimmerman— Bigmouth Buffalo, Mooneye, Redeye Bass, River Carpsucker, Sea Lamprey, *Ictiobus* spp.

Illustrations were obtained from the following sources with permission by the publishers or were in the public domain:

American Society of Ichthyologists and Herpetologists
American Fisheries Society
Food and Agriculture Organization of the United Nations
North Carolina Department of Environmental Quality—Division of Energy, Minerals, and Land Resources
North Carolina Wildlife Resources Commission
The University of South Carolina Press

We are indebted to the following friends and colleagues who aided us in the collecting of fishes to be photographed or who provided us with specimens to be photographed: Robert Aguilar, Tim Aldridge, Caitlin Bourner, Cody Cromer, Luke Etchison, Michael Fisk, Ryan Heise, James Henne, Chip Hildreth,

Stephen Johnson, Shay Kashey, Ray Katula, Shannon Kemp, David Mathews, Jason Mays, Kelly McDonald, Dylan Owensby, Michael Perkins, Benjamin J. Ricks, Chantelle Rondel, Brandi Salmon, Rick Sanchez, C. J. Schlick, Konrad Schmidt, David Smith, Dustin Smith, Fletcher Stone, J. Michael Swing, Todd D. VanMiddlesworth, Derek Wheaton, Jason Yates, and Parker Yates. We are especially thankful to the following staff of the South Carolina Department of Natural Resources who obtained big-river fishes from the Cooper and Congaree Rivers: Jason Bettinger, Dani Carty, Rhett Casey, Jeremy Grigsby, Emily Klimczak, Corbert Norwood, Conner Owens, Bill Post, Greg Song, and Ellen Waldrop.

Identification keys were adopted from multiple sources (see Literature Consulted). We are indebted to the following colleagues for reviewing them: Robert Adams, Sierra B. Benfield, J. Michael Fisk, Andrew Glen, Ryan Heise, Lily C. Hughes, Thomas Russ, Nick Wahl, Michael Walter, and David Werneke.

Thank you to Chris Scharpf for reviewing the appendix, which contains, with full permission, the meanings of the scientific names of North Carolina's freshwater fishes. This information was adapted from the ETYFish Project (etyfish.org).

We are grateful for receiving two grants from the North Carolina Chapter of the American Fisheries Society, which funded trips to photograph specimens.

We are very appreciative of financial support for publication of this book received from the North Carolina Wildlife Resources Commission. Its generosity directly impacted the final cost of this book and allows it to be accessible for all.

The Meanings of the Scientific Names of North Carolina's Freshwater Fishes

The following meanings of scientific names have been adopted, with full permission, from the ETYFish Project (etyfish.org), created by Christopher Scharpf and Kenneth J. Lazara and last accessed by the authors on June 1, 2023.

Family Petromyzontidae Bonaparte, 1831, Lampreys

- *Ichthyomyzon* Girard, 1858—*ichthyos*, fish; *myzon*, to suck (borrowed from *Petromyzon*), i.e., a sucking fish, referring to their suctorial behavior
    - a. *Ichthyomyzon bdellium* (Jordan, 1885)—*ium*, adjectival suffix; *bdella*, leech, referring to its leech-like suctorial and/or parasitic feeding behavior (as adults)
    - b. *Ichthyomyzon greeleyi* Hubbs and Trautman, 1937—patronym, in honor of fisheries scientist John R. Greeley (1904–64), who collected holotype and granted the authors permission to describe it

- *Lampetra* Bonaterre, 1878—*lambo*, to lick; *petra*, rock, referring to their suctorial behavior (adults attach to rocks during nest building and mating)
    - a. *Lampetra aepyptera* (Abbott, 1860)—*aepy*, from *aipys*, high; *pteron*, fin, referring to enlarged dorsal fins of nuptial males

- *Lethenteron* Creaser and Hubbs, 1922—etymology not explained, perhaps *lethalis* (Latin), lethal, or *lethe* (Greek), forgetfulness; *enteron* (Greek), presumably referring to "degenerate and nonfunctional" intestine of adult *L. appendix*
    - a. *Lethenteron appendix* (DeKay, 1842)—Latin for appendage, referring to "thread-like appendix" (genital papilla) on anterior portion of nuptial males (DeKay believed this papilla was appended to the anal fin)

- *Petromyzon* Linnaeus, 1758—*petra*, rock or stone; *myzon*, to suck, referring to the suctorial behavior (adults attach to rocks during nest building and mating)
    - a. *Petromyzon marinus* Linnaeus, 1758—of the sea, referring to marine habitat (as a nonbreeding adult). See Scharpf (2014a).

Family Acipenseridae Bonaparte, 1831, Sturgeons

- *Acipenser* Linnaeus, 1758—Latin for sturgeon, derived from *akis*, point; *pente*, five, referring to five rows of body scutes, but see Scharpf (2021a) for an alternative explanation
  - a. *Acipenser brevirostrum* Lesueur, 1818—*brevis*, short; *rostrum*, nose, referring to shorter snout compared with *A. oxyrinchus*
  - b. *Acipenser fulvescens* Rafinesque, 1817—*fulvous*, yellowish brown; *escens*, becoming, referring to olive-brown coloration on upper half of body, Latin for yellowish or tawny, referring to its "dark fulvus color"
  - c. *Acipenser oxyrinchus* Mitchell, 1815—*oxy*, sharp; *rhynchus*, snout, referring to sharply V-shaped snout

Family Polyodontidae Bonaparte, 1835, Paddlefishes

- *Polyodon* Lacepède, 1797—*poly*, many; *odon*, tooth, referring to its many teeth compared with an absence of teeth in *Acipenser* (sturgeons). However, only juvenile Paddlefish have teeth, and the type specimen is a small specimen. The genus name, *Polyodon*, does not allude to the numerous gill rakers possessed by the adults, as reported by many authors, but to the numerous teeth possessed by the juveniles. See Scharpf (2014d).
  - a. *Polyodon spathula* (Walbaum, 1792)—*spatula*, Latin for a paddle, spoon, or broad blade used for stirring, derived from the Greek *spathe*, referring to paddle-shaped rostrum

Family Lepisosteidae Bonaparte, 1835, Gars

- *Lepisosteus* Lacepède, 1803—*lepis*, scale; *osteus*, bony, referring to bone-like (ganoid) scales
  - a. *Lepisosteus osseus* (Linnaeus, 1758)—*osseus*, Latin for bone or bony, referring to bone-like (ganoid) scales on head and body

Family Amiidae Bonaparte, 1831, Bowfins

- *Amia* Linnaeus, 1766—meaning unknown, presumably derived from *amia*, ancient name for an unknown fish, long held to be a bonito or a kind of "tunny" (e.g., Atlantic Bonito, *Sarda sarda*) since the Renaissance but now identified as the Bluefish, *Pomatomus saltatrix*
  - a. *Amia calva* Linnaeus, 1766—*calva*, smooth or bald, probably referring to its scaleless head

Family Anguillidae Rafinesque, 1810, Freshwater Eels

- *Anguilla* Schrank, 1798—Latin for eel, derived from *anguis* (Latin), snake, referring to snakelike body shape (tautonymous with *Muraena anguilla* Linnaeus, 1758)
  - a. *Anguilla rostrata* (Lesueur, 1817)—Latin for beaked, possibly referring to the "elongated, pointed and strait" snout on the specimen Lesueur examined

Family Hiodontidae Valenciennes, 1847, Mooneyes

- *Hiodon* Lesueur, 1818—*hio*, from either *hyoides* (New Latin) or *huoeidēs* (Greek), shaped like the letter Ypsilon, referring to Y-shaped hyoid bone forming base of tongue of *H. tergisus*; *odon* (Greek), tooth, referring to "strong teeth" on tongue
  a. *Hiodon tergisus* Lesueur, 1818—Latin for polished, probably referring to its silvery sides (i.e., like polished silver)

Family Alosidae Svetovidov, 1953, Shads and Sardines

- *Alosa* Linck, 1790—from *alausa*, Latin name for *A. alosa* from the river Moselle in Germany, but in this case presumably tautonymous with *Clupea alosa* Linnaeus, 1758 (no species mentioned)
  a. *Alosa aestivalis* (Mitchill, 1814)—Latin for "of the summer," presumably referring to its later spawning run compared with *A. pseudoharengus* (Mitchill called it the "Summer Herring")
  b. *Alosa mediocris* (Mitchill, 1814)—Latin for middling or ordinary, referring to its taste or food value compared with *A. sapidissima*
  c. *Alosa pseudoharengus* (Wilson, 1811)—*pseudo*, from *pseúdēs* (Greek), false; *harengus*, Medieval Latin for herring, described as an American herring "well tasted, either fresh or salted, [but] not so fat as European Herring" (*A. alosa* or *A. fallax*)
  d. *Alosa sapidissima* (Wilson, 1811)—Latin for "most delicious," the two words Wilson used to describe the palatability of this shad

- *Brevoortia* Gill, 1861—*ia* (Latin suffix), pertaining to James Carson Brevoort (1818–87), "the well known ichthyologist of New York" (Brevoort was a businessman and philanthropist who supported various literary and scientific societies and institutions and was himself a fine amateur naturalist; his zoological library was then reputed to be the finest in the country)
  a. *Brevoortia tyrannus* (Latrobe, 1802)—Latin for tryrant, named for the Tyranni rulers of ancient Rome who forced food-tasters to test their food for poison, referring to this fish's relationship with the parasitic isopod *Cymothoa praegustator* (a "pretaster," described by Latrobe in the same publication), which lives in the mouths of many specimens (see also *B. patronus*) and thus represents "the minion of a tyrant . . . for he is not without those who suck him" (i.e., like all tyrants, this fish has parasites or hangers-on, perhaps reflecting Latrobe's enthusiasm for American independence from England). See Scharpf (2013).

Family Dorosomatidae Bleeker, 1872, Thread Herrings and Sardinellas

- *Dorosoma* Rafinesque, 1820—*dóratos* (Greek), genitive of *dóry*, lnce or spear; *sóma* (Greek), referring to "lanceolate" body of *D. notata* (= *cepedianum*), "tapering gradually towards the tail"
  a. *Dorosoma cepedianum* (Lesueur, 1818)—*anum* (Latin, neuter), belonging to Bernard-Germain-Étienne de La Ville-sur-Illon, comte de La Cepède (also

spelled as La Cépède, Lacépède, or Lacepède, 1756–1825), whose five-volume *Histoire Naturelle des Poissons* (1798–1803) was the standard ichthyological reference of his day; Lesueur may have honored his fellow countryman for having proposed the tarpon genus *Megalops*, into which this species was originally assigned

    b. *Dorosoma petenense* (Günther, 1867)—*ense*, Latin suffix (neuter) denoting place: Lake Petén (now Petén Itzá), Guatemala, type locality

Family Engraulidae Gill, 1861, Anchovies

• *Anchoa* Jordan and Evermann, 1927—anchovy-like, indicating a "transition to *Anchovia*"

    a. *Anchoa mitchilli* (Valenciennes, 1848)—in honor of Samuel Latham Mitchill (1764–1831), naturalist, physician, and US senator, who studied the fishes of New York Harbor

Family Catostomidae Agassiz, 1850, Suckers

• *Carpiodes* Rafinesque, 1820—*oides*, Latinized suffix adopted from *eidos* (Greek) form or shape: Common Carp, *Cyprinus carpio*

    a. *Carpiodes carpio* (Rafinesque, 1820)—derived from the Latin *carpa*, carp, probably originally from an unknown Slavic language (Danube region) used by Germanic tribes and then the Romans, referring to its resemblance to the Common Carp, *Cyprinus carpio*

    b. *Carpiodes cyprinus* (Lesueur, 1817)—from *kyprīnos* (Greek), carp, referring to resemblance to *Cyprinus carpio*

• *Catostomus* Lesueur, 1817—tautonymous with *Cyprinus catostomus* Forster, 1773, *cato*, from *katá* (Greek), downwards, beneath, below, or under; *stomus*, Latinization of *stóma* (Greek), mouth, referring to its ventral position compared with terminal mouth of Common Carp, *Cyprinus carpio*, its presumed congener at the time

    a. *Catostomus commersonii* (Lacepède, 1803)—in honor of French naturalist Philibert Commerçon (also spelled Commerson, 1727–73), whose collections where studied by Lacepède

• *Erimyzon* Jordan, 1876—*eri* (Greek), much or very; *myzon*, from *mýzō* (Greek), to suck, referring to fleshy, papillose lips that suck up food, a "free translation" of the vernacular name chubsucker

    a. *Erimyzon oblongus* (Mitchill, 1814)—Latin for oblong (longer than broad), referring to its more elongate shape compared with *E. sucetta*

    b. *Erimyzon sucetta* (Lacepède, 1803)—Latinization of the French *sucet*, meaning sucker

• *Hypentelium* Rafinesque, 1818—*hypó* (Greek), under, beneath, or less than; *pénte* (Greek), five, referring to "lower jaw shorter with five lobes" (a character that does not fit the genus)

a. *Hypentelium nigricans* (Lesueur, 1817)—Latin for blackish, referring to color of back

b. *Hypentelium roanokense* Raney and Lachner, 1947—*ensis*, Latin suffix denoting place: Roanoke River drainage, Virginia and North Carolina, where it is endemic

- *Ictiobus* Rafinesque, 1820—*icti*, presumably from *ichthýs* (Greek), fish; bus, from *boûs* (Greek), bull, ox, or buffalo, referring to humpbacked nape of *I. bubalus* and *I. niger*

    a. *Ictiobus bubalus* (Rafinesque, 1818)—possibly a Latin loanword from *boûs* (Greek), bull, ox, or buffalo, referring to its humpbacked nape

    b. *Ictiobus cyprinellus* (Valenciennes, 1844)—diminutive of *cyprinus*, from *kyprīnos* (Greek), carp, referring to its resemblance to the Common Carp, *Cyprinus carpio*

    c. *Ictiobus niger* (Rafinesque, 1819)—Latin for black, referring to its blackish fins

- *Minytrema* Jordan, 1878—*minýs* (Greek), less, small, or reduced; *trêma* (Greek), hole or aperture, referring to its lateral line, absent in juveniles and incomplete (consisting of only four unpored scales) in adults

    a. *Minytrema melanops* (Rafinesque, 1820)—*mélanos* (Greek), genitive of *mélas*, black; *óps* (Greek), eye, face, or countenance, referring to its head, "flat above, blackish there and in the fore part" (Rafinesque called it the "Black-face Sucker")

- *Moxostoma* Rafinesque, 1820—*moxo*, probably a variant spelling of *mýzō* (Greek), to suck; *stoma*, mouth, referring to "fleshy, thick, or lobed sucking lips"

    a. *Moxostoma anisurum* (Rafinesque, 1820)—*ánisos* (Greek), unequal; *oura* (Greek), tail, but treated as an adjective (tailed), referring to upper lobe of tail being narrower and longer than lower (which it is not)

    b. *Moxostoma ariommum* Robins and Raney, 1956—*arí* (Greek), much or very; *ómma* (Greek), eye, but treated as an adjective (eyed), referring to its large eyes

    c. *Moxostoma breviceps* (Cope, 1870)—*brevis* (Latin), short, *ceps* (Neo Latin), headed, referring to small head and mouth

    d. *Moxostoma carinatum* (Cope, 1870)—Latin (neuter) for keeled, referring to low ridges on roof of skull

    e. *Moxostoma cervinum* (Cope, 1868)—Latin (neuter) for "of or relating to deer," referring to tawny or fawn-like coloration of lateral stripe on *Thoburnia rhothoeca*, which Cope confused with this species

    f. *Moxostoma collapsum* (Cope, 1870)—Latin (neuter) for flattened sidewise, referring to its compressed body

    g. *Moxostoma duquesnei* (Lesueur, 1817)—of Fort Duquesne (now Pittsburgh, PA) on Ohio River, type locality

    h. *Moxostoma erythrurum* (Rafinesque, 1818)—*erythrós* (Greek), red; *ourá* (Greek), tail, but treated here as an adjective (tailed), which accurately describes color of lower fins in some adults but not the yellowish tail

      i. *Moxostoma macrolepidotum* (Lesueur, 1817)—*macro*, from *makrós* (Greek), long or large; *lepidōtós* (Greek), scaly, allusion not explained, probably referring to how dark scale bases make scales appear larger than they are

      j. *Moxostoma pappillosum* (Cope, 1870)—Latin (neuter) for papillose, referring to pimple-like surface of its lips

      k. *Moxostoma robustum* (Cope, 1870)—Latin (neuter) for full-bodied, referring to its large size

      l. *Moxostoma rupiscartes* Jordan and Jenkins, 1889—*rupes* (Latin), rock; *skarthmós* (Greek), leaping or leap, i.e., a jumprock (its common name), inspired by *Rupiscartes* Swainson, 1839 for a genus of blenniids (=*Alticus*) that "jump on rocks, like a lizard," probably referring to the proclivity of some specimens to jump or break surface of water while spawning

- *Thoburnia* Jordan and Snyder, 1917—*ia* (Latin), belonging to American biologist Wilbur Wilson Thoburn (1859–99), who described *T. rhothoeca* and taught bionomics (ecology) at Stanford University (where Jordan had been president and chancellor)

      a. *Thoburnia hamiltoni* Raney and Lachner, 1946—in honor of vertebrate zoologist William J. Hamilton Jr. (1902–90), Cornell University, the author's friend and teacher, "whose stimulating suggestions and assistance over a period of ten years have been invaluable"

Family Cobitidae Swainson, 1838, Spined Loaches

- *Misgurnus* Lacepède, 1803—Latinization of either *mis'gurn*, *misgurne*, or *misgurnos*, Old English, French, and Spanish vernaculars, respectively, for *M. fossilis*

      a. *Misgurnus anguillicaudatus* (Cantor, 1842)—*anguilla* (Latin), eel, presumably referring to its long, eel-like body; *caudatus* (Latin), tailed, probably referring to its "much elongated" caudal fin

Family Cyprinidae Rafinesque, 1815, Carps

- *Carassius* Jarocki, 1822—tautonymous with *Cyprinus carassius* Linnaeus, 1758, from the French *carassin*, Golden or Crucian Carp

      a. *Carassius auratus* (Linnaeus, 1758)—Latin for gilded, referring to golden color

- *Cyprinus* Linnaeus, 1758—Latinization of *kypríños* (Greek) for carp, possibly derived from Kypris, also known as Venus (or Aphrodite), goddess of love, referring to fecundity of *C. carpio*, or from *Kýpros* (Greek), henna-like, referring to bronze color of *C. carpio*. See Scharpf (2022).

      a. *Cyprinus carpio* Linnaeus, 1758—derived from the Latin *carpa*, carp, probably originally from an unknown Slavic language (Danube region) used by Germanic tribes and then the Romans. See Scharpf (2022).

      b. *Cyprinus rubrofuscus* Lacepède, 1803—*rubro* from *ruber* (Latin), red; *fuscus* (Latin), dusky or dark, referring to its golden-brown coloration

Family Xenocyprididae Günther, 1868, Sharpbellies

- *Ctenopharyngodon* Steindachner, 1866—*cteno*, from *ktenós* (Greek), genitive of *kteís*, comb; *phárynx* (Greek), throat; *odon*, Latinized and grammatically adjusted from the Greek nominative *odoús*, tooth, referring to its comblike pharyngeal teeth
  a. *Ctenopharyngodon idella* (Valenciennes, 1844)—*ella* (Latin), diminutive suffix, etymology not explained nor evident, possibly referring to Ide or *idus*, from the Eurasian *Leuciscus idus* (Cypriniformes: Leuciscidae), in whose genus it was originally placed

Family Leuciscidae Bonaparte, 1835, Minnows

- *Alburnops* Girard, 1856—*ops*, appearance, referring to "striking external resemblance" to minnows Girard had placed in the European genus *Alburnus* (Leuciscinae)
  a. *Alburnops chalybaeus* (Cope, 1867)—steel-colored, referring to dark lateral stripe
  b. *Alburnops petersoni* (Fowler, 1942)—in honor of C. Bernard Peterson (1906–63), Fowler's editor at the Academy of Natural Sciences of Philadelphia, who helped collect type

- *Campostoma* Agassiz, 1855—*campo*, bent or curved; *stoma*, mouth, referring to U-shaped mouth of *C. anomalum*
  a. *Campostoma anomalum* (Rafinesque, 1820)—anomalous, differing from all other Ohio minnows by its "unequal bilobed tail" (not different or abnormal appearance of ridge on lower jaw, as reported by some authorities)

- *Chrosomus* Rafinesque, 1820—*chroma*, color; *soma*, body, referring to vibrant coloration of *Chrosomus erythrogaster*
  a. *Chrosomus oreas* Cope, 1868—of the mountains, referring to occurrence in montane and upland regions

- *Clinostomus* Girard, 1856—*clino*, latinization of the Greek *klino*, bend, slope, or slant; *stoma*, mouth, referring to mouth shape of *C. elongatus*, the "lower jaw longer than the upper, beyond which it protrudes, giving to the cleft an oblique direction upwards"
  a. *Clinostomus funduloides* Girard, 1856—*oides*, having the form of, referring to superficial resemblance to topminnows (Cyprinodontiformes—Fundulidae, *Fundulus*)

- *Coccotis* Jordan, 1882—*coccom*, berry; *otis*, presumably from *otos*, genitive of *ous*, ear, referring to prominent crimson slash on opercle of *C. coccogenis*
  a. *Coccotis coccogenis* (Cope, 1868)—*coccom*, berry; *geneion*, cheek, referring to prominent crimson slash on opercle

- *Cyprinella* Girard, 1856—diminutive of *cypris*, carp, i.e., a small carp or minnow
  a. *Cyprinella analostana* Girard, 1859—*ana*, belonging to—Analostan (now Theodore Roosevelt) Island, Potomac River, Washington, DC, type locality

    b. *Cyprinella chloristia* (Jordan and Brayton, 1878)—*chloros*, green; *histia*, sail, referring to green dorsal fin

    c. *Cyprinella galactura* (Cope, 1868)—*galactos*, milk; *oura*, tailed, referring to two clear-to-white areas on caudal fin base

    d. *Cyprinella labrosa* (Cope, 1870)—thick-lipped, referring to its "prominent" lips

    e. *Cyprinella lutrensis* (Baird and Girard, 1853)—*ensis*, suffix denoting place; *lutra*, otter, referring to Otter Creek, Oklahoma (erroneously reported as Arkansas), type locality

    f. *Cyprinella nivea* (Cope, 1870)—snow, referring to white fins of breeding males

    g. *Cyprinella pyrrhomelas* (Cope, 1870)—*pyrrhos*, flame; *melas*, black, referring to red-black caudal fin of males

    h. *Cyprinella spiloptera* (Cope, 1867)—*spilos*, spot; *ptera*, finned, referring to black spot on dorsal fin

    i. *Cyprinella zanema* (Jordan and Brayton, 1878)—*za*, very; *nemus*, thread, referring to "extremely long [barbels], probably longer than in any other of our Cyprinoids"

• *Erimonax* Jordan, 1924—*eri*, very; *monax*, solitary, referring to isolated suite of characters of *E. monachus*

    a. *Erimonax monachus* (Cope, 1868)—solitary, referring to its isolated suite of characters and the fact that Cope saw it only "singly or in pairs"

• *Erimystax* Jordan, 1882—*eri*, very; *mystax*, mustached, presumably referring to small maxillary barbel of *E. dissimilis*

    a. *Erimystax insignis* (Hubbs and Crowe, 1956)—conspicuous, referring to blotches on sides

• *Exoglossum* Rafinesque, 1818—*ex*, outside; *glossa*, tongue, referring to bony tongue-like extension of lower jaw

    a. *Exoglossum laurae* (Hubbs, 1931)—in honor of Carl L. Hubbs's wife, Laura (1893–1988)

    b. *Exoglossum maxillingua* (Lesueur, 1817)—*maxilla*, jawbone; *lingua*, tongue, referring to bony tongue-like extension of lower jaw

• *Hudsonius* Girard, 1856—named for type species *Clupea hudsonia*, referring to its type locality, Hudson River, New York (not tautonymous because Girard unnecessarily renamed *H. hudsonia* as *H. fluviatilis*)

    a. *Hudsonius altipinnis* (Cope, 1870)—*altus*, high; *pinna*, fin, referring to "much elevated" dorsal fin compared with other minnows Cope grouped with this species

    b. *Hudsonius hudsonius* (Clinton, 1824)—*ius*, adjectival suffix; referring to Hudson River, New York, type locality

• *Hybognathus* Agassiz, 1855—*hybos*, gibbous or blunt; *gnathus*, jaw, referring to "slight tubercle" at symphysis of lower jaw of *H. nuchalis*

a. *Hybognathus regius* Girard, 1856—royal, a "large and beautiful species, the largest [member of genus] that has, so far, come to our knowledge, some of the specimens measuring seven inches in length"

- *Hybopsis* Agassiz, 1854—*hybos* (Greek), gibbous or blunt; *ops*, from *opsis*, Greek for appearance or face, allusion not explained, presumably referring to "obtuse prominent snout" of *H. gracilis* (= *amblops*)
    a. *Hybopsis amblops* (Rafinesque, 1820)—*amblys*, blunt; *ops*, face or appearance, referring to "round" snout
    b. *Hybopsis hypsinotus* (Cope, 1870)—*hypselos*, high; *notos*, back, referring to strongly arched back
    c. *Hybopsis rubrifrons* (Jordan, 1877)—*rubrum*, red; *frons*, forehead, referring to rosy red color of anterior portion of body of breeding males

- *Hydrophlox* Jordan, 1878—*hydro*, water; *phlox*, flame, referring to red or orange colors of breeding males
    a. *Hydrophlox chiliticus* (Cope, 1870)—lipped, referring to vermilion lips (and snout) on males
    b. *Hydrophlox chlorocephalus* (Cope, 1870)—*chloros*, green; *cephalus*, head, referring to head coloration of breeding males
    c. *Hydrophlox lutipinnis* (Jordan and Brayton, 1878)—*luteus*, yellow; *pinnis*, fin, referring to color of fins on breeding males
    d. *Hydrophlox rubricroceus* (Cope, 1868)—*ruber*, red; *croceus*, saffron, referring to dominant colors of body and fins, respectively, of breeding males

- *Luxilus* Rafinesque, 1820—*lux*, light, connoting the American vernacular *shiner*; *illus*, diminutive suffix, i.e., a small, shiny fish
    a. *Luxilus albeolus* (Jordan, 1889)—whitish, the sides and fins a "pure silvery white"
    b. *Luxilus cerasinus* (Cope, 1868)—cherry red, referring to body color of breeding males
    c. *Luxilus chrysocephalus* Rafinesque, 1820—*chryso*, golden; *cephalus*, head, referring to "gilt" head

- *Lythrurus* Jordan, 1876—*lythrum*, gore; *oura*, tailed, referring to blood-red caudal fin often seen on males of *L. ardens* and *L. diploemius* (= *Chrosomus erythrogaster*)
    a. *Lythrurus ardens* (Cope, 1868)—ardent, referring to bright colors of breeding males
    b. *Lythrurus matutinus* (Cope, 1870)—of the morning, or rosy, referring to "rufous" muzzle and chin

- *Miniellus* Jordan, 1882—etymology not explained but almost certainly *iellus*, a diminutive, i.e., a small minnow (or "minnie" in American vernacular), described as "small, plain" fishes, the "smallest and most insignificant of American Cyprinidae" (date of authorship often incorrectly given as 1888; not to be confused with *Minnilus* Rafinesque, 1820, a junior synonym of *Notropis*)

a. *Miniellus alborus* (Hubbs and Raney, 1947)—*albus*, white; *oris*, mouth, referring to unpigmented lips and mouth

b. *Miniellus mekistocholas* (Snelson, 1971)—*mekisto*, longest; *cholas*, intestine, referring to elongate, convoluted intestine, an adaptation to its herbivorous diet

c. *Miniellus procne* (Cope, 1865)—*Prokne*, from Greek mythology, whom the gods transformed into a swallow, alluding to its deeply forked tail

d. *Miniellus scabriceps* (Cope, 1868)—*scaber*, rough; *ceps*, head, referring to abrasive tubercles on heads of breeding males

- *Nocomis* Girard, 1856—a Native American word, one of several that Girard arbitrarily chose for North American leuciscid genera "as being more euphonic than any one I might have framed from the Greek" (there is no evidence that Girard named the genus after Nookomis, the name of a grandmother in traditional stories among the Indigenous Ojibwe people of Michigan, Wisconsin, Minnesota, North Dakota, and Ontario and made famous in Longfellow's 1855 epic poem *The Song of Hiawatha*, in which a major female character named Nokomis falls from the moon)

  a. *Nocomis leptocephalus* (Girard, 1856)—*leptos*, small or slender; *cephalus*, head, referring to smaller head compared to *Ceratichthys* (= *Hybopsis*) *amblops*

  b. *Nocomis micropogon* (Cope, 1865)—*micro*, small; *pogon*, beard, referring to very small barbels on holotype (which was later discovered to be a *Luxilus cornutus* × *N. micropogon* hybrid; name validated by substituting holotype with a neotype). See Scharpf (2017c).

  c. *Nocomis platyrhynchus* Lachner and Jenkins, 1971—*platy*, wide; *rhynchus*, snout, referring to large gape width

  d. *Nocomis raneyi* Lachner and Jenkins, 1971—in honor of ichthyologist Edward C. Raney (1909–84), Cornell University, "whose enthusiasm and guidance placed many American students on the professional pathway to ichthyology"

- *Notemigonus* Rafinesque 1819—*notos*, back; [*h*]*emi-*, from *hemisys*, half; *gonia*, corner or angle, referring to obtusely angled, or carinated, back, in which it differs from the superficially similar genus *Clupea*.

  a. *Notemigonus crysoleucas* (Mitchell 1814)—*chrysos*, golden, referring to color of eyes and gill cover and "tinge of the same along the belly"; *leukos*, white, referring to its "shining white scales"

- *Notropis* Rafinesque, 1818—*notos*, back; *tropis*, keel, referring to ridged or keeled back of *N. atherinoides*, possibly due to shrinkage of the specimen that Rafinesque examined. See Scharpf (2017c).

  a. *Notropis amoenus* (Abbott, 1874)—pleasing, or "beautiful," as Abbott described it

  b. *Notropis bifrenatus* (Cope, 1867)—*bi*, two; *frenatus*, bridled, referring to black bars across snout

    c. *Notropis maculatus* (Hay, 1881)—spotted, referring to large caudal spot

    d. *Notropis micropteryx* (Cope, 1868)—*micro*, small; *pteryx*, fin, referring to smaller fins compared with *Alburnellus jaculus* (= *Notropis rubellus*)

    e. *Notropis photogenis* (Cope, 1865)—*photo*, light; *genis*, cheek, referring to its "bright silvery" sides, "especially brilliant" on the operculum

    f. *Notropis scepticus* (Jordan and Gilbert, 1883)—observant, referring to its large eyes

    g. *Notropis telescopus* (Cope, 1868)—far-seeing, referring to its large eyes

- *Paranotropis* Fowler, 1904—*para*, near, proposed as a subgenus of *Notropis*
  - a. *Paranotropis leuciodus* (Cope, 1868)—*leucos*, white; *eidus*, form or resemblance, i.e., whitish, presumably referring to silver sides
  - b. *Paranotropis spectrunculus* (Cope, 1868)—*specca*, spot; *trunculus*, stem, referring to spot at end of caudal peduncle
  - c. *Paranotropis volucellus* (Cope, 1865)—diminutive of *volucer*, flying or swift, probably referring to its "elongate fins, especially the dorsal"

- *Phenacobius* Cope, 1867—*phenax*, imposter; *bios*, life, referring to *P. teretulus* and *P. uranops*, which look like herbivores and superficially like suckers (Catostomidae: *Catostomus*) but are neither
  - a. *Phenacobius crassilabrum* Minckley and Craddock, 1962—*crassus*, fat; *labrum*, lip, referring to large, fleshy lips
  - b. *Phenacobius teretulus* Cope, 1867—referring to terete body form

- *Pimephales* Rafinesque, 1820—*pimele*, fat; *cephales*, head, the head of *P. promelas* being "soft and fat all over," a clear reference to fleshy growth on nape of breeding males (Rafinesque twice incorrectly translated name as "Flat-head" in description of genus, possibly a typesetting error, but correctly translated it as "Fat-head" in description of *P. promelas*)
  - a. *Pimephales notatus* (Rafinesque, 1820)—marked, probably referring to caudal fin spot
  - b. *Pimephales promelas* Rafinesque, 1820— *pro*, in front of; *melas*, black, referring to black head of breeding males

- *Pteronotropis* Fowler, 1935—*ptero*, winged, i.e., *Notropis* species with enlarged dorsal fin on breeding males
  - a. *Pteronotropis cummingsae* (Myers, 1925)—in honor of Mrs. J. H. Cummings (1885–?), amateur naturalist, for her "investigation of the Wilmington [NC] fauna and flora" (she and her husband also hosted Myers in their houseboat during his North Carolina fieldwork)

- *Rhinichthys* Agassiz, 1849—*rhinos*, nose, referring to prominent snout of *R. atronasus* (= *atratulus*); *ichthys*, fish
  - a. *Rhinichthys atratulus* (Hermann, 1804)—dressed in black, referring to stripe on body and around snout

b. *Rhinichthys cataractae* (Valenciennes, 1842)—of cataracts, referring to area around Niagara Falls, North America, type locality

c. *Rhinichthys obtusus* Agassiz, 1854—blunt, referring to more blunt body compared with *R. marmoratus* (= *cataractae*)

• *Semotilus* Rafinesque, 1820— etymology not explained, probably derived from *sēmeíon* (Greek), banner, i.e., dorsal fin, referring to spot at dorsal-fin base of *S. atromaculatus*; Jordan (1878) offered this explanation for *"tilus"*: *teleis*, or some similar word, supposed by Rafinesque to mean "spotted." But the closest word to *tilus* (*téleios*) does not mean "spotted" but quite the opposite: "perfect" or "without spot or blemish" (Christopher Scharpf, personal communication)

a. *Semotilus atromaculatus* (Mitchill, 1818)—*atro*, black; *maculatus*, spotted, referring to large black spot at front of dorsal fin base

b. *Semotilus lumbee* Snelson and Suttkus, 1978—referring to Lumbee Indians who inhabited Lumber River system in North Carolina, type locality

Family Loricariidae Rafinesque, 1815, Suckermouth Armored Catfishes

• *Pterygoplichthys* Gill, 1858—*pterygion*, diminutive of *pteryx* and fin; *hoplon*, weapon, referring to sail-like dorsal fin with single large spine; *ichthys*, fish

a. *Pterygoplichthys pardalis* (Castelnau, 1855)—*pardalis*, like a leopard, referring to round spots or points of dark brown on a light-yellow background

Family Ictaluridae Gill, 1861, North American Catfishes

• *Ameiurus* Rafinesque, 1820—*a*, without; *meiosis*, to reduce; *urus*, tailed, literally "not curtailed," referring to absence of deep notch in caudal fin compared to forked tail of *Ictalurus*

a. *Ameiurus brunneus* Jordan, 1877—brown, referring to brownish color of young and juveniles

b. *Ameiurus catus* (Linnaeus, 1758)—Latin for cat, referring to its catlike whiskers

c. *Ameiurus melas* (Rafinesque, 1820)—black, referring to color (which varies to yellowish and brown)

d. *Ameiurus natalis* (Lesueur, 1819)—Latin for "of or belonging to birth," often applied to Christmas (Noël in French), as reflected in Lesueur's vernacular name for this catfish, "Pimelode Noël," allusion not explained but almost certainly in honor of Simon Barthélemy Joseph Noël de la Morinière (1765–1822), French naturalist, journalist, author, and fisheries inspector who devoted twenty years to a projected six-volume history of fisheries of which only one volume (1815) appeared (Lesueur mentioned Noël in his 1817 description of the American Eel, *Anguilla rostrata*); most sources claim name means "having large nates or buttocks," referring to either a swollen and elevated caudal peduncle, a large adipose fin, or the swollen head and nape muscles of breeding males, an etymological error apparently based on the assumption that *natalis* was the adjectival form of the Latin noun *natis* (rump or buttocks) (Scharpf, 2020a)

    e. *Ameiurus nebulosus* (Lesueur, 1819)—Latin for cloudy, referring to olivaceous body color, "clouded with irregular brown spots"

    f. *Ameiurus platycephalus* (Girard, 1859)—*platys*, flat; *cephalus*, head, referring to "very much depressed" head

- *Ictalurus* Rafinesque, 1820—*ichthys*, fish; *aelurus*, cat, i.e., "catfish"
  a. *Ictalurus furcatus* (Valenciennes, 1840)—forked, referring to forked tail (authorship often credited to Lesueur, 1840, whose 1829 description of *Pimelodis caudofurcatus* [perhaps a synonym of *I. punctatus*] may have been unnecessarily renamed by Valenciennes)
  b. *Ictalurus punctatus* (Rafinesque, 1818)—spotted, referring to small, dark spots on body

- *Noturus* Rafinesque, 1818—*noton*, back; *oura*, tail, i.e., tail over the back, referring to connected caudal and adipose fins
  a. *Noturus eleutherus* Jordan, 1877—free, referring to "free adipose fin," i.e., incomplete fusion of adipose and caudal fins
  b. *Noturus flavus* Rafinesque, 1818—yellow, referring to the Kentucky specimens Rafinesque examined, "entirely of rufous yellow"
  c. *Noturus furiosus* Jordan and Meek, 1889—mad, "the poison of its axillary gland is more virulent than that of" its congeners
  d. *Noturus gilberti* Jordan and Evermann, 1889—in honor of friend and colleague Charles H. Gilbert (1859–1928), ichthyologist and fisheries biologist
  e. *Noturus gyrinus* (Mitchill, 1817)—latinization of *gyrinos*, tadpole, referring to tadpole-like shape
  f. *Noturus insignis* (Richardson, 1836)—remarkable or extraordinary, allusion not evident since Richardson did not provide a description; Taylor, in his 1969 revision of the genus, said the "probable intention [of the name] was to emphasize the [yellowish] color and the long adipose fin, features which were at one time considered unique"

- *Pylodictis* Rafinesque, 1819—*pelos*, mud; *ictis*, variant spelling of *ichthys*, fish, with the *d* likely inserted for euphony, reflecting Rafinesque's belief that *P. limosus* (an imaginary fish, based on a drawing by James Audubon, presumably presented to Rafinesque as a prank) lives on muddy bottoms and buries itself in the mud in the winter (Jordan, 1877 synonymized *P. limosus* with *P. olivaris*, not realizing that the latter fish was imaginary). See Scharpf (2021b).
  a. *Pylodictis olivaris* (Rafinesque, 1818)—Latin for olive-colored, referring to its body color, "olivaceous, shaded with brown"

Family Esocidae Rafinesque, 1815, Pikes

- *Esox* Linnaeus, 1758—latinized Gaulish word for a large fish from the Rhine, possibly originally applied to salmons, now applied to pikes
    a. *Esox americanus* Gmelin, 1789—American, distinguishing it from the circumpolar *E. lucius*
    b. *Esox masquinongy* Mitchill, 1824—Native American name for this species, from the Ojibway (Chippewa) mask, ugly; *kinongé*, fish (due to a bibliographic error, Mitchill's description had been "lost" since its publication until 2015, when it was rediscovered by Ronald Fricke, upon which it was revealed that Mitchill used a vernacular name instead of proposing a new binomial; Jordan, who searched for Mitchill's description but never found it, nevertheless treated the name as valid in 1885, a decision accepted by every fish taxonomist ever since; technically, name and/or author and/or date should change depending on first available name [not researched], but prevailing usage may apply). See Scharpf (2016b, 2020b).
    c. *Esox niger* Lesueur, 1818—black or dark, referring to its juvenile coloration

Family Umbridae Bonaparte, 1845, Mudminnows

- *Umbra* Kramer, 1777—shade or shadow, allusion not explained; according to Valenciennes (1855), name refers to belief among early naturalists that *U. krameri* is rarely seen because it "preferably lives in underground caves where light does not penetrate" (translation) (name first published in 1756 but not available until 1777)
    a. *Umbra pygmaea* (DeKay, 1842)—*pygmaea*, dwarflike, referring to small size (ca. 25.4 mm) of type specimens (now lost), described as a "pigmy dace"

Family Salmonidae Jarocki or Schinz, 1822, Trouts and Salmons

- *Oncorhynchus* Suckley, 1861—*onkos*, hook; *rhynchus*, snout, referring to hooked lower jaw, or kype, of breeding males
    a. *Oncorhynchus mykiss* (Walbaum, 1792)—derived from *mykizha*, vernacular name for this species used in the Kamchatka Peninsula in the sixteenth century
    b. *Oncorhynchus nerka* (Walbaum, 1792)—Russian name for this species

- *Salmo* Linnaeus, 1758—ancient word for salmon, probably either of Celtic origin or from the pre-Indo-European Iberian of Aquitania (Andrews, 1955)
    a. *Salmo trutta* Linnaeus, 1758—Latin word for trout, possibly derived from the Greek *troktes*, meaning nibbler

- *Salvelinus* Richardson, 1836—latinization of *säibling*, an old German word for char (Richardson credited the name to Nilsson, 1832, who used it for the group "Salvelini")

a. *Salvelinus fontinalis* (Mitchill, 1814)—living in or near springs; "He lives in running waters only," Mitchill wrote, "and not in stagnant ponds; and, therefore, the lively streams, descending north and south from their sources on Long–Island [NY], exactly suit the constitution of this fish"

Family Aphredoderidae Bonaparte, 1845, Pirate Perches

• *Aphredoderus* Lesueur, 1833—*aphodos*, excrement; *dere*, neck or throat, referring to anterior placement of anus, just under head in front of pelvic fins (note: vernacular name "Pirate Perch" originated with "The Pirate," coined by New Jersey archaeologist and naturalist Charles C. Abbott [1843–1919] in 1872, who observed a specimen trying to steal a minnow from the mouth of another in an aquarium)

    a. *Aphredoderus sayanus* (Gilliams, 1824)—*anus*, belonging to; eponym not identified but almost certainly in honor of Gilliams's good friend and colleague, naturalist Thomas Say (1787–1834). See Scharpf (2014b).

Family Amblyopsidae Bonaparte, 1845, Cavefishes

• *Chologaster* Agassiz, 1853—*cholos*, maimed; *gaster*, belly, referring to its lack of ventral fins

    a. *Chologaster cornuta* Agassiz, 1853—*cornuta*, horned, referring to its tubular, hornlike nostrils

Family Eleotridae Bonaparte, 1835, Sleepers

• *Dormitator* Gill, 1861—Latin for "one who sleeps," derived from "dormeur," vernacular used in the early nineteenth-century French colonies of South America and source of the English vernacular "sleeper," presumably referring to their seemingly lethargic behavior (see *Eleotris*)

    a. *Dormitator maculatus* (Bloch, 1792)—spotted, described as having brown spots on head and sides and black spots on belly

• *Eleotris* Bloch and Schneider, 1801—according to Cuvier and Valenciennes (1837), derived from a Greek name for an unidentified fish from the Nile, whereas Jordan and Gilbert (1883) claim that name derives from the Greek *eleos* (bewildered), perhaps alluding to the vernacular "sleeper," which appears to be an English translation of the vernacular "dormeur" used in the early nineteenth-century French colonies of South America; Valenciennes stated that dormeurs are "usually lazy fish, which stand quietly in the mud or rocks holes" (translation). See Scharpf (2015).

    a. *Eleotris amblyopsis* (Cope, 1871)—*amblys*, blunt; *opsis*, face, referring to its prominent chin

Family Gobiidae Cuvier, 1816, Gobies

• *Awaous* Valenciennes, 1837—latinization of *awao* or *awaou*, local name of *A. ocellaris* in Tahiti

    a. *Awaous banana* (Valenciennes, 1837)—latinization of *banane*, its local name in the Dominican Republic

- *Ctenogobius* Gill, 1858—*cteno*, comb, referring to its "pectinated" scales; *gobius*, goby
    a. *Ctenogobius boleosoma* (Jordan and Gilbert, 1882)—*bole*, dart; *soma*, body, but here referring to its "remarkable resemblance" to the North American percid *Etheostoma* (*Boleosoma*) *olmstedi*
    b. *Ctenogobius shufeldti* (Jordan and Eigenmann, 1887)—in honor of American surgeon-zoologist (and later outspoken white supremacist) Robert Wilson Shufeldt (1850–1934), who collected type. See Scharpf (2017b).

- *Evorthodus* Gill, 1859—*eu*, well; *orthos*, straight; *don*, tooth, allusion not explained, possibly referring to nearly horizontal teeth on lower jaw of *E. breviceps* (= *lyricus*)
    a. *Evorthodus lyricus* (Girard, 1858)—pertaining to a lyre, allusion not explained, perhaps referring to filiform middle rays of first dorsal fin

- *Gobiosoma* Girard, 1858—*gobius*, goby; *soma*, body, allusion not explained; Girard established this genus for *Gobius* that are "deprived of scales"
    a. *Gobiosoma bosc* (Lacepède, 1800)—in honor of French naturalist Louis-Augustin Bosc d'Antic (1759–1828), whose manuscript provided the basis of Lacepède's description (presumably a noun in apposition, without the patronymic *i*)
    b. *Gobiosoma ginsburgi* Hildebrand and Schroeder, 1928—in honor of colleague Isaac Ginsburg (1886–1975), goby taxonomist, US National Museum, who identified many gobies for the authors' monograph on fishes of Chesapeake Bay and called attention to how this species differed from *G. bosc*

- *Microgobius* Poey, 1876—*micro*, small, referring to size of *M. signatus* (described at thirty-five to forty mm but reaches sixty mm TL); *gobius*, goby
    a. *Microgobius thalassinus* (Jordan and Gilbert, 1883)—sea-green, referring to translucent body "overlaid by brilliant green luster, which is formed by exceedingly minute close-set green points"

Family Paralichthyidae Regan, 1910, Sand Flounders

- *Paralichthys* Girard, 1858—*parallens*, parallel, allusion not explained, perhaps referring to oblong body of *P. maculosus* (= *californicus*)
    a. *Paralichthys albigutta* Jordan and Gilbert, 1882—*albus*, white; *gutta*, spot, referring to "very small pale spots" on dark greenish body (eyed side)
    b. *Paralichthys dentatus* (Linnaeus, 1766)—toothed, referring to large canine teeth
    c. *Paralichthys lethostigma* Jordan and Gilbert, 1884—*lethos*, to forget; *stigma*, mark or spot, referring to absence of spots compared to the spotted *P. dentatus*, which it otherwise resembles

Family Achiridae Rafinesque, 1815, American Soles

- *Trinectes* Rafinesque, 1832—*tri*, three; *nectes*, swimmer, referring to "only three fins, dorsal, anal and caudal" of *T. scabra* (= *maculatus*) (rudimentary pectoral fin sometimes present on eyed side of body)
  - a. *Trinectes maculatus* (Bloch and Schneider, 1801)—spotted, referring to black spots on both sides of body (usually fainter and sometimes absent on blind side)

Family Cichlidae Bonaparte, 1835, Cichlids

- *Coptodon* Gervais, 1853—*copto*, split or divided; *odon*, tooth, presumably referring to bifid maxillary teeth of *C. zillii*
  - a. *Coptodon zillii* (Gervais, 1848)—in honor of M. (probably Monsieur) Zill, the "distinguished naturalist" (translation) who collected type and sent it to Gervais

- *Oreochromis* Günther, 1889—*oreos*, mountains, referring to Mt. Kilimanjaro (Kilimanjaro District, Tanzania), location of *O. hunteri*, type species; *chromis*, a name dating to Aristotle, possibly derived from *chroemo* (to neigh), referring to a drum (Sciaenidae) and its ability to make noise, later expanded to embrace cichlids, damselfishes, dottybacks, and wrasses (all perch-like fishes once thought to be related), often used in the names of African cichlid genera following *Chromis* (now *Oreochromis*) *mossambicus* Peters, 1852
  - a. *Oreochromis aureus* (Steindachner, 1864)—golden, referring to golden-yellow body color

Family Atherinopsidae Fitzinger, 1873, New World Silversides

- *Labidesthes* Cope, 1870—*labidos*, forceps; *esthio*, eat, referring to prolonged jaws, which form a short, depressed beak
  - a. *Labidesthes sicculus* (Cope, 1865)—etymology not explained, perhaps an adjectival form of *sicula*, a small dagger, referring to its sharp snout and daggerlike shape; Jordan and Evermann (1896–1900) posit that the name derives from *siccus*, dry or desiccated, referring to Cope having found his specimens in dried ponds, but these specimens were collected in 1869 from Tennessee, whereas the type specimen came from Michigan ca. 1864. See also Scharpf (2014c).
  - b. *Labidesthes vanhyningi* Bean and Reid, 1930—in honor of herpetologist Oather C. Van Hyning (1901–73), who collected type (and not his father Thompson H. Van Hyning, first director of the Florida Museum of Natural History, as is sometimes reported)

- *Membras* Bonaparte, 1836—Greek word for a kind of herring or anchovy (i.e., a small silvery fish that lives in the sea), dating to at least Aristotle
  - a. *Membras martinica* (Valenciennes, 1835)—*ica*, belonging to; Martinique Island, West Indies, type locality (although record of it occurring there is doubtful, per Chernoff, 1986)

- *Menidia* Bonaparte, 1836—presumably tautonymous with *Atherina menidia* (no species mentioned), diminutive of *mene*, moon, ancient name of some small silvery fish, referring to silver-metallic white of its scales
  - a. *Menidia beryllina* (Cope, 1867)—like the mineral beryl (e.g., emerald), presumably referring to "bright pale olive" body color
  - b. *Menidia extensa* Hubbs and Raney, 1946—*extensa*, stretched out, referring to its more slender body compared with congeners
  - c. *Menidia menidia* (Linnaeus, 1766)—diminutive of *mene*, moon, ancient name of some small silvery fish, referring to silver-metallic white of its scales

Family Fundulidae Günther, 1866, Topminnows

- *Fundulus* Lacepède, 1803—*fundus*, bottom; *ulus*, a diminutive suffix, i.e., a "small burrower," referring to "mudfish," local name for *F. heteroclitus* in South Carolina, perhaps referring to its occurrence in muddy pools, creeks, and ditches and/or to how they bury themselves fifteen to twenty cm into the mud during winter
  - a. *Fundulus chrysotus* (Günther, 1866)—based on manuscript name coined by physician-naturalist John E. Holbrook (1796–1871); scholars have offered two etymologies: gilded, referring to gold flecks on sides, and *chrysos*, gold, and *otos*, ear, referring to gold iridescence on opercle (neither character mentioned by Günther, who remarked "it is impossible to know whether the specimens described are identical with those for which Holbrook intended this name")
  - b. *Fundulus confluentus* Goode and Bean, 1879—flowing together, allusion not explained, perhaps referring to confluence of salt and fresh water at type locality (Lake Monroe, FL), which is 161 miles from the sea; Wildekamp and Watters (1996) say name refers to "partial interconnection of the cross-bars on the sides of the body" but provides no source for this explanation
  - c. *Fundulus diaphanus* (Lesueur, 1817)—transparent, referring to its semitranslucent ("diaphanous") body (probably a male)
  - d. *Fundulus heteroclitus* (Linnaeus, 1766)—*heteros*, different; *clinus*, leaning or inclining, i.e., deviating, abnormal, or different, allusion not explained, perhaps referring to Linnaeus's uncertainty ("Genus *nondun certam*") in placing it in the loach genus *Cobitis*, from which it clearly differs; Wildekamp and Watters (1996) states that name refers to "differences between the sexes," but sexual dimorphism is not included in Linnaeus's brief description (based on notes from South Carolina naturalist Andrew Garden, who sent Linnaeus right half-skins of two specimens, pressed in a botanical press, varnished, and glued to a sheet of herbarium paper). See Scharpf (2019).
  - e. *Fundulus lineolatus* (Agassiz, 1854)—lined, presumably referring to black stripes on sides of females (vertical bars on males)
  - f. *Fundulus luciae* (Baird, 1855)—in honor of Baird's daughter, Lucy Hunter Baird (1848–1913)

   g. *Fundulus majalis* (Walbaum, 1792)—pertaining to May, based on "Mayfish," local name recorded by Schöpf (1788), who collected specimens from New York City's East River

   h. *Fundulus rathbuni* Jordan and Meek, 1889—in honor of Richard Rathbun (1852–1918), chief of the Division of Scientific Inquiry, US Fish Commission

   i. *Fundulus waccamensis* Hubbs and Raney, 1946—*ensis*, suffix denoting place: Lake Waccamaw, North Carolina, where it is endemic

- *Lucania* Girard, 1859—a Native American word chosen presumably because Girard liked the sound of it

   a. *Lucania goodei* Jordan, 1880—in honor of ichthyologist George Brown Goode (1851–96), who collected type

   b. *Lucania parva* (Baird and Girard, 1855)—small, referring to its "diminutive size" (up to 6.2 cm TL)

Family Cyprinodontidae Wagner, 1828, Pupfishes

- *Cyprinodon* Lacepède, 1803—*cyprinus*, carp or minnow; *odon*, tooth, i.e., a carp- or cyprinid-like fish but with teeth (hence "tooth carps," another name for the order)

   a. *Cyprinodon variegatus* Lacepède, 1803—*variegatus*, variegated, referring to variable color patterns of brown spots and bands on sides

Family Poeciliidae Bonaparte, 1831, Livebearers

- *Gambusia* Poey, 1854—latinization of the Cuban *gambusino*, signifying "nothing," referring to their inconsequential size; "to fish for gambusinos" is an idiomatic way of saying one has caught nothing

   a. *Gambusia affinis* (Baird and Girard, 1853)—related, allusion not explained, but probably referring to its close relationship with *G. holbrooki*, whose name (i.e., *G. holbrooki*), although not formally published until 1859, was first used in an unpublished 1853 manuscript by Louis Agassiz (Agassiz, 1853).

   b. *Gambusia holbrooki* Girard, 1859—manuscript name coined by Louis Agassiz in 1853 but first made available by Girard; patronym not identified but certainly in honor of physician-naturalist John E. Holbrook (1796–1871), who lived in South Carolina, where Girard said this species was "very abundant in ponds and ditches of fresh water"

- *Heterandria* Agassiz, 1853—*heteros*, different; *andros*, male, referring to sexually dimorphic anal fin, i.e., the gonopodium

   a. *Heterandria formosa* (Girard, 1859)—comely or attractive, allusion not explained but probably referring to its appearance [name coined by Louis Agassiz in 1855 who mentioned only its diminutive size; workers who believe this mention constitutes a distinguishing feature assign authorship to Agassiz, 1855; Girard's authorship often appears without parentheses, but he originally placed species in *Girardinus*]

- *Poecilia* Bloch and Schneider, 1801—from the Greek *poikilos*, variegated or speckled, referring to color pattern of *P. vivipara* and other cyprinodontiform fishes (e.g., *Fundulus heteroclitus*) that Bloch and Schneider placed in the genus
    a. *Poecilia latipinna* (Lesueur, 1821)—*latus*, broad; *pinna*, fin, referring to enlarged dorsal fin of male

Family Belonidae Bonaparte, 1835, Needlefishes

- *Strongylura* van Hasselt, 1824—*strongylos*, round; *oura*, tailed, allusion not explained but probably referring to conspicuous round spot on tail of *Belone* (now *Strongylura*) *strongylura* (not strictly a tautonym since van Hasselt, 1824 is a French translation of van Hasselt, 1823, which is in Dutch; for reasons unexplained, the French version renamed *Belone strongylura* as *Strongylura caudimaculata*)
    a. *Strongylura marina* (Walbaum, 1792)—of the sea, at the time presumed to be a marine inhabitant of the pike genus *Esox* (Esociformes: Esocidae)

Family Mugilidae Jarocki, 1822, Mullets

- *Dajaus* Valenciennes, 1836—latinization of *dajao*, local name for this mullet in Puerto Rico
    a. *Dajaus monticola* (Bancroft, 1834)—of the mountains, referring to its occurrence in steep forested streams as high as 1500 m

- *Mugil* Linnaeus, 1758—Latin for mullet, possibly derived from *mulgeo*, to suck, referring to how *M. cephalus* feeds by sucking up sediment
    a. *Mugil cephalus* Linnaeus, 1758—*kephalos*, a name dating to Aristotle (it is not clear if the name relates to *cephalus*, meaning head)
    b. *Mugil curema* Valenciennes, 1836—*curema*, Portuguese vernacular used by Dutch naturalist Georg Marcgraf in his 1648 *Historia naturalis brasiliae*, doubtless corresponding to the Spanish vernacular *querimana* (or *queriman*)

Family Moronidae Jordan and Evermann, 1896, Temperate Basses

- *Morone* Mitchill, 1814—etymology not explained nor evident, making it perhaps the most enigmatic name for such a well-known group of fishes; our best guess is that it is from *morone*, an archaic version of maroon, possibly referring to the red, ruddy, or rusty colors Mitchill described on all four taxa he included in the genus (less than a year later, Mitchill discarded *Morone* for the labrid name *Bodianus* and never mentioned *Morone* again). See Scharpf (2016a).
    a. *Morone americana* (Gmelin, 1789)—American, then believed to be an American representative of the largely European genus *Perca* (Perciformes: Percidae)
    b. *Morone chrysops* (Rafinesque, 1820)—*chrysos*, gold; *ops*, eye, referring to gold or yellow cast of iris
    c. *Morone saxatilis* (Walbaum, 1792)—living among rocks, presumably derived from its common name in New York, Rockfish, as reported in Schöpf (1788), possibly referring to its often being caught near coastal rocky ledges

Family Sciaenidae Cuvier, 1829, Drums and Croakers

- *Aplodinotus* Rafinesque, 1819—etymology not explained, perhaps (*h*)*aplous*, single, and *notos*, back, referring to confluent spinous and soft dorsal fins. But see Scharpf (2020d).
  - a. *Aplodinotus grunniens* Rafinesque, 1819—Latin for grunting, referring to drumlike sounds that resonate from swim bladder of mature males (hence the common names drum and croaker)

- *Bairdiella* Gill, 1861—*ella*, diminutive connoting endearment; patronym not identified but almost certainly in honor of Spencer Fullerton Baird (1823–87), director, US National Museum (where Gill worked)
  - a. *Bairdiella chrysoura* (Lacepède, 1802)—*chrysos*, gold; *oura*, tailed, referring to yellow caudal fin

- *Leiostomus* Lacepède, 1802—*leios*, smooth; *stomus*, mouth, referring to lack of teeth on lower jaw of adults (upper jaw with minute teeth)
  - a. *Leiostomus xanthurus* Lacepède, 1802—*xanthus*, yellow; *oura*, tailed, referring to yellow caudal fin, a misnomer since fin is actually dusky or olivaceous (description based on notes provided by naturalist Louis-Augustin Bosc d'Antic, who may have confused this species with *Bairdiella chrysoura*)

- *Micropogonias* Bonaparte, 1831—*micro*, small; *pogonias*, bearded, replacement name for *Micropogon* Cuvier, 1830 (preoccupied by *Micropogon* Boie, 1826 in birds), referring to three to five pairs of small barbels or "whiskers" on chin of *M. lineatus* (= *undulatus*)
  - a. *Micropogonias undulatus* (Linnaeus, 1766)—wavy, referring to dark, wavy streaks on sides

- *Sciaenops* Gill, 1863—*ops*, appearance, presumably similar to (and previously recognized as) *Sciaena*
  - a. *Sciaenops ocellatus* (Linnaeus, 1766)—with eyelike spots, referring to distinctive black spot near base of caudal fin (some individuals exhibit several spots)

Family Percidae Rafinesque, 1815, Perches and Darters

- *Etheostoma* Rafinesque, 1819—etymology not explained, perhaps *etheo*, filter or strain, and *stoma*, mouth, which does not seem to match Rafinesque's 1820 explanation ("The name means different mouths," in which case name should be "heterostoma"), perhaps referring to how the three darter species known to him (*Percina caprodes, Etheostoma blennioides,* and *E. flabellare*) are so different in respect of the form of the mouth that he conceived that they might belong to different subgenera
  - a. *Etheostoma blennioides* Rafinesque, 1819—*oides*, having the form of; *blennius*, blenny, having the "appearance, head, and spots of many Blennies" (per Rafinesque, 1820)

b. *Etheostoma brevispinum* (Coker, 1926)—*brevis*, short; *spinus*, spined, allusion not explained, perhaps referring to "short stiff spines" of first dorsal fin, lower than second dorsal fin

c. *Etheostoma collis* (Hubbs and Cannon, 1935)—*collis*, of the high ground, referring to how it "seems to occur at a higher elevation than most species" of darter (above the Fall Line)

d. *Etheostoma flabellare* Rafinesque, 1819—like a *flabellum*, fan, referring to fanlike pattern on caudal fin of adult males

e. *Etheostoma fusiforme* (Girard, 1854)—*fusus*, spindle; *forma*, shape, referring to "slender and fusiform" body shape

f. *Etheostoma gutselli* (Hildebrand, 1932)—*gutselli*, in honor of James S. Gutsell (1887–1976), US Bureau of Fisheries, who collected type while studying the effects of tannery and factory wastes into western North Carolina streams

g. *Etheostoma inscriptum* (Jordan and Brayton, 1878)—*inscriptum*, inscribed, presumably referring to an "orange spot on each scale, these forming continuous lines along the rows of scales" of mature males

h. *Etheostoma kanawhae* (Raney, 1941)—*kanawhae*, of the Kanawha River system, Virginia and North Carolina, of which the New River (where this darter is endemic) is a main tributary

i. *Etheostoma maculaticeps* (Cope, 1870)—*maculatus*, spotted; *ceps*, head, referring to large brown spots on top of nape, head, and snout

j. *Etheostoma mariae* (Fowler, 1947)—*mariae*, in honor of Maria Darlington (d. 1951), wife of entomologist Emlen P. Darlington, Fowler's colleague at the Academy of Natural Sciences of Philadelphia; Mrs. Darlington was the "generous sponsor whose kindly interest has made possible the discovery of this interesting little fish"

k. *Etheostoma perlongum* (Hubbs and Raney, 1946)—*per*, very; *longus*, long, i.e., "extremely long" according to the authors, referring to its length compared with other *Boleosoma*

l. *Etheostoma podostemone* Jordan and Jenkins, 1889—named for its occurrence in swift water, especially among rocks covered by the riverweed *Podostemum ceratophyllum.*

m. *Etheostoma serrifer* (Hubbs and Cannon, 1935)—*serra*, saw; *fero*, to bear, referring to "diagnostically serrate preopercle"

n. *Etheostoma simoterum* (Cope, 1868)—*simus*, flat-nosed; *teres*, rubbed off, i.e., snub-nosed, referring to its "obtuse muzzle"

o. *Etheostoma swannanoa* Jordan and Evermann, 1889—*swannanoa*, named for the South Fork of the Swannanoa River, Black Mountain, North Carolina, paratype locality (also occurs in Virginia and Tennessee)

p. *Etheostoma thalassinum* (Jordan and Brayton, 1878)—*thalassinum*, sea-green, presumably referring to male coloration in life: dark-green or olive body with six to nine dark blue-green vertical bars, "grass-green" opercular region and middle of pectoral fins, and "brilliant" blue-green color at base of anal fins

q. *Etheostoma vitreum* (Cope, 1870)—*vitreum*, glassy, referring to the "transparency of its muscles" in life

r. *Etheostoma zonale* (Cope, 1868)—*zonale*, banded, referring to "broad brown lateral band [usually a series of unconnected spots], from which eight narrower bands more or less completely encircle the belly," these bands are bright green in breeding males

- *Nothonotus* Putnam, 1863—*nothos*, false; *notos*, back, meaning unknown; Jordan and Evermann (1896–1900) state that *nothos* means "prominent," in which case it may be a misspelled abridgment of *notablis*, presumably referring to "longer and higher" first dorsal fin compared with *Catonotus* (subgenus of *Etheostoma*); Boschung and Mayden (2004) suggest *nothos* means "spotted," but this is doubtful since Agassiz, who coined *Nothonotus* in an unpublished manuscript, also coined *Catonotus*, in which *notus* clearly means "back" and the two taxa are distinguished by the size of their first dorsal fins

  a. *Nothonotus acuticeps* (Bailey, 1959)—*acutus*, pointed or sharp; *ceps*, head, referring to its "extremely sharp" snout

  b. *Nothonotus chlorobranchius* (Zorach, 1972)—*chloro*, green; *branchos*, fin, referring to prominent, broad, dark-green bands in median fins of males in life

  c. *Nothonotus rufilineatus* (Cope, 1870)—*rufus*, reddish; *lineatus*, lined, referring to narrow longitudinal lines of males, which contain "quadrate spots of a mahogany or brick red color"

  d. *Nothonotus vulneratus* (Cope, 1870)—wounded, referring to red spots on males

- *Perca* Linnaeus, 1758—from *perke*, ancient Greek name for *P. fluviatilis* of Europe, dating to at least Aristotle

  a. *Perca flavescens* (Mitchill, 1814)—*flavescens*, yellowish or golden-yellow, described as "brown or olive on the back, turning to yellow on the sides"

- *Percina* Haldeman, 1842—*percina*, diminutive of *Perca*, meaning "little perch"; *ina*, Latin feminine suffix with diminutive implications, proposed as a subgenus of *Perca* (name often translated as "small perch," but Haldeman did not reference size in his description)

  a. *Percina aurantiaca* (Cope, 1868)—*aurantiaca*, orange-colored, referring to "bright yellow" color in life (probably referring to yellow and tangerine middle of lower body of breeding males)

  b. *Percina burtoni* Fowler, 1945—*burtoni*, in honor of E. Milby Burton (1898–1977), director of the Charleston Museum (SC), who collected local fishes for his museum, including type of this one, and invited Fowler to study them; Fowler called him a friend to whom he was "indebted for many favors"

  c. *Percina caprodes* (Rafinesque, 1818)—*caprodes* means "like a pig," from the prominent snout that often is fleshy, upturned, and blunt

  d. *Percina crassa* (Jordan and Brayton, 1878)—*crassa*, thick, referring to its stocky appearance, "more heavily built" than *Alvordius aspro* (= *P. maculata*), "the form being less graceful than that of the other members of the genus"

  e. *Percina evides* (Jordan and Copeland, 1877)—*evides*, comely, being "one of the most beautiful of all the darters"

  f. *Percina gymnocephala* Beckham, 1980—*gymno*, naked or lightly clad; *cephala*, head, referring to reduced scalation of head and nape compared with other members of the *P. maculata* species group

  g. *Percina nevisensis* (Cope, 1870) (*Percina nevisense* [Cope, 1870])—in the book *American Darters*, Kuehne and Barbour (1983) stated that *nevisense* means "birthmark," probably referring to lateral blotches that Edward D. Cope, the species' author, described as "dark chestnut quadrate spots" on sides. However, this is likely a mistranslation. Since the species was originally described in the genus *Etheostoma*, which is neuter in gender, it seems more likely that *nevisense* is the neuter version of *nevisensis* (-*ensis* is an adjectival suffix that connotes place). Cope described the species from one specimen from the Neuse River in Wake County, so maybe Cope latinized the *u* to *v* (in Classical Latin the *v* shape was used for both the vowel *u* and the consonant *v*); the *i* may have been added for euphony or is a printer's error (Christopher Scharpf, personal communication).

  h. *Percina oxyrhynchus* (Hubbs and Raney, 1939)—*oxy*, sharp; *rhynchus*, nose, referring to its "extraordinarily sharp snout"

  i. *Percina rex* (Jordan and Evermann, 1889)—*rex*, king, referring to its being the largest known darter at the time (that distinction belongs to *P. lenticula*)

  j. *Percina roanoka* (Jordan and Jenkins, 1889)—*roanoka*, named for the Roanoke River system of Virginia, type locality

  k. *Percina squamata* (Gilbert and Swain, 1887)—*squamata*, scaled, referring to any or all of the following: body covered with very small scales, uniform in size, covering belly (instead of belly scutes as in presumed congeners in *Etheostoma*); cheeks, breast, and nuchal region covered with "still finer" scales; larger, spinous scales on opercle; an enlarged black humeral scale

  l. *Percina westfalli* (Fowler, 1942)—*westfalli*, named after Minter J. Westfall Jr. (1916–2003), collector of the species

• *Sander* Oiken, 1817—latinization of Cuvier's "Les Sandres," from *zander*, German name for *S. lucioperca* (treated as an invalid name by some workers, replaced by *Stizostedion* Rafinesque, 1820 [*stizo*, prick; *stithios*, a little breast, referring to "pungent throat" or spiny opercle at pectoral-fin base of *S. vitreus*]); consulting with Ronald Fricke of *Eschmeyer's Catalog of Fishes*, we believe *Sander* was validly published and should be retained (Scharpf, 2021c; Scharpf and Fricke, 2022).

  a. *Sander canadensis* (Griffith and Smith, 1834)—*ensis*, suffix denoting place; described from Canada (but no types known)

  b. *Sander vitreus* (Mitchill, 1818)—*vitreus*, glassy, referring to the tapetum lucidum of its eyes, "appearing like the semi-globes of glass in the decks of vessels, when illuminated on the opposite sides"

Family Gasterosteidae Bonaparte, 1831, Sticklebacks

- *Apeltes* DeKay, 1842—*a*, without; *peltes*, shield, referring to lack of lateral plates
  a. *Apeltes quadracus* (Mitchill, 1815)— *quadra*, four; *acus*, needle, referring to four (sometimes two or three) free dorsal spines

Family Cottidae Bonaparte, 1831, Sculpins

- *Cottus* Linnaeus, 1758—Latinization of the Greek *kottos, kothos*, and a few similar words, denoting a bulging head, used as a name for small freshwater fishes with a large head
  a. *Cottus bairdii* Girard, 1850—in honor of Spencer Fullerton Baird (1823–87), assistant secretary of the Smithsonian Institution, director of the US National Museum, and US Commissioner of Fish and Fisheries, who collected type
  b. *Cottus caeruleomentum* Kinziger, Raesly and Neely, 2000 —*caeruleus*, blue; *mentum*, chin, referring to blue chin of spawning males
  c. *Cottus carolinae* (Gill, 1861)—in honor of Gill's "estimable young friend" Miss Caroline Henry (1839–1920); she was the daughter of Joseph Henry (1797–1878), first secretary of the Smithsonian Institution, who took Gill under his wing when Gill was beginning his career

Family Centrarchidae Bleeker, 1859, Sunfishes

- *Acantharchus* Gill, 1864—*acanthus*, thorn or spine; *archos*, anus, allusion not explained, presumably referring to five anal fin spines
  a. *Acantharchus pomotis* (Baird, 1855)—*poma*, lid or gill cover; *otos*, ear, allusion not explained, possibly referring to scaly operculum common to all centrarchids (*Pomotis* is also a junior synonym of *Lepomis* and may have been used here as a generic term for sunfish)

- *Ambloplites* Rafinesque, 1820—*amblys*, blunt; *hoplites*, armed, defined by Rafinesque to mean "obtuse weapons," referring to "two flat, broad and obtuse spines above the opercula" of *A. ictheloides* (= *rupestris*)
  a. *Ambloplites cavifrons* Cope, 1868—*cavus*, concave; *frons*, forehead, referring to concave profile over eyes
  b. *Ambloplites rupestris* (Rafinesque, 1817)—living among rocks, a "permanent fish living generally in rocky bottoms" (hence the common name Rock Bass)

- *Centrarchus* Cuvier, 1829—*kentron*, spine; *archos*, anus, similar to *Pomotis* (= *Lepomis*) but with more (seven to nine versus three) anal fin spines
  a. *Centrarchus macropterus* (Lacepède, 1801)—*macro*, long or large; *pterus*, fin, presumably referring to long rays of dorsal and anal fins, nearly all of which Lacepède said were "furnished with filaments" (translation), which they are not

- *Enneacanthus* Gill, 1864—*ennea*, nine; *acanthus*, thorn or spine, referring to nine dorsal fin spines of *E. obesus*
  a. *Enneacanthus chaetodon* (Baird, 1855)—etymology not explained, possibly referring to marine butterflyfishes, *Chaetodon* (Acanthuriformes: Chaetodontidae), some of which have a similarly banded appearance
  b. *Enneacanthus gloriosus* (Holbrook, 1855)—splendid, a "beautiful little fish," its head, body, and dorsal and anal fins marked with silver to blue spots
  c. *Enneacanthus obesus* (Girard, 1854)—plump or fat, allusion not explained, perhaps referring to occasional chubby appearance of adult males

- *Lepomis* Rafinesque, 1819—*lepis*, scale; *poma*, lid or cover, referring to scaly operculum ("opercules ecailleux")
  a. *Lepomis auritus* (Linnaeus, 1758)—eared, referring to long opercular flap, especially in adults. See Scharpf (2018).
  b. *Lepomis cyanellus* Rafinesque, 1819—diminutive of *cyano*, blue, allusion not explained, probably referring to sky-blue spot on each scale and blue flexuous lines on cheek
  c. *Lepomis gibbosus* (Linnaeus, 1758)—humpbacked, allusion not explained, perhaps referring to steep, convex profile of large males
  d. *Lepomis gulosus* (Cuvier, 1829)—large-mouthed, alluding to its common name among the French colonists of Louisiana, *de grande gueule*, referring to its large mouth, the maxillary reaching posterior border of eye
  e. *Lepomis macrochirus* Rafinesque, 1819—*macro*, long or large; *cheiros*, hand, referring to long pectoral fins, reaching anal fin
  f. *Lepomis marginatus* (Holbrook, 1855)—bordered, referring to green border on opercular appendix
  g. *Lepomis microlophus* (Günther, 1859)—*micro*, small; *lophus*, crest or nape, allusion not explained; replacement name for *Pomotis speciosus* Holbrook, 1855, preoccupied by *P. speciosus* Baird and Girard, 1854, now a subspecies of *L. macrochirus*, perhaps referring to "fleshy appendix" of opercle (per Holbrook) or "small" opercular lobe (per Günther) (Pflieger, 1997, says name means "small nape," an accurate translation, but nape is not mentioned in either description, nor does it seem small to us)
  h. *Lepomis punctatus* (Valenciennes, 1831)—spotted, referring to small black (actually black to reddish) spots on cheeks and lower part of body

- *Micropterus* Lacepède, 1802—*micro*, small; *pterus*, fin, referring to detached posterior rays of damaged dorsal fin on type specimen of *M. dolomieu*, which Lacepède mistakenly believed was a separate but small fin
  a. *Micropterus coosae* Hubbs and Bailey, 1940—of the Coosa River system, Alabama and Georgia, where it occurs (a separate population in the Savannah River basin of Georgia may represent a separate species known as "Bartram's Bass")

b. *Micropterus dolomieu* Lacepède, 1802—in honor of geologist Déodat de Dolomieu (1750–1801, for whom "dolomite" is named), expressing joy and relief that Dolomieu had just been released by Italy after twenty-one months in solitary confinement during a France-Italy border dispute (Dolomieu was sick when he was imprisoned, and his health deteriorated in the horrible conditions of his captivity; by the time the description of *M. dolomieu* appeared in print, he was dead). See Scharpf (2020c).

c. *Micropterus henshalli* Hubbs and Bailey, 1940—in honor of angler-naturalist James A. Henshall (1836–1925) for "raising the black basses to their position of high esteem in the minds of sportsmen" and for "determining their proper nomenclature"

d. *Micropterus punctulatus* (Rafinesque, 1819)—diminutive of *punctum*, spot, referring to olivaceous body covered with blackish dots (per Rafinesque, 1820)

e. *Micropterus salmoides* (Lacepède, 1802)—*-oides*, having the form of; *salmo*, salmon; Lacepède based name on a specimen labeled "trout," reflecting local name for this fish in Charleston, South Carolina, type locality

- *Pomoxis* Rafinesque, 1818—*poma*, lid or cover; *oxys*, sharp, referring to opercles ending in two flat points, as opposed to extended into a flap as in some other sunfishes

    a. *Pomoxis annularis* Rafinesque, 1818—ringed or ring-shaped, presumably referring to "golden ring at the base of the tail"

    b. *Pomoxis nigromaculatus* (Lesueur, 1829)—*nigro*, black; *maculatus*, spotted, referring to blackish speckles and mottles on silver body and to blackish spots between rays of fins

Family Elassomatidae Jordan, 1877, Pygmy Sunfishes

- *Elassoma* Jordan, 1877—*elasson*, smaller; *soma*, body, referring to their diminutive size, or "a being reduced or diminished" (per Jordan)

    a. *Elassoma boehlkei* Rohde and Arndt, 1987—in honor of James E. Böhlke (1930–82), late Curator of Fishes, Academy of Natural Sciences of Philadelphia, for his contributions to ichthyology and his interest in *Elassoma*. See Scharpf (2017a).

    b. *Elassoma evergladei* Jordan, 1884—of the Florida Everglades, where type locality (Lake Jessup) is situated (occurs in coastal streams of North Carolina south to the Everglades and west to Alabama)

    c. *Elassoma zonatum* Jordan, 1877—banded, described as having approximately eleven parallel, vertical, dark-olive bands on body

GLOSSARY

Adapted from the glossaries of Froese and Pauly (2022), Jenkins and Burkhead (1994), and Rohde et al. (2009)

abbreviate heterocercal caudal fin— A caudal fin type in which the posterior end of the vertebral column is distinctly upturned but short.

Abbreviate heterocercal caudal fin of Bowfin, *Amia calva*.

adipose eyelids—Translucent tissue that totally or partially covers the eye in some fishes.

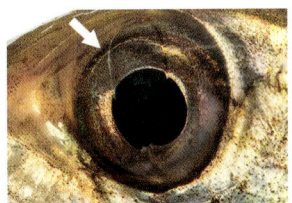

Adipose eyelid in Blueback Herring, *Alosa aestivalis*.

adipose fin—A small or medium-sized fleshy fin lacking rays and spines and occurring on the dorsum between the dorsal and caudal fins.

Adipose fin of Rainbow Trout, *Oncorhynchus mykiss*.

anteromedially—Toward the anterior middle region of the breast between the pectoral fins.

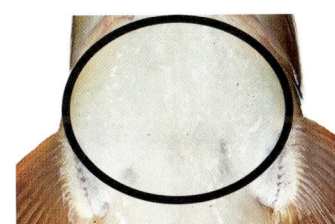

Anteromedial region of Black Redhorse, *Moxostoma duquesnei*.

axillary process (axillary scale) — Narrow flap of flesh located just above the outside base of the pectoral or the pelvic fin.

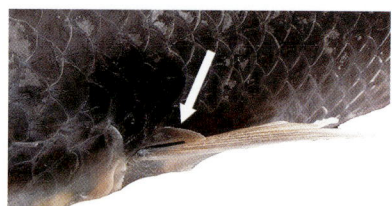

Axillary process (axillary scale) of Striped Mullet, *Mugil cephalus*.

barbel—Fleshy sensory projections around the mouth, nostrils, or chin; it may be large and obvious or small and inconspicuous.

basicaudal spot—Spot at the base of the caudal fin.

Basicaudal spot of Fatlips Minnow, *Phenacobius crassilabrum*.

bicuspid—A tooth with two points (cusps) on its upper surface.

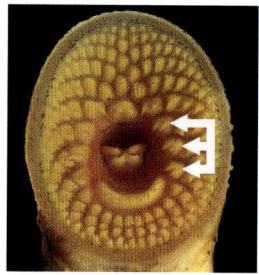

Bicuspid lateral teeth of Ohio Lamprey, *Ichthyomyzon bdellium*.

branchial region—Region posterior to the oral disk and ventral to the gill openings in lampreys.

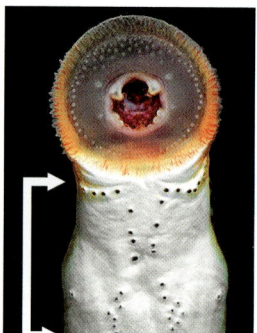

Branchial region of Least Brook Lamprey, *Lampetra aepyptera*.

branchiostegal rays—Raylike bones that support the gill membranes ventral to the operculum.

Branchiostegal rays of Walleye, *Sander vitreus*.

ca.—An abbreviation of the Latin word *circa*, meaning approximately.

caudal peduncle—Part of the body between the posterior insertion of the anal fin and the hypural plate.

Caudal peduncle of Redear Sunfish, *Lepomis microlophus*.

cephalic lateralis (lateral cephalic sensory canal; supratemporal canal)—The system of canals on the head, which are a sonar-like division of the nervous system; at intervals the canals have pores that open to the exterior and contain sensory organs that detect displacement of water.

Cephalic lateralis (lateral cephalic sensory canal; supratemporal canal) of Golden Redhorse, *Moxostoma erythrurum*.

cf.—Used to express a possible identity, or at least a significant resemblance, such as between an observed specimen and a known species or taxon. Such a usage might suggest a specimen's membership of the same genus or possibly of a shared higher taxon.

chin bar—Pigmented rectanglar-shaped bar between the halves of the lower lip.

Chin bar of Piedmont Darter, *Percina crassa*.

circumpeduncle scale count—A count of the number of scales encircling the caudal peduncle.

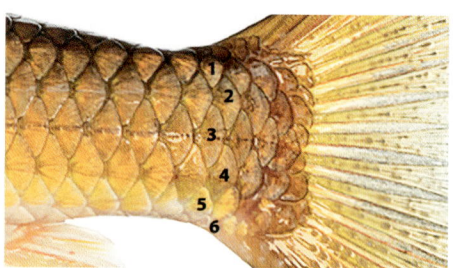

Circumpeduncle scale count in Golden Redhorse, *Moxostoma erythrurum*.

corrugate—Appearing as wrinkled, folded, or with shallow creases.

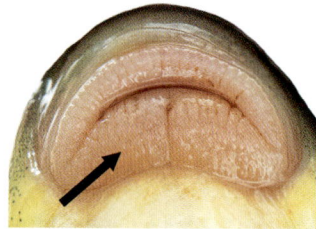

Corrugated plicae of the lower lip of Shorthead Redhorse, *Moxostoma macrolepidotum*.

ctenoid scale—Type of thin scale with numerous small, backward-pointing "teeth" on the outer edge.

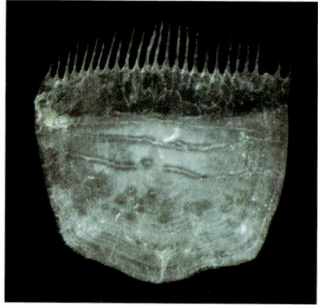

Ctenoid scale from an Atlantic Menhaden, *Brevoortia tyrannus*.

cycloid scale—Thin, light, flexible scale with smooth outer edge.

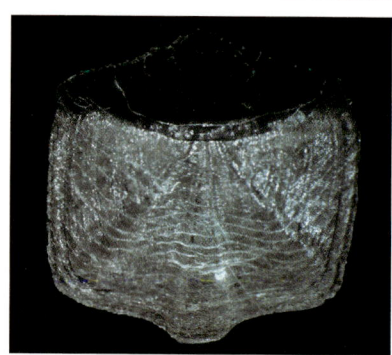

Cycloid scale from an American Shad, *Alosa sapidissima*.

emarginate—Usually referring to the caudal fin having a notched or slight mid-distal concavity of the fin margin.

Emarginate caudal fin of Redbreast Sunfish, *Lepomis auritus*.

embedded scales—Scales that are not obvious owing to deep embedment or full covering by skin.

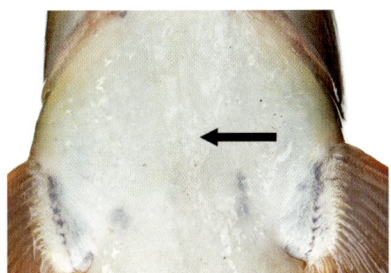

Embedded breast scales in Black Redhorse, *Moxostoma duquesnei*.

falcate—Fins with a markedly concave or sickle-shaped distal margin.

Falcate dorsal fin of Smallmouth Redhorse, *Moxostoma breviceps*.

frenum—Fleshy bridge or connection between snout and upper lip.

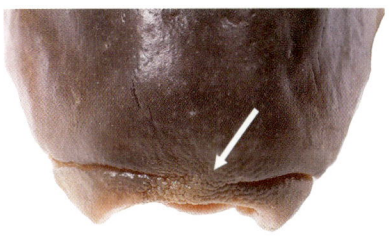

Frenum of Tonguetied Minnow, *Exoglossum laurae*.

ganoid scales—Thick scale covered with hard enamel.

Ganoid scales in Longnose Gar, *Lepisosteus osseus.*

gill rakers—Projections along the anterior edge of the gill arch.

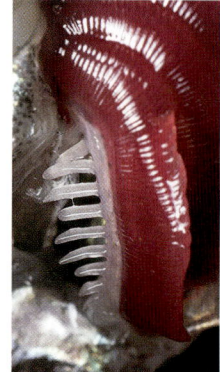

Gill rakers of Bluegill, *Lepomis macrochirus.*

gonopodium—Modified front rays of anal fin of male livebearers that serve as intromittent organs.

Gonopodium of a melanistic male Eastern Mosquitofish, *Gambusia holbrooki.*

gular plate—Large, hard bony plate found on throat.

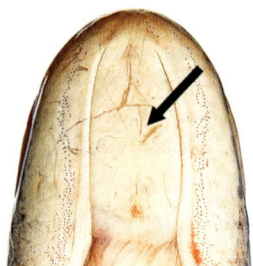

Gular plate of Bowfin, *Amia calva.*

heterocercal caudal fin—Type of caudal fin in which the vertebral column extends into the upper lobe.

Heterocercal caudal fin of Atlantic Sturgeon, *Acipenser oxyrinchus*. South Carolina. Photograph taken under the authority of National Marine Fisheries Service Endangered Species Permit No. 21198.

homocercal caudal fin—Type of caudal fin in which the vertebral column ends at the caudal fin base.

Homocercal caudal fin of Mountain Redbelly Dace, *Chrosomus oreas*.

humeral bar—Pigmented bar in the shoulder scapular area, laterally just behind the head.

Humeral bar of Fieryblack Shiner, *Cyprinella pyrrhomelas*.

hypural plate—Expanded terminal bones that form a fanlike support for the caudal fin rays. The end of the plate usually appears as a crease across the caudal peduncle.

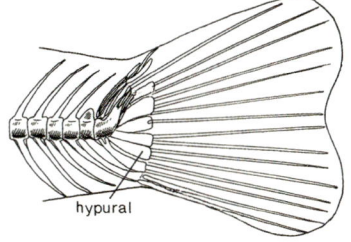

Hypural plate. Illustration adapted from Jenkins and Burkhead (1994).

infraorbital canal—The pored canal passing just below the eye; part of the cephalic lateralis system.

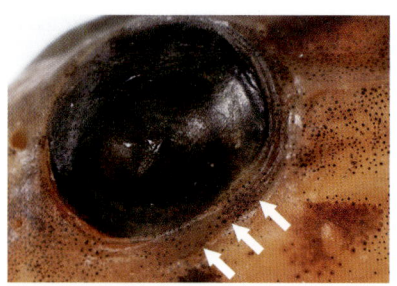

Infraorbital canal in "Tessellated" Darter, *Etheostoma* sp.

interorbital area—Top of the head between the eyes.

Interorbital area of Sharpnose Darter, *Percina oxyrhynchus*.

interradial membranes—Membranes between rays in the fins.

Interradial membranes—pigmented in the dorsal fin of Fieryblack Shiner, *Cyprinella pyrrhomelas*.

lachrymal groove—A small groove on the side of snout, caused by folding of skin under edge of lachrymal bone just above upper lip.

Lachrymal groove of Eastern Silvery Minnow, *Hybognathus regius*.

lunula (pl. **lunulae**)—The visible, exposed posterior part of a scale when in its natural position.

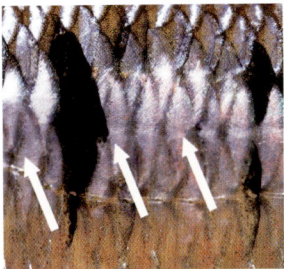

Lunulae of Crescent Shiner, *Luxilus cerasinus*.

maxilla—Bone in the upper jaw that lies immediately above or behind the premaxilla. The premaxilla and maxilla are the most anterior bones in the upper jaw.

Maxilla of Largemouth Bass, *Micropterus salmoides*.

maxillary barbel—A slender fleshy protuberance, short to long, usually tapered to a point, found dorsolaterally at the corner of the mouth on the maxillary bone.

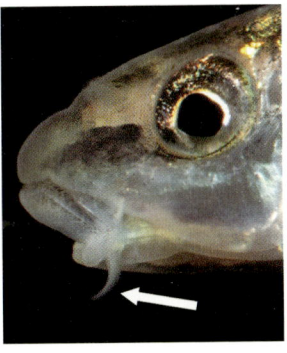

Maxillary barbel of Thicklip Chub, *Cyprinella labrosa*.

melanistic—Quite dark or black.

Melanistic male Eastern Mosquitofish, *Gambusia holbrooki*.

melanophore—A cell-bearing melanin or dark pigment, which produces shades of gray to black depending upon the concentration.

Melanophores in the dorsal fin of Fieryblack Shiner, *Cyprinella pyrrhomelas*.

molariform—Molar-like; teeth with flattened or broadly rounded crowns adapted for grinding.

Molariform teeth on the pharyngeal arch of Robust Redhorse, *Moxostoma robustum*.

myomeres—Bundles of muscle, dorso-ventrally or obliquely oriented, arranged in a series along the body.

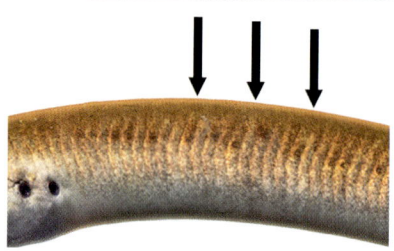

Myomeres of American Brook Lamprey, *Lethenteron appendix*.

nape—The dorsal area between the posterior end of the head occiput and the dorsal fin.

Nape region of Silver Shiner, *Notropis photogenis*.

neuromasts—Tiny pit-like sensory structures.

Neuromasts (shown as faint depressions) on the lateral surface of the head of Mimic Shiner, *Paranotropis volucellus*.

occiput (occipital area)—The postero-dorsal portion of the head, immediately anterior to the nape.

Occipital area (O) and nape (N) of Southern Tessellated Darter, *Etheostoma maculaticeps*.

ocelli—Eyelike spots in which the central color is surrounded by one or more differently colored rings.

Ocelli on the posterior half of Red Drum, *Sciaenops ocellatus*.

opercle (sing. operculum)—Uppermost and largest bone of the gill cover, sometimes a synonym for "gill cover."

Opercle (A) and preopercle (B), completely scaled, of Redfin Pickerel, *Esox americanus*.

oral disk—The circular mouth area of lampreys, bearing horny teeth in juveniles and adults. Preservation may cause disk to become noncircular.

Oral disk and lateral line neuromasts (black dots) on the ventral surface of the branchial region of Least Brook Lamprey, *Lampetra aepyptera*.

orbit—Eye socket.

Orbit of Striped Jumprock, *Moxostoma rupiscartes*.

palatine tooth patch—A patch of small teeth on the paired palatine bones in the roof of the mouth, just posterior or lateral to medial vomer or prevomer bone.

Ovals encircling the palatine tooth patches in Mottled Sculpin, *Cottus bairdii*.

papillose lips—Lips bearing papillae, which are small, rounded, fleshy protuberances that are knob-like or elongate.

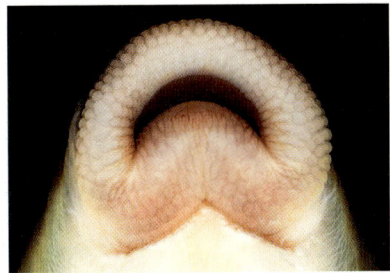

Papillose lips of Northern Hog Sucker, *Hypentelium nigricans*.

peritoneum—The membrane lining the abdominal cavity.

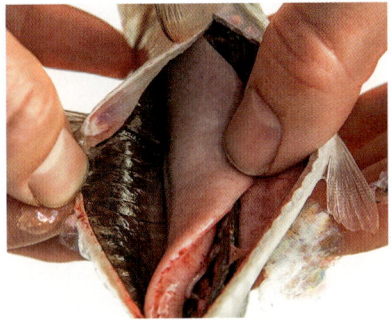

Peritoneum—black in Blueback Herring, *Alosa aestivalis*.

pharyngeal arch—The bony modified last posterior gill arch. This term applies when this arch bears definitive teeth, as in "minnows" (families Cyprinidae, Leuciscidae, and Xenocyprididae) and suckers (family Catostomidae).

Pharyngeal arch of Warpaint Shiner, *Coccotis coccogenis*.

pharyngeal teeth—Teeth on the pharyngeal arch.

phylogenetic—Based upon evolutionary history and relationships.

plicate lips (pl. **plicae**)—Lips bearing plicae, which are parallel ridges and grooves, appearing pleated or folded.

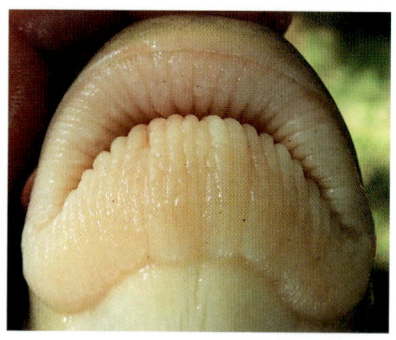

Plicate lips of River Redhorse, *Moxostoma carinatum*. Photograph courtesy of Luke Etchison.

predorsal circumferential scale count—A count, done in a zigzag pattern, of the number of scales encircling the body anterior to the dorsal fin.

Predorsal circumferential scale count in Bluehead Chub, *Nocomis leptocephalus*.

predorsal length—The distance from the tip of the snout to the structural base of the first dorsal fin, ray, or spine.

predorsal profile—Profile of the body anterior to the dorsal fin.

Predorsal profile of Comely Shiner, *Notropis amoenus*.

predorsal scales—Scales anterior to the dorsal fin.

Predorsal scales of Swallowtail Shiner, *Miniellus procne*.

premaxilla—One of the most anterior paired dermal bones of the upper jaw proximal or anterior to the maxillaries.

Premaxilla in Redbreast Sunfish, *Lepomis auritus*.

premaxillary band of teeth (tooth patch)—Tooth patch located at the anterior portion on the roof of the mouth, best observed by placing the fish on its back and prying open the mouth.

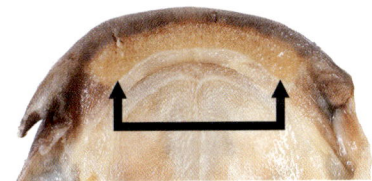

Premaxillary band of teeth of Flat Bullhead, *Ameiurus platycephalus*.

prenasal—Anterior to the nares (nostril openings).

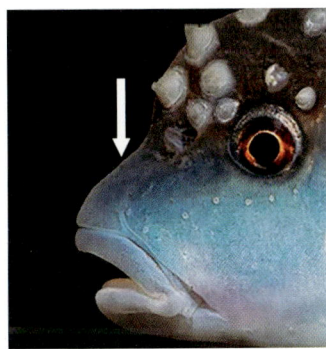

Prenasal area, with absence of tubercles, of Bluehead Chub, *Nocomis leptocephalus*.

preopercle—Sickle-shaped bone located on the front portion of the gill cover and forming the posterior boundary of the cheek; see figure under "opercle."

preopercular spine—Spine located on the preopercle bone, which is the bone just anterior to the opercle, forming the posterior boundary of the cheek.

Preopercular spine (beneath the tissue) of Banded Sculpin, *Cottus carolinae*.

preorbital area (preorbital stripe)—Area of the head anterior to the eye.

Preorbital area and preorbital stripe of Swallowtail Shiner, *Miniellus procne*.

principal ray—A fin ray that extends to the distal margin of a median fin.

Principal rays (11) in the dorsal fin of Striped Jumprock, *Moxostoma rupiscartes*.

protractile mouth (protrusile jaw)—An upper jaw able to protract from the snout tip when the mouth is opened.

Protractile mouth of Rough Silverside, *Membras martinica*.

scute—A large, modified, often thick scale or plate.

Scutes, dorsal, lateral, and ventral, of Atlantic Sturgeon, *Acipenser oxyrinchus*. South Carolina. Photograph taken under the authority of National Marine Fisheries Service Endangered Species Permit No. 21198.

serrate—Sawtooth-like, bearing a series of serrae or jagged edge.

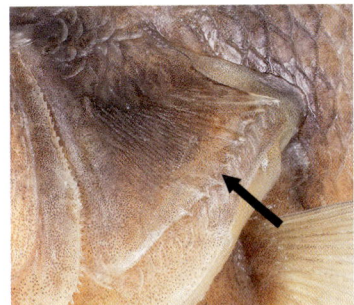

Serrated operculum of Yellow Perch, *Perca flavescens*.

soft dorsal fin—The posterior, soft-ray dorsal fin (A).

spinous dorsal fin—The anterior, spine-supported dorsal fin (B).

Soft (A) and spinous (B) dorsal fins of Roanoke Darter, *Percina roanoka*.

standard length (SL)—Straight-line distance from the anteriormost point on the head to the posterior end on the caudal fin base (hypural plate).

Standard length (SL) of Fieryblack Shiner, *Cyprinella pyrrhomelas*.

subnasal—Below the nostrils.

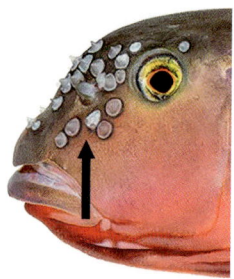

Subnasal area with cephalic (head) tubercles of River Chub, *Nocomis micropogon*.

subocular (suborbital) bar—A mostly vertical dark bar beneath the eye.

Subocular bar of Chainback Darter, *Percina nevisensis*.

subopercle—A bone of the gill cover that is below the opercle bone, which is the largest bone of the gill cover.

Subopercle of River Redhorse, *Moxostoma carinatum*.

supraoral lamina—A horny ridge bearing teeth above (anterior to) the mouth opening in lampreys.

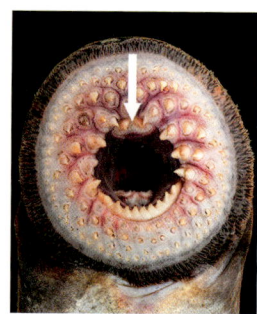

Supraoral lamina of Sea Lamprey, *Petromyzon marinus*. New Hampshire. Photograph courtesy of Brian Zimmerman.

supratemporal canal—A lateral line canal extending across the occiput, or behind that point, of a fish's head that connects the lateral canals of the two sides; often incomplete.

Supratemporal canal in Golden Redhorse, *Moxostoma erythrurum*.

total length (TL)—Straight-line distance from the anteriormost part of the head to the posteriormost part of the caudal fin with the caudal fin tips compressed together.

Total length (TL) in Mountain Redbelly Dace, *Chrosomus oreas*.

truncate—Squared-off or straight; for example, the distal margin of the caudal fin.

Truncate caudal fin of Waccamaw Killifish, *Fundulus waccamensis*.

tubercles—Small, white, hard (keratin) protuberances on the skin or fins, usually present only on a breeding male.

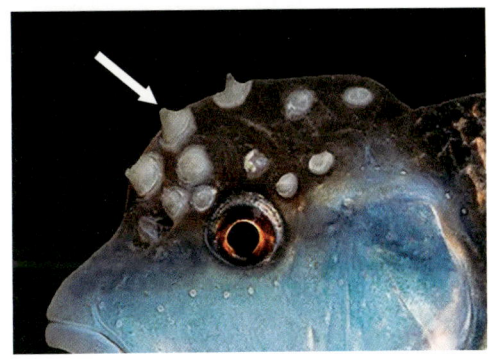

Tubercles of Bluehead Chub, *Nocomis leptocephalus*.

unicuspid—A tooth with a single point cusp on its upper surface.

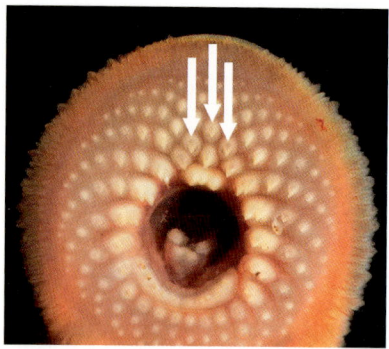

Unicuspid anterior teeth of Mountain Brook Lamprey, *Ichthyomyzon greeleyi*.

urogenital opening—The exterior opening of both the urinary and reproductive tracts (A).

villi—Small, narrow, fleshy growths (B).

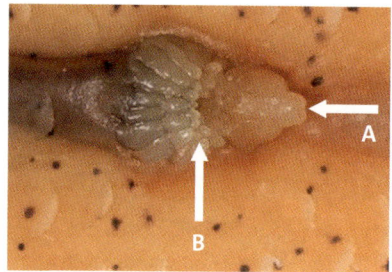

Urogenital opening (A) and villi (B) of male Glassy Darter, *Etheostoma vitreum*.

vomerine tooth patch—A patch of teeth on the vomer or prevomer bone, centered on the roof of the mouth anteromedially.

Vomerine tooth patch of Mottled Sculpin, *Cottus bairdii*.

LITERATURE CITED

Anonymous. 1873. "A New Fish." *Asheville (NC) Weekly Citizen*. July 24, 1873.
——. 1874. "An Odd Fish." *Asheville (NC) Weekly Pioneer*. June 27, 1874.
Agassiz, L. 1853. "9. Recent Researches of Prof. Agassiz," in "Scientific Intelligence:
IV. Botany and Zoology." *American Journal of Science and Arts*, Second Series,
vol. 16 (46):134–36.
——. 1855. Remarks on B. Dowler, "Discovery of Viviparous Fish in Louisiana."
American Journal of Science and Arts, Second Series, vol. 19 (55):133–36.
Andrews, A. 1955. "Greek and Latin Terms for Salmon and Trout." *Transactions and
Proceedings of the American Philological Association* 86:308–18.
Ash, T. 1682. *Carolina; or a Description of the Present State of That Country, and
the Natural Excellencies Thereof, Viz. the Healthfulness of the Air, Pleasantness of
the Place, Advantage and Usefulness of Those Rich Commodities There Plentifully
Abounding, Which Much Increase and Flourish by the Industry of the Planters That
Daily Enlarge That Colony*. London.
Bailey, J. R. 1949. *Yadkin River Survey Report*. Fish Division. Raleigh: North Caro-
lina Wildlife Resources Commission.
Baker, W. D., and W. B. Smith. 1965a. *Survey and Classification of the Perquimans-
Pasquotank-North Rivers and Tributaries, North Carolina*. Final Report, Federal
Aid in Fish Restoration, Job I-R, Project F-14-R. Raleigh: North Carolina Wildlife
Resources Commission.
——. 1965b. *Survey and Classification of the Scuppernong-Alligator Rivers and
Tributaries, North Carolina*. Final Report, Federal Aid in Fish Restoration, Job
I-S, Project F-14-R. Raleigh: North Carolina Wildlife Resources Commission.
Bayless, J. D. 1963. *Survey and Classification of the Northeast Cape Fear River and
Tributaries, North Carolina*. Final Report, Federal Aid in Fish Restoration, Job
I-E, Project F-14-R. Raleigh: North Carolina Wildlife Resources Commission.
——. 1966. *Coastal Lakes I. 1965 Surveys*. Raleigh: North Carolina Wildlife Re-
sources Commission.
Bayless, J. D., and E. H. Shannon. 1965. *Survey and Classification of the Pamlico River
and Tributaries, North Carolina*. Final Report, Federal Aid in Fish Restoration,
Job I-K, Project F-14-R. Raleigh: North Carolina Wildlife Resources Commission.
Bayless, J. D., and W. B. Smith. 1963. *Survey and Classification of the Neuse River and
Tributaries, North Carolina*. Final Report, Federal Aid in Fish Restoration, Job
I-A, Project F-14-R. Raleigh: North Carolina Wildlife Resources Commission.
Beane, J. 2017. "Meet Our Superb Suckers!" *Wildlife in North Carolina*. January–
February 2017.

Birdsong, R. S. 1981. "A Review of the Gobiid Fish Genus *Microgobius* Poey." *Bulletin of Marine Science* 31:267–306.

Boschung, H. T., and R. L. Mayden. 2004. *Fishes of Alabama*. Washington, DC: Smithsonian Institution Press.

Brickell, J. 1737. *The Natural History of North-Carolina. With an Account of the Trade, Manners, and Customs of the Christian and Indian Inhabitants. Illustrated with Copper-Plates, Whereon Are Curiously Engraved on the Map of the Country, Several Strange Beasts, Birds, Fishes, Snakes, Insects, Trees, and Plants, &c.* Dublin.

Brownstein, C. D., D. Kim, O. D. Orr, G. M. Hogue, B. H. Tracy, M. W. Pugh, R. Singer, C. Myles-McBurney, J. M. Mollish, J. W. Simmons, S. R. David, G. Watkins-Colwell, E. A. Hoffman, and T. J. Near. 2022. "Hidden Species Diversity in an Iconic Living Fossil Vertebrate." *Biology Letters* 18 (11): no. 20220395.

Carnes, W. C. 1965. *Survey and Classification of the Roanoke River and Tributaries, North Carolina*. Final Report, Federal Aid in Fish Restoration, Job I-Q, Project F-14-R. Raleigh: North Carolina Wildlife Resources Commission.

Carnes, W. C., J. R. Davis, and B. L. Tatum. 1964. *Survey and Classification of the Deep-Haw Rivers and Tributaries, North Carolina*. Final Report, Federal Aid in Fish Restoration, Job I-G, Project F-14-R. Raleigh: North Carolina Wildlife Resources Commission.

Chernoff, B. 1986. "Phylogenetic Relationships and Reclassification of Menidiine Silverside Fishes with Emphasis on the Tribe Membradini." *Proceedings of the Academy of Natural Sciences of Philadelphia* 138:189–249.

Cloutman, D. G., and L. L. Olmsted. 1978. *The Fishes of Mecklenburg County, North Carolina*. Final report to Charlotte Nature Museum, Charlotte, NC.

———. 1983. "Vernacular Names of Freshwater Fishes of the Southeastern United States." *Fisheries* 8 (2): 7–11.

Cope, E. D. 1870a. "On Some Etheostomine Perch from Tennessee and North Carolina." *Proceedings of the American Philosophical Society* 11 (81): 261–70.

———. 1870b. "A Partial Synopsis of the Fishes of the Freshwaters of North Carolina." *Proceedings of the American Philosophical Society* 11 (81): 448–95.

Crossman, E. J. 1962. "The Redfin Pickerel, *Esox a. americanus* in North Carolina." *Copeia* 1962:114–23.

Crowell, T. E. 1965. *Survey and Classification of the Toe River and Tributaries, North Carolina*. Final Report, Federal Aid in Fish Restoration, Job I-U, Project F-14-R. Raleigh: North Carolina Wildlife Resources Commission.

———. 1966. *Coastal Lakes II, 1965 Surveys*. Raleigh: North Carolina Wildlife Resources Commission.

Cuvier, G., and A. Valenciennes. 1837. *Histoire naturelle des poissons*. Vol. 12. Paris: F. G. Levrault.

Davis, J. R., and E. G. McCoy. 1965. *Survey and Classification of the New-Whiteoak-Newport Rivers and Tributaries, North Carolina*. Final Report, Federal Aid in Fish Restoration, Job I-T, Project F-14-R. Raleigh: North Carolina Wildlife Resources Commission.

Eddy, S. 1969. *How to Know the Freshwater Fishes*. 2nd ed. Dubuque, IA: W. C. Brown.

Engels, W. L. 1941a. "A Survey of the Fish Resources of the New River System in North Carolina." *North Carolina Wildlife Conservation* 5 (11): 4–6, 12–13.

———. 1941b. "A Survey of the Fish Resources of the Watauga River System in North Carolina." *North Carolina Wildlife Conservation* 5 (12): 8–9, 14–15.

———. 1942. "Survey of the Fish Resources of the Nolichucky River System in North Carolina." *North Carolina Wildlife Conservation* 6 (1): 6–7, 14–15; 6 (2): 125–13.

Etnier, D. A., and W. C. Starnes. 1993. *The Fishes of Tennessee*. Knoxville: University of Tennessee Press.

Evermann, B. W. 1916. "Notes on the Fishes of the Lumbee River." *Copeia* 1916:77–80.

Evermann, B. W., and U. O. Cox. 1896. "The Fishes of the Neuse River Basin." *Bulletin of the U. S. Fisheries Commission* 15:303–10.

Fowler, H. W. 1945. *A Study of the Fishes of the Southern Piedmont and Coastal Plain*. Monograph No. 7. Philadelphia: Academy of Natural Sciences of Philadelphia.

Frey, D. G. 1951. "The Fishes of North Carolina's Bay Lakes and Their Intraspecific Variation." *Journal of the Elisha Mitchell Scientific Society* 67:1–44.

Fricke, R., W. N. Eschmeyer, and R. van der Laan, eds. 2022. *Eschmeyer's Catalog of Fishes: Genera, Species, References*. http://researcharchive.calacademy.org/research/ichthyology/catalog/fishcatmain.asp.

Froese, R., and D. Pauly, eds. 2022. *FishBase*. Electronic publication. www.fishbase.org, version 06/2022.

Fuller, P. L., L. G. Nico, and J. D. Williams. 1999. *Nonindigenous Fishes Introduced into Inland Waters of the United States*. American Fisheries Society Special Publication 27. Bethesda, MD: American Fisheries Society.

Gilbert, C. R. 2004. "Family Elassomatidae Jordan 1877—Pygmy Sunfishes." *Annotated Checklists of Fishes* 33:1–5.

Girard, C. 1856. "Researches upon the Cyprinoid Fishes Inhabiting the Fresh Waters of the United States of America, West of the Mississippi Valley, from Specimens in the Museum of the Smithsonian Institution." *Proceedings of the Academy of the Natural Sciences of Philadelphia* 8:165–213.

Goodfred, D., K. Hodges, and S. Loftis. 2023. "Alabama Bass." *Wildlife in North Carolina*. May–June 2023: 46–47.

Hardy, J. D. 1980. "*Fundulus confluentus* Goode and Bean, Marsh Killifish." In *Atlas of North American Freshwater Fishes*, edited by D. S. Lee, C. R. Gilbert, C. H. Hocutt, R. E. Jenkins, D. E. McAllister, and J. R. Stauffer Jr., 512. Raleigh: North Carolina State Museum of Natural History.

Harrison, I. J. 2002. "Mugilidae. Mullets." In Carpenter 2002a, 1071–85.

Henson, M. N., D. D. Aday, and J. A. Rice. 2018. "Thermal Tolerance and Survival of Nile Tilapia and Blue Tilapia under Rapid and Natural Temperature Declination Rates." *Transactions of the American Fisheries Society* 147:278–86.

Hildebrand, S. F. 1932. "On a Collection of Fishes from the Tuckaseegee and Upper Catawba River Basins, N.C., with a Description of a New Darter." *Journal of the Elisha Mitchell Scientific Society* 48:50–82.

Hodson, R. G. 1989. *Hybrid Striped Bass. Biology and Life History.* Southern Regional Aquaculture Center Publication No. 300. Stoneville, MS: Southern Regional Aquaculture Center.

Hogue, G. M., and B. H. Tracy. 2014. "It's All in the Lips." *Newsletter of the North Carolina Chapter of the American Fisheries Society,* Fall 2014. https://nc.fisheries.org/newsletters/.

Hubbs, C. L., and E. C. Raney. 1946. "Endemic Fish Fauna of Lake Waccamaw, North Carolina." *Miscellaneous Publications.* No. 65. Ann Arbor: Museum of Zoology, University of Michigan.

Hubbs, C. L., and K. F. Lagler. 1964. *Fishes of the Great Lakes Region.* Ann Arbor: University of Michigan Press.

Jenkins, R. E., and N. M. Burkhead. 1994. *Freshwater Fishes of Virginia.* Bethesda, MD: American Fisheries Society.

Jenkins, R. E., and D. S. Sorensen. 1980. "*Notropis procne* (Cope), Swallowtail Shiner." In *Atlas of North American Freshwater Fishes,* edited by D. S. Lee, C. R. Gilbert, C. H. Hocutt, R. E. Jenkins, D. E. McAllister, and J. R. Stauffer Jr., 298. Raleigh: North Carolina State Museum of Natural History.

Jones, B. K., and T. D. Ewing. 2019. *Elassoma boehlkei, Carolina Pygmy Sunfish, Including Information from the North Carolina Status Assessment 2014–2015 and North and South Carolina Surveys 2018.* Raleigh: North Carolina Wildlife Resources Commission.

Jordan, D. S. 1877. "Synopsis of the Siluridae of the Fresh Waters of North America." *Bulletin of the United States National Museum* 1 (10):69–103.

———. 1878. *Manual of the Vertebrates of the Northern United States, Including the District East of the Mississippi River, and North of North Carolina and Tennessee, Exclusive of Marine Species.* Chicago: Jansen, McClurg & Company.

———. 1889a. "Descriptions of Fourteen Species of Fresh-Water Fishes Collected by the U.S. Fish Commission in the Summer of 1888." *Proceedings of the United States National Museum* 11:351–62.

———. 1889b. "Report of Explorations Made during the Summer and Autumn of 1888, in the Alleghany Region of Virginia, North Carolina and Tennessee, and in Western Indiana, with an Account of the Fishes Found in Each of the River Basins in Those Regions." *Bulletin of the United States Fish Commission* 8:97–173.

Jordan, D. S., and H. W. Clark. 1930. *Check List of the Fishes and Fishlike Vertebrates of North and Middle America North of the Northern Boundary of Venezuela and Colombia.* Report to the United States Commissioner of Fisheries for 1928. Appendix X. Washington, DC: United States Commission of Fish and Fisheries.

Jordan, D. S., and B. W. Evermann. 1896–1900. "The Fishes of North and Middle America: A Descriptive Catalog of the Species of Fish-Like Vertebrates Found in the Waters of North America, North of the Isthmus of Panama." *Bulletin of the United States National Museum* 47 (parts 1–4): 1–3313.

Jordan, D. S., and C. H. Gilbert. 1883. "Synopsis of the Fishes of North America." *Bulletin of the United State National Museum* 16:1–1018.

Kassler, T. W., J. B. Koppelman, T. J. Near, C. B. Dillman, J. M. Levengood, D. L. Swofford, J. L. VanOrman, J. E. Claussen, and D. P. Philipp. 2002. "Molecular and Morphological Analyses of the Black Basses: Implications for Taxonomy and Conservation." *American Fisheries Society Symposium* 31:291–322.

Kells, V. A., and K. Carpenter. 2011. *A Field Guide to Coastal Fishes: From Maine to Texas*. Baltimore: Johns Hopkins University Press.

Kim, D., A. T. Taylor, and T. J. Near. 2022. "Phylogenomics and Species Delimitation of the Economically Important Black Basses (*Micropterus*)." *Scientific Reports* 12 (9113): 1–14. https://doi.org/10.1038/s41598-022-11743-2.

King, J., and D. Grove. 1942a. "A Survey of the Fish Resources of the Upper Waters of the Catawba River." *North Carolina Wildlife Conservation* 6 (12): 11–14.

———. 1942b. "A Survey of the Fish Resources of the Upper Waters of the Yadkin River." *North Carolina Wildlife Conservation* 6 (11): 8–11.

Krabbenhoft, T. J., F. C. Rohde, and J. M. Quattro. 2005. "Threatened Fishes of the World: *Menidia extensa* (Hubbs and Raney 1946) (Atherinopsidae)." *Environmental Biology of Fishes* 73:48.

———. 2006. "Threatened Fishes of the World: *Etheostoma perlongum* (Hubbs and Raney 1946) (Percidae)." *Environmental Biology of Fishes* 76:411–12.

———. 2009. "Threatened Fishes of the World: *Fundulus waccamensis* (Hubbs and Raney 1946) (Fundulidae)." *Environmental Biology of Fishes* 84:173–74.

Kuehne, R. A., and R. W. Barbour. 1983. *The American Darters*. Lexington: University of Kentucky Press.

Lachner, E. A., and E. E. Deubler Jr. 1960. "*Clinostomus funduloides* Girard to Replace *Clinostomus vandoisulus* (Valenciennes) as the Name of the Rosyside Dace of Eastern North America." *Copeia* 1960:358–60.

Lanteigne, J. 1988. "Identification of Lamprey Larvae of the Genus *Ichthyomyzon* (Petromyzontidae)." *Environmental Biology of Fishes* 23:261–65.

Lawson, J. 1709. *A New Voyage to Carolina; Containing the Exact Description and Natural History of That Country: Together with the Present State Thereof. And a Journal of a Thousand Miles, Travel'd thro' Several Nations of Indians. Giving a Particular Account of Their Customs, Manners &c.* London.

Lee, D. S. 1980. "*Fundulus rathbuni* Jordan and Meek, Speckled Killifish." In *Atlas of North American Freshwater Fishes*, edited by D. S. Lee, C. R. Gilbert, C. H. Hocutt, R. E. Jenkins, D. E. McAllister, and J. R. Stauffer Jr., 526. Raleigh: North Carolina State Museum of Natural History.

Lee, D. S., C. R. Gilbert, C. H. Hocutt, R. E. Jenkins, D. E. McAllister, and J. R. Stauffer Jr., eds. 1980. *Atlas of North American Freshwater Fishes*. Raleigh: North Carolina State Museum of Natural History.

Louder, D. E. 1962a. "An Annotated Check List of the North Carolina Bay Lakes Fishes." *Journal of the Elisha Mitchell Scientific Society* 78:68–73.

———. 1962b. *Survey and Classification of the Lumber River and Shallotte River, North Carolina*. Final Report, Federal Aid in Fish Restoration, Job I-C, Project F-14-R. Raleigh: North Carolina Wildlife Resources Commission.

———. 1963. *Survey and Classification of the Cape Fear River and Tributaries, North Carolina*. Final Report, Federal Aid in Fish Restoration, Job I-G, Project F-14-R. Raleigh: North Carolina Wildlife Resources Commission.

———. 1964. *Survey and Classification of the Catawba River and Tributaries, North Carolina*. Final Report, Federal Aid in Fish Restoration, Job I-H, Project F-14-R. Raleigh: North Carolina Wildlife Resources Commission.

MacGuigan, D. J., O. D. Orr, and T. J. Near. 2023. "Phylogeography, Hybridization, and Species Discovery in the *Etheostoma nigrum* Complex (Percidae: *Etheostoma: Boleosoma*)." *Molecular Phylogenetics and Evolution* 178: no. 107645.

Manooch, C. S., III. 1984. *Fisherman's Guide: Fishes of the Southeastern United States*. Raleigh: North Carolina State Museum of Natural History.

Marcgraf, G., and W. Piso. 1648. *Historia naturalis brasiliae*. Amsterdam: Elzevir.

McCargo, J. 2020. "Methuselah Striped Bass Recovered in Roanoke River." *Newsletter of the North Carolina Chapter of the American Fisheries Society*, Summer 2020, p. 6. https://nc.fisheries.org/newsletters/.

Menhinick, E. F. 1991. *The Freshwater Fishes of North Carolina*. Raleigh: North Carolina Wildlife Resources Commission.

———. 2010. *Fishes of Gaston County, Fishes by Species*. Charlotte: Department of Biology, University of North Carolina–Charlotte.

———. N.d. *Studies of Streams of the Uwharrie National Forest, Parts 1 and 2*. Uwharrie National Forest, Troy, NC.

Menhinick, E. F., T. M. Burton, and J. R. Bailey. 1974. "An Annotated Checklist of the Freshwater Fishes of North Carolina." *Journal of the Elisha Mitchell Scientific Society* 90:24–50.

Messer, J. B. 1964a. *Survey and Classification of the Pigeon River and Tributaries, North Carolina*. Final Report, Federal Aid in Fish Restoration, Job I-N, Project F-14-R. Raleigh: North Carolina Wildlife Resources Commission.

———. 1964b. *Survey and Classification of the Savannah River and Tributaries, North Carolina*. Final Report, Federal Aid in Fish Restoration, Job I-M, Project F-14-R. Raleigh: North Carolina Wildlife Resources Commission.

———. 1965. *Survey and Classification of the Hiwassee River and Tributaries, North Carolina*. Final Report, Federal Aid in Fish Restoration, Job I-W, Project F-14-R. Raleigh: North Carolina Wildlife Resources Commission.

———. 1966. *Mountain Reservoirs—1965 Surveys*. Raleigh: North Carolina Wildlife Resources Commission.

Messer, J. B., J. R. Davis, T. E. Crowell, and W. C. Carnes. 1965. *Survey and Classification of the Broad River and Tributaries, North Carolina*. Final Report, Federal Aid in Fish Restoration, Job I-V, Project F-14-R. Raleigh: North Carolina Wildlife Resources Commission.

Messer, J. B., and H. M. Ratledge. 1963. *Survey and Classification of the Little Tennessee River and Tributaries, North Carolina*. Final Report, Federal Aid in Fish Restoration, Job I-D, Project F-14-R. Raleigh: North Carolina Wildlife Resources Commission.

Moser, M. L., F. C. Rohde, R. G. Arndt, and K. W. Ashley. 1998. "Occurrence of the Brook Silverside, *Labidesthes sicculus* (Atheriniformes: Atherinidae), in North Carolina." *Brimleyana* 25:135–39.

Murdy, E. O., and D. F. Hoese. 2002. "Gobiidae. Gobies." In *The Living Marine Resources of the Western Central Atlantic. Volume 3. Bony Fishes Part 2 (Opistognathidae to Molidae), Sea Turtles and Marine Mammals*, edited by K. E. Carpenter, 1781–96. Rome, Italy: Food and Agriculture Organization of the United Nations. https://www.fao.org/publications/card/en/c/bdf3d5ff-704c-5a41-82fd-82dc2049cd99/.

NCAC (North Carolina Administrative Code). 2023. *Subchapters 10I. 0103–10I. 0105: Endangered, Threatened, and Special Concern Species Listed*. Amended effective February 1, 2023. Raleigh: North Carolina Administrative Code.

NCDMF (North Carolina Division of Marine Fisheries). 2020. *North Carolina Recreational Coastal Waters Guide for Sports Fishermen—December 2020 and Subsequent Versions*. Morehead City: North Carolina Division of Marine Fisheries.

NCNHP (North Carolina Natural Heritage Program). 2022. *Natural Heritage Program List of Rare Animal Species of North Carolina*. Raleigh: North Carolina Natural Heritage Program, North Carolina Department of Natural and Cultural Resources.

NCWRC (North Carolina Wildlife Resources Commission). 1961. *Inventory of Fish Population in Lentic Waters*. Job completion report. Job No. 1, Projects F-5-R and F-6-R. Raleigh: North Carolina Wildlife Resources Commission.

———. 2010. *Bodie Bass*. North Carolina Sport Fish Profiles. Updated May 2010. Raleigh: North Carolina Wildlife Resources Commission.

———. 2019a. *Discontinued: Red Drum Fishery in Hyco Lake*. Fisheries Research Fact Sheet, April 2019. Raleigh: North Carolina Wildlife Resources Commission.

———. 2019b. *Herring and Shad Identification Tips*. Raleigh: North Carolina Wildlife Resources Commission.

———. 2020. *Warmwater Stocking List*. Raleigh: North Carolina Wildlife Resources Commission.

———. 2021. *Protected Wildlife Species of North Carolina*. Raleigh: North Carolina Wildlife Resources Commission.

———. 2022. *North Carolina Inland Fishing, Hunting and Trapping Regulations Digest, 2022–2023*. Raleigh: North Carolina Wildlife Resources Commission. https://www.ncwildlife.org/licensing/regulations#7126766-fishing.

———. N.d. (a). *Catfishes of North Carolina*. Raleigh: Division of Inland Fisheries, North Carolina Wildlife Resources Commission.

———. N.d. (b). *Brook Trout*. North Carolina Wildlife Profiles. Raleigh: North Carolina Wildlife Resources Commission.

Near, T. J., M. Sandel, K. L. Kuhn, P. J. Unmack, C. Wainwright, and W. L. Smith. 2012. "Nuclear Gene–Inferred Phylogenies Resolve the Relationships of the Enigmatic Pygmy Sunfishes, *Elassoma* (Teleostei: Percomorpha)." *Molecular Phylogenetics and Evolution* 63:388–95.

Nico, L. G., P. J. Schofield, and M. E. Neilson. 2021. "*Oreochromis niloticus* (Linnaeus, 1758)." U.S. Geological Survey, Nonindigenous Aquatic Species Database, Gainesville, FL. https://nas.er.usgs.gov/queries/FactSheet.aspx?SpeciesID=468.

Nilsson, S. 1832. *Prodromus ichthyologiae Scandinavicae.* Lund, Sweden.

Outten, L. M. 1956. "Studies of the Life History of the Cyprinid Fishes *Notropis coccogenis, galacturus,* and *rubricroceus.*" PhD diss., Cornell University, Ithaca.

———. 1957. "A Study of the Life History of the Cyprinid Fish *Notropis coccogenis.*" *Journal of the Elisha Mitchell Scientific Society* 73:68–84.

———. 1958. "Studies of the Life History of the Cyprinid Fishes *Notropis galacturus* and *rubricroceus.*" *Journal of the Elisha Mitchell Scientific Society* 74:122–34.

Page, L. M., K. E. Bemis, T. E. Dowling, H. Espinosa-Pérez, L. T. Findley, C. R. Gilbert, K. E. Hartel, R. N. Lea, N. E. Mandrak, M. A. Neighbors, J. J. Schmitter-Soto, and H. J. Walker Jr. 2023. *Common and Scientific Names of Fishes from the United States, Canada, and Mexico.* American Fisheries Society Special Publication 37. Bethesda, MD: American Fisheries Society.

Page, L. M., and B. M. Burr. 2011. *Peterson Field Guide to Freshwater Fishes of North America North of Mexico.* 2nd ed. New York: Houghton Mifflin.

Pezold, F. L., and R. J. Edwards. 1983. "Additions to the Texas Marine Ichthyofauna, with Notes on the Rio Grande Estuary." *Southwestern Naturalist* 28:102–5.

Pflieger, W. L. 1997. *The Fishes of Missouri.* Jefferson City: Missouri Department of Conservation.

Powers, S. L. 2020. "Mugilidae: Mullets." In *Freshwater Fishes of North America,* edited by M. L. Warren Jr. and B. M. Burr, 367–83. Baltimore: Johns Hopkins University Press.

Quattro, J. M., W. J. Jones, and F. C. Rohde. 2001. "Evolutionary Significant Units of Rare Pygmy Sunfishes (Genus *Elassoma*)." *Copeia* 2001:514–520.

Rafinesque, C. S. 1820. *Ichthyologia Ohiensis, or Natural History of the Fishes Inhabiting the River Ohio and Its Tributary Streams, Preceded by a Physical Description of the Ohio and Its Branches.* Lexington, KY: W. G. Hunt [for C. S. Rafinesque].

Ramsey, J. S. 1965. "Zoogeographic Studies of the Freshwater Fish Fauna of Rivers Draining the Southern Appalachian Region." PhD diss., Tulane University, New Orleans.

Randall, J. F. 1957. "The Distribution of the Fishes of the Catawba-Wateree River Drainage Basin, North Carolina and South Carolina." PhD diss., University of South Carolina, Columbia.

Raney, E. C. 1947. "Subspecies and Breeding Behavior of the Cyprinid Fish *Notropis procne* (Cope)." *Copeia* 1947:103–9.

Ratledge, H. M., W. C. Carnes, and E. D. Collins. 1966. *Checklist of Fishes Collected from the Lotic Waters of North Carolina, 1960–1964.* Raleigh: North Carolina Wildlife Resources Commission.

Renaud, C. B. 2011. *Lampreys of the World: An Annotated and Illustrated Catalogue of Lamprey Species Known to Date.* FAO Species Catalogue for Fishery Purposes. No. 5. Rome, Italy: Food and Agriculture Organization of the United Nations.

Richardson, F. R. 1964. *Survey and Classification of the Watauga River and Tributaries, North Carolina.* Final Report, Federal Aid in Fish Restoration, Job I-J, Project F-14-R. Raleigh: North Carolina Wildlife Resources Commission.

Richardson, F. R., and W. C. Carnes. 1964. *Survey and Classification of the New River and Tributaries, North Carolina.* Final Report, Federal Aid in Fish Restoration, Job I-O, Project F-14-R. Raleigh: North Carolina Wildlife Resources Commission.

Richardson, F. R., J. B. Messer, and H. M. Ratledge. 1963. *Survey and Classification of the French Broad River and Tributaries, North Carolina.* Final Report, Federal Aid in Fish Restoration, Job I-I, Project F-14-R. Raleigh: North Carolina Wildlife Resources Commission.

Roberts, J. H., and A. E. Rosenberger. 2008. "Threatened Fishes of the World: *Percina rex* (Jordan and Evermann 1889) (Percidae)." *Environmental Biology of Fishes* 83:439–40.

Robins, R. H., L. M. Page, J. D. Williams, Z. R. Randall, and G. E. Sheehy. 2018. *Fishes in the Fresh Waters of Florida: An Identification Guide and Atlas.* Gainesville: University of Florida Press.

Rohde, F. C., and R. G. Arndt. 1987. "Two New Species of Pygmy Sunfishes (Elassomatidae, *Elassoma*) from the Carolinas." *Proceedings of the Academy of Natural Sciences of Philadelphia* 139:65–85.

Rohde, F. C., R. G. Arndt, D. J. Coughlan, and S. M. Smith. 2003. "An Annotated List of the Fishes Known from the Dan River in Virginia and North Carolina (Blue Ridge/Piedmont Provinces)." *Southeastern Fishes Council Proceedings* 45:1–10.

Rohde, F. C., R. G. Arndt, J. W. Foltz, and J. M. Quattro. 2009. *Freshwater Fishes of South Carolina.* Columbia: University of South Carolina Press.

Rohde, F. C., R. G. Arndt, D. G. Lindquist, and J. F. Parnell. 1994. *Freshwater Fishes of the Carolinas, Virginia, Maryland and Delaware.* Chapel Hill: University of North Carolina Press.

Rohde, F. C., R. G. Arndt, and S. M. Smith. 2001. "Longitudinal Succession of Fishes in the Dan River in Virginia and North Carolina (Blue Ridge/Piedmont Provinces)." *Southeastern Fishes Council Proceedings* 42:1–13.

Rohde, F. C., G. H. Burgess, and G. W. Link Jr. 1979. "Freshwater Fishes of Croatan National Forest, North Carolina, with Comments on the Zoogeography of Coastal Plain Fishes." *Brimleyana* 2:97–118.

Rohde, F. C., M. L. Moser, and R. G. Arndt. 1998. "Distribution and Status of Selected Fishes in North Carolina with a New State Record." *Brimleyana* 25:43–68.

Ross, S. W., and F. C. Rohde. 2004. "The Gobioid Fishes of North Carolina (Pisces: Gobioidei)." *Bulletin of Marine Science* 74:287–323.

Sammons, S. M., L. G. Dorsey, C. S. Loftis, P. Chrisman, M. Scott, J. Hammonds, M. Jolley, H. Hatcher, J. Odenkirk, J. Damer, M. R. Lewis, and E. J. Peatman. 2023. "Alabama Bass Alter Reservoir Black Bass Species Assemblages When Introduced outside Their Native Range." *North American Journal of Fisheries Management* 43:384–99.

Sandel, M., and P. M. Harris. 2007. "Threatened Fishes of the World: *Elassoma boehlkei* (Rohde and Arndt 1987) (Elassomatidae)." *Environmental Biology of Fishes* 78:289–90.

Scharpf, C. 2013. "*Brevoortia tyrannus* (Latrobe 1802)." ETYFish Project, Names of the Week, December 11, 2013. https://etyfish.org/name-of-the-week2013/.

———. 2014a. "*Petromyzon marinus* Linnaeus 1758." ETYFish Project, Names of the Week, January 1, 2014. https://etyfish.org/name-of-the-week2014/.

———. 2014b. "*Aphredoderus sayanus* (Gilliams 1824)." ETYFish Project, Names of the Week, May 28, 2014. https://etyfish.org/name-of-the-week2014/.

———. 2014c. "Cope Headscratcher #6: *Labidesthes sicculus*." ETYFish Project, Names of the Week, August 27, 2014. https://etyfish.org/name-of-the-week2014/.

———. 2014d. "*Polyodon* Lacépède 1797." ETYFish Project, Names of the Week, November 26, 2014. https://etyfish.org/name-of-the-week2014/.

———. 2015. "*Eleotris* Bloch & Schneider 1801." ETYFish Project, Names of the Week, March 11, 2015. https://etyfish.org/name-of-the-week2014/.

———. 2016a. "The Mystery of *Morone*: Solved at Last?" ETYFish Project, Names of the Week, April 20, 2016. https://etyfish.org/name-of-the-week2016/.

———. 2016b. "*Esox masquinongy* Mitchill 1824." ETYFish Project, Names of the Week, April 27, 2016. https://etyfish.org/name-of-the-week2016/.

———. 2017a. "*Elassoma boehlkei* Rohde & Arndt 1987." ETYFish Project, Names of the Week, January 11, 2017. https://etyfish.org/name-of-the-week2017/.

———. 2017b. "*Ctenogobius shufeldti* (Jordan & Eigenmann 1887)." ETYFish Project, Names of the Week, July 26, 2017. https://etyfish.org/name-of-the-week2017/.

———. 2017c. "*Nocomis micropogon* and *Notropis rubricroceus*." ETYFish Project, Names of the Week, December 24, 2017. https://etyfish.org/name-of-the-week 2017/.

———. 2018. "America's First Fish: *Lepomis auritus* (Linnaeus 1758)." ETYFish Project, Names of the Week, July 4, 2018. https://etyfish.org/name-of-the -week2018/.

———. 2019. "*Fundulus heteroclitus* and the 'Mudfish' Muddle." ETYFish Project, Names of the Week, April 10, 2019. https://etyfish.org/name-of-the-week2019/.

———. 2020a. "Lost in Translation: The True Meaning of 'Natalis' in the Name of the Yellow Bullhead *Ameiurus natalis*." *American Currents* 45 (2): 11–17.

———. 2020b. "The Authorial Mystery of *Esox masquinongy*." *American Currents* 45 (1): 33.

———. 2020c. "*Micropterus dolomieu* Lacépède 1802." ETYFish Project, Names of the Week, February 19, 2020.

———. 2020d. "*Aplodinotus grunniens* Rafinesque 1819." ETYFish Project, Names of the Week, February 26, 2020. https://etyfish.org/name-of-the-week2020/.

———. 2021a. "*Acipenser* Linnaeus 1758." ETYFish Project, Names of the Week, March 31, 2021. https://etyfish.org/name-of-the-week2021/.

———. 2021b. "*Pylodictis*—Real and Imagined." ETYFish Project, Names of the Week, July 14, 2021. https://etyfish.org/name-of-the-week2021/.

———. 2021c. "*Stizostedion* vs. *Sander*." ETYFish Project, Names of the Week, November 17, 2021. https://etyfish.org/name-of-the-week2021/.

———. 2022. "Carpe Diem." ETYFish Project, Names of the Week, May 25, 2022. https://etyfish.org/name-of-the-week2022/.

Scharpf, C., and R. Fricke. 2022. "*Sander* Oken 1817 (Percidae) is the Valid Generic Name for Walleye, Sauger, and European Pikeperches: A Response to Bruner (2021)." *Fisheries* 47:151–53.

Schöpf, J. D. 1788. "Beschreibungen einiger Nord-Amerikanischer fische, vorzüglich aus den Neu-Yorkischen Gewässern." *Schriften der Berlinischen Gesellschaft Naturforschender Freunde* 8:138–94.

Shute, J. R., P. W. Shute, and D. G. Lindquist. 1981. "Fishes of the Waccamaw River Drainage." *Brimleyana* 6:1–24.

Shute, P. W., D. G. Lindquist, and J. R. Shute. 1983. "Breeding Behavior and Early Life History of the Waccamaw Killifish, *Fundulus waccamensis*." *Environmental Biology of Fishes* 8:293–300.

Silliman, K., H. Zhao, M. Justice, W. Thongda, B. Bowen, and E. Peatman. 2021. "Complex Introgression among Three Diverged Largemouth Bass Lineages." *Evolutionary Applications* 14:2815–2830.

Smith, H. M. 1907. *The Fishes of North Carolina*. Vol. 2. Raleigh: North Carolina Geological and Economic Survey.

Smith, W. B. 1963. *Survey and Classification of the Chowan River and Tributaries, North Carolina*. Final Report, Federal Aid in Fish Restoration, Job I-F, Project F-14-R. Raleigh: North Carolina Wildlife Resources Commission.

Smith, W. B., and J. D. Bayless. 1964. *Survey and Classification of the Tar River and Tributaries, North Carolina*. Final Report, Federal Aid in Fish Restoration, Job I-L, Project F-14-R. Raleigh: North Carolina Wildlife Resources Commission.

Snelson, F. F., Jr. 1971. "*Notropis mekistocholas*, a New Herbivorous Cyprinid Fish Endemic to the Cape Fear River Basin, North Carolina." *Copeia* 1971:449–62.

Snelson, F. F., Jr., T. J. Krabbenhoft, and J. M. Quattro. 2009. "*Elassoma gilberti*: A New Species of Pygmy Sunfish (Elassomatidae) from Florida and Georgia." *Bulletin of the Florida Museum of Natural History* 48:119–44.

Starnes, W. C., and G. M. Hogue. 2011. *Curation and Databasing of Voucher Collections from the North Carolina Wildlife Resources Commission 1960s Statewide Surveys of Fishes*. Final Report, Federal Aid in Sport Fish Restoration, Project F-91: Curate Fish Collection July 2008–December 2010. Raleigh: North Carolina Wildlife Resources Commission.

Tan, M., and J. W. Armbruster. 2018. "Phylogenetic Classification of Extant Genera of Fishes of the Order Cypriniformes (Teleostei: Ostariophysi)." *Zootaxa* 4476:6–39.

Tatum, B. L., W. C. Carnes, and F. Richardson. 1963. *Survey and Classification of the Yadkin River and Tributaries, North Carolina*. Final Report, Federal Aid in Fish Restoration, Job I-B, Project F-14-R. Raleigh: North Carolina Wildlife Resources Commission.

Taylor, W. R. 1969. "A Revision of the Catfish Genus *Noturus* Rafinesque, with an Analysis of Higher Groups in the Ictaluridae." *Bulletin of the United States National Museum* 282:1–315.

Tracy, B. H. 2013. "Bluehead Chub, *Nocomis leptocephalus* (Girard 1856), the First Species of Freshwater Fish Scientifically Described from North Carolina." *American Currents* 38 (4): 2–8.

Tracy, B. H., and R. E. Jenkins. 2021. "Professor Edward Drinker Cope's Travels through North Carolina, August–December 1869: Insights from the Transcriptions and Annotations of Letters to His Father and His Contributions to North Carolina Ichthyology." *Southeastern Fishes Council Proceedings* 61:72–135.

Tracy, B. H., R. E. Jenkins, and W. C. Starnes. 2013. "History of Fish Investigations in the Yadkin–Pee Dee River Drainage of North Carolina and Virginia with an Analysis of Nonindigenous Species and Invasion Dynamics of Three Species Suckers (Catostomidae)." *Journal of the North Carolina Academy of Science* 129:82–106.

Tracy, B. H., F. C. Rohde, and G. M. Hogue. 2020. "An Annotated Atlas of the Freshwater Fishes of North Carolina." *Southeastern Fishes Council Proceedings* No. 60. https://trace.tennessee.edu/sfcproceedings/vol1/iss60/1.

Tracy, B. H., S. A. Smith, J. L. Bissette, and F. C. Rohde. 2021. "Ahead by a Whisker: Freshwater Catfish (Family Ictaluridae) Diversity in North Carolina." *American Currents* 46 (3): 17–25.

———. 2022. "Just Below the Surface: Topminnow (Family Fundulidae) Diversity in North Carolina." *American Currents* 47 (1): 19–26, 28.

Tracy, B. H., S. A. Smith, and F. C. Rohde. 2021. "Identifying North Carolina's Suckers May Not Be as Hard as You Think." *American Currents* 46 (1): 3–14.

USGS (US Geological Survey). 2021. *Nonindigenous Aquatic Species Database*. Gainesville, FL: US Geological Survey. http://nas.er.usgs.gov.

Valenciennes, A. 1855. "Ichthyologie." In A. du Petit-Thouars, *Voyage autour du monde sur la frégate "Vénus," pendant les années 1836–1839*. Paris: Gide et J. Baudry.

Wang, Q., L. P. Dizaj, J. Huang, K. K. Sarker, C. Kevrekidis, B. Reichenbacher, H. R. Esmaeili, N. Straube, T. Moritz, and C. Li. 2022. "Molecular Phylogenetics of the Clupeiformes Based on Exon-Capture Data and a New Classification of the Order." *Molecular Phylogenetics and Evolution* 175:107590.

Werneke, D. C., and J. W. Armbruster. 2015. "Silversides of the Genus *Labidesthes* (Atheriniformes: Atherinopsidae)." *Zootaxa* 4032:535–50.

Wildekamp, R. H., and B. R. Watters. 1996. *A World of Killies: Atlas of the Oviparous Cyprinodontiform Fishes of the World*. Mishikawa, IN: American Killifish Association.

LITERATURE CONSULTED

Adams, G. L., B. M. Burr, and M. L. Warren Jr. 2020. "Amblyopsidae: Cavefishes." In *Freshwater Fishes of North America,* edited by M. L. Warren Jr. and B. M. Burr, 281–321. Baltimore: Johns Hopkins University Press.

Bailey, J. R., and C. L. Hubbs. 1949. "The Black Basses (*Micropterus*) of Florida, with Description of a New Species." *Occasional Papers of the Museum of Zoology,* University of Michigan, 516:1–44.

Baker, W. H., R. E. Blanton, and C. E. Johnston. 2013. "Diversity within the Redeye Bass, *Micropterus coosae* (Perciformes: Centrarchidae) Species Group, with Descriptions of Four New Species." *Zootaxa* 3635: 379–401.

Baker, W. H., C. E. Johnston, and G. W. Folkerts. 2008. "The Alabama Bass, *Micropterus henshalli* (Teleostei: Centrarchidae) from the Mobile River Basin." *Zootaxa* 1861:57–67.

Blanton, R. E., and G. A. Schuster. 2008. "Taxonomic Status of *Etheostoma brevispinum*, the Carolina Fantail Darter (Percidae: *Catonotus*)." *Copeia* 2008: 844–57.

Burkhead, N. M., R. E. Jenkins, and E. G. Maurakis. 1980. "New Records, Distribution and Diagnostic Characters of Virginia Ictalurid Catfishes with an Adnexed Adipose Fin." *Brimleyana* 4:75–93.

Carpenter, K. E. 2002a. *The Living Marine Resources of the Western Central Atlantic. Volume 2. Bony Fishes Part 1 (Acipenseridae to Grammatidae).* Rome, Italy: Food and Agriculture Organization of the United Nations. https://www.fao.org /documents/card/en/c/eb133bb9-b358-5887-b82f-23c9f8ee94fd/.

———. 2002b. *The Living Marine Resources of the Western Central Atlantic. Volume 3. Bony Fishes Part 2 (Opistognathidae to Molidae), Sea Turtles and Marine Mammals.* Rome, Italy: Food and Agriculture Organization of the United Nations. https://www.fao.org/publications/card/en/c/bdf3d5ff-704c-5a41-82fd-82dc2049 cd99/.

Chao, N. L. 2002. "Sciaenidae. Croakers (Drums)." In Carpenter 2002b, 1583–1653.

Chernoff, B. 2002. "Atherinopsidae. New World Silversides." In. Carpenter 2002a, 1090–1103.

Collette, B. B. 2002. "Belonidae. Needlefishes." In Carpenter 2002a, 1104–13.

Dimmick, W. W., K. L. Fiorino, and B. M. Burr. 1996. "Reevaluation of the *Lythrurus ardens* (Cypriniformes: Cyprinidae) Complex with Recognition of Three Evolutionary Species." *Copeia* 1996:813–23.

Fuller, P. 2020. "*Membras martinica* (Valenciennes in Cuvier and Valenciennes, 1835)." US Geological Survey, Nonindigenous Aquatic Species Database, Gainesville, FL. https://nas.er.usgs.gov/queries/FactSheet.aspx?SpeciesID=320.

Fuller, P., L. G. Nico, and M. Neilson. 2020. "*Menidia beryllina* (Cope, 1867)." US Geological Survey, Nonindigenous Aquatic Species Database, Gainesville, FL. https://nas.er.usgs.gov/queries/FactSheet.aspx?SpeciesID=321.

Jenkins, R. E. 1995. "Workshop on Suckers" handout (unpublished). November 2, 1995. Duke Power Company, Huntersville, NC.

Jenkins, R. E., and N. M. Burkhead. 1994. *Freshwater Fishes of Virginia*. Bethesda, MD: American Fisheries Society.

Kinziger, A. P., R. L. Raesly, and D. A. Neely. 2000. "New Species of *Cottus* (Teleostei: Cottidae) from the Middle Atlantic Eastern United States." *Copeia* 2000: 1007–18.

Kraczkowski, M. L., and B. Chernoff. 2014. "Molecular Phylogenetics of the Eastern and Western Blacknose Dace, *Rhinichthys atratulus* and *R. obtusus* (Teleostei: Cyprinidae)." *Copeia* 2014:325–38.

Lachner, E. A., and R. E. Jenkins. 1971. *Systematics, Distribution, and Evolution of the Chub Genus* Nocomis Girard *(Pisces, Cyprinidae) of Eastern United States, with Descriptions of New Species*. Smithsonian Contributions to Zoology No. 85. Washington, DC. Smithsonian Institution Press.

Lanteigne, J. 1988. "Identification of Lamprey Larvae of the Genus *Ichthyomyzon* (Petromyzontidae)." *Environmental Biology of Fishes* 23:261–65.

Mansueti, A. J. 1963. "Some Changes in Morphology during Ontogeny in the Pirate Perch, *Aphredoderus s. sayanus*." *Copeia* 1963: 546–57.

Miller, R. E. 1968. "A Systematic Study of the Greenside Darter, *Etheostoma blennioides* Rafinesque (Pisces: Percidae)." *Copeia* 1968: 1–40.

Munroe, T. A. 2002. "Paralichthyidae. Sand Flounders." In Carpenter 2002b, 1898–1921.

Munroe, T. A., and M. S. Nizinski (FAO 2002). 2002. "Clupeidae. Herrings (Shad, Menhadens)." In Carpenter 2002a, 804–30.

Murdy, E. O., and D. F. Hoese. 2002a. "Eleotridae. Sleepers." In Carpenter 2002b, 1778–80.

———. 2002b. "Gobiidae. Gobies." In Carpenter 2002b, 1781–96.

Pflieger, W. L. 1975. *The Fishes of Missouri*. Jefferson City: Missouri Department of Conservation.

Piller, K. R., and H. L. Bart Jr. 2017. "Rediagnosis of the Tuckasegee Darter, *Etheostoma gutselli* (Hildebrand), a Blue Ridge Endemic." *Copeia* 2017: 569–74.

Ross, S. W., and F. C. Rohde. 2003. "Life History of the Swampfish from a North Carolina Stream." *Southeastern Naturalist* 2:105–20.

Schofield, P. J., J. D. Williams, L. G. Nico, P. Fuller, and M. R. Thomas. 2005. *Foreign Nonindigenous Carps and Minnows (Cyprinidae) in the United States— A Guide to Their Identification, Distribution, and Biology*. US Geological Survey Scientific Investigations Report 2005–5041. Reston, VA: US Geological Survey.

Starnes, W. C. 2006a. "*Procne* and Friends . . . or Aid for Identification of Five Similar and Sympatric Shiners in the Cape Fear River Basin, NC: Spottail Shiner (*Notropis hudsonius*), Swallowtail Shiner (*N. procne*), Cape Fear Shiner (*N. mekisto-*

cholas), Coastal Shiner (*N. petersoni*) and Whitemouth Shiner (*N. alborus*).” Revised March 31, 2006 (unpublished). North Carolina Museum of Natural Sciences Raleigh.

———. 2006b. “Aid to Identification of Highfin Shiner (*Notropis altipinnis*) versus Dusky Shiner (*Notropis cummingsae*).” July 3, 2006 (unpublished). North Carolina Museum of Natural Sciences, Raleigh.

———. 2006c. “Aid to Identification of Chubsuckers, *Erimyzon* spp., in North Carolina.” December 29, 2006 (unpublished). North Carolina Museum of Natural Sciences, Raleigh.

Stout C., S. Schönhuth, R. Mayden, N. L. Garrison, and J. W. Armbruster. 2022. “Phylogenomics and Classification of *Notropis* and Related Shiners (Cypriniformes: Leuciscidae) and the Utility of Exon Capture on Lower Taxonomic Groups.” *PeerJ* 10:e14072. DOI 10.7717/peerj.14072.

Weyand, C. A., and K. R. Piller. 2020. “Assessing Phylogeographic Variation in the Rosyside Dace (Teleostei, Leuciscidae), a Widespread Morphologically Variable Taxon.” *Zoologica Scripta* 49:563–74.

Wood, R. M., R. L. Mayden, R. H. Matson, B. R. Kuhajda, and S. R. Layman. 2002. “Systematics and Biogeography of the *Notropis rubellus* Species Group (Teleostei: Cyprinidae).” *Bulletin of the Alabama Museum of Natural History* 22:37–80.

Yerger, R. W., and K. Relyea. 1968. “The Flathead Bullheads (Pisces: Ictaluridae) of the Southeastern United States, and a New Species of *Ictalurus* from the Gulf Coast.” *Copeia* 1968: 361–84.

GENERAL INDEX

Bold italic page numbers refer to illustrations. For definitions of meristic and morphological terms, consult the glossary (pp. 401–20).

INDEX OF COMMON AND SCIENTIFIC NAMES

Bold italic page numbers refer to illustrations.

Trinectes maculatus, 22, *60*, 73, *83*, 389
Trout
    Brook, 231, 232–34, *234*
    Brown, 21, *63*, 231, 232, 234, *235*
    Rainbow, 20, 231, 232, 234, *234*, 235,
     401
Tuckasegee Darter, 26, 46, 300, 323, *324*,
   325, 326
Turquoise Darter, 27, *69*, 300, 330, *330*, 331

Umbra pygmaea, 20, *63*, 73, 80, *81*, 386
Umbridae, 2, 20, 63, 73, 80, 386

V-lip Redhorse, 9, 44, 112, *115*, 127, *127*,
  128

Waccamaw Darter, 27, 46, 48, 300, 319,
  332, *332*, 333
Waccamaw Killifish, 23, 43, 46, 48, 261,
  262, 264, *267*, 272, *272*, 273, *419*
Waccamaw Silverside, 23, 43, 45, 48, 255,
  256, 259, *259*
Walleye, 30, 299, 301, 302, 304, *304*, *403*
Warmouth, 32, 341, 342, 349. *350*
Warpaint Shiner, 11, 38, 146, 149, 150, 184,
  184, 185, *413*

Western Blacknose Dace, 17, 148, 157, *158*,
  159
Western Mosquitofish, 24, *64*, 278, 281,
  282
White Bass, 25, 287, 288, *290*, 291, *291*
White Catfish, 18, 208, 209, 218, *218*, 221,
  221, 222
White Crappie, 33, 341, 342, 363, *363*, 364
Whitefin Shiner, 12, 44, 146, 180, *180*
Whitemouth Shiner, 14, 45, 147, 149, *195*,
  196, *196*
White Mullet, 283–85, *286*
White Perch, 25, *71*, 287–89, *290*
White Shiner, 14, 147, 186, *186*, 187
White Sucker, 7, 111, 112, 113, *116*, 120, *120*
Whitetail Shiner, 11, 38, 146, 149, 182, *182*
Wounded Darter, 28, 47, 300, 301, 322,
  322, 323

Xenocyprididae, 1, 2, 11, 55, 67, 72, 73, 145,
  146, 149, 152, 154, 210, 379, 413

Yellow Bullhead, 18, 208, 209, 221, *221*, 222
Yellowfin Shiner, 13, 147, 149, 150, 202,
  205, *205*
Yellow Perch, 29, 299–303, *303*, *417*

ABOUT THE AUTHORS

Bryn H. Tracy is a retired fish biologist residing in Apex, North Carolina. **Fred C. (Fritz) Rohde** is a fish biologist with the National Marine Fisheries Service. **Scott A. Smith** and **Jesse L. Bissette** are fish biologists with the North Carolina Division of Marine Fisheries. **Gabriela M. Hogue** is the Collections Manager of Fishes in the Ichthyology Unit at the North Carolina Museum of Natural Sciences. All authors are members of the NCfishes.com team.

Other **Southern Gateways Guides** you might enjoy

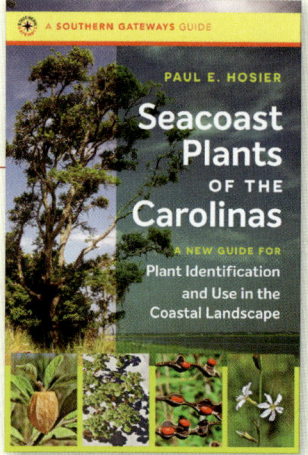

Seacoast Plants of the Carolinas

A New Guide for Plant Identification and
Use in the Coastal Landscape

PAUL E. HOSIER

Published in association with North Carolina Sea Grant

The must-have guide for plant lovers along the North Carolina coast

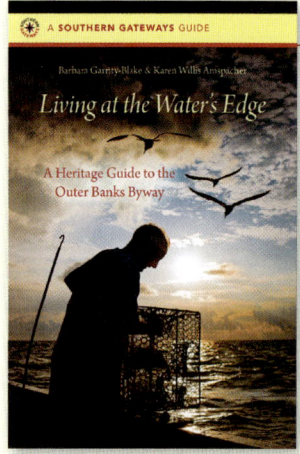

Living at the Water's Edge

A Heritage Guide to the Outer Banks Byway

**BARBARA GARRITY-BLAKE
AND KAREN WILLIS AMSPACHER**

A unique guide to the byway's people and places

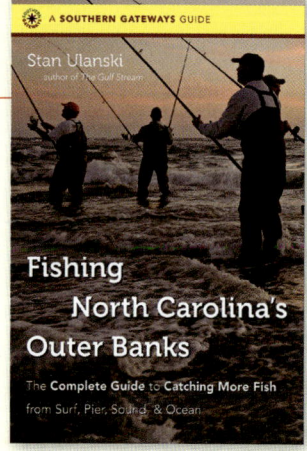

Fishing North Carolina's Outer Banks

The Complete Guide to Catching More Fish
from Surf, Pier, Sound, and Ocean

STAN ULANSKI

*Improve your fishing techniques (and success) by learning
the science of the sea*